Communications
in Computer and Information Science   1099

Commenced Publication in 2007
Founding and Former Series Editors:
Simone Diniz Junqueira Barbosa, Phoebe Chen, Alfredo Cuzzocrea,
Xiaoyong Du, Orhun Kara, Ting Liu, Krishna M. Sivalingam,
Dominik Ślęzak, Takashi Washio, Xiaokang Yang, and Junsong Yuan

More information about this series at http://www.springer.com/series/7899

Henry Han · Tie Wei · Wenbin Liu ·
Fei Han (Eds.)

Recent Advances in Data Science

Third International Conference
on Data Science, Medicine, and Bioinformatics, IDMB 2019
Nanning, China, June 22–24, 2019
Revised Selected Papers

 Springer

Editors
Henry Han (ID)
Fordham University
New York, NY, USA

Tie Wei
Guangxi University
Nanning, China

Wenbin Liu
Guangzhou University
Guangzhou, China

Fei Han
Jiangsu University
Zhenjiang, China

ISSN 1865-0929   ISSN 1865-0937   (electronic)
Communications in Computer and Information Science
ISBN 978-981-15-8759-7   ISBN 978-981-15-8760-3   (eBook)
https://doi.org/10.1007/978-981-15-8760-3

This Springer imprint is published by the registered company Springer Nature Singapore Pte Ltd.
The registered company address is: 152 Beach Road, #21-01/04 Gateway East, Singapore 189721, Singapore

Preface

With the surge of big data and massive data, data science becomes a new exciting interdisciplinary field that is bringing challenges and opportunities to science, engineering, health, and business. The huge amount of data provide more available information and demand for more complicated and explainable knowledge discovery methods to decipher the information. For example, high-frequency trading data from algorithmic trading needs special algorithms and computing approaches to disclose the latent information encoded in the structured big data.

Data science is seeing the trend of problem solving from model-driven to data-driven. The model-driven was widely used in almost all fields because limited data is available. Researchers have to build different models from a theoretical standing point and apply it to 'fit' the limited data. It is likely that the models can have unrealistic assumptions that do not match real data well in modeling. Usually one or few types of data are considered in the models. Thus, the model-driven approaches may need to build different models to handle different types of data. As a result, they can have a hard time generalizing one result to another. Furthermore, the model-driven approaches cannot take advantage of a large amount of available data mainly because the modeling process is somewhat independent from data. On the other hand, the data-driven approaches derive and develop customized but more generalized models or algorithms from data. The data-driven approaches take advantage of the large amount of data and let data talk in decision making. In other words, data-driven approaches are driven and adjusted by data dynamically. The state-of-the-art artificial intelligence (AI) approaches (e.g. machine learning) can be viewed as a typical data-driven approach in data science to exploit the large amount of data.

The future of data science relies on how well AI methods are developed for different subfields. It is expected that different machine learning and deep learning methods are developed for specific type of data. For example, financial data science need their own deep learning methods to dig knowledge from financial data that generally has few independent variables but a huge amount of observations from time series. On the other hand, biological and health data science may want to derive the AI models that work well for high-dimensional data. Quite a lot of pioneering studies have been conducted in biological data science fields in recent years while other data science fields are starting to catch up.

This volume serves the proceeding of the International Conference on Data science, Medicine, and Bioinformatics (IDMB 2019), Nanning, China. It aims to report recent advances in data science, business data science, health and biological data science, and other data science fields. It will be a good guide for data scientists and practitioners in the fields. IDMB 2019 only accepted 35 high-quality papers among 93 submissions through a very rigorous reviewing process. There were 15 papers among the accepted papers recommended to the 3 IDMB 2019 associative SCI-indexed journals: BMC Bioinformatics, Computational Biology & Chemistry, and the Chinese Electronic

Journal, and published in two special journal issues. The volume consists of 20 cutting-edge papers devoted to state-of-the-art data science research in business data science, biological and health data science, and novel data science theory and applications.

We thank all authors, committee members, and session chairs involved in IDMB 2019, as well as folks that kindly provided their support to IDMB 2019. In particular, we thank the business school of Guangxi University and National Science Foundation of China (No. 71562001) for their valuable support alongside Dr. Tie Wei's great leadership.

Henry Han
Tie Wei
Wenbin Liu
Fei Han

Organization

IDMB 2019 Chairs

General Chair

Henry Han   Fordham University, USA

Co-chair

Tie Wei   Guangxi University, China

IDMB 2019 Steering Committee

Henry Han	Fordham University, USA
Fei Han	Jiangsu University, China
Wentian Li	Feinstein Institute for Medical Research, USA
Xiangrong Liu	Xiamen University, China
Wenbin Liu	Guangzhou University, China
Ke Men	Xi'an Medical University, China

IDMB 2019 Program Committee

Yongsheng Bai	University of Michigan, USA
Yaodong Chen	Northwest University, China
Jeff Forest	Slippery Rock University, USA
Qiannong Gu	Ball State University, USA
Deqing Li	University of Maryland, USA
Junyi Li	Harbin Institute of Technology, China
Zhen Li	Texas Woman's University, USA
Zhou Ji	Columbia University, USA
Ye Tian	Case Western Reserve University, USA
Guimin Qin	Xidian University, China
Chaoyong Qin	Guangxi University, China
Stellar Tao	UT Health Center, USA
Hongling Wang	ACT, USA
John Wu	Nanjing Tech University, China
Xiaoguang Tian	Purdue University, USA
Zhenning Xu	University of Southern Maine, USA
Yan Yu	St John's University, USA
Honggang Zhang	University of Massachusetts, USA
Jeff Zhang	California State University, USA
Yanbang Zhang	Northwestern Polytechnical University, China

IDMB 2019 Reviewers

Wei Chen	Fordham University, USA
Zhihua Chen	Guangzhou University, China
Gang Fang	Guangzhou University, China
Chunmei Feng	Harbin Institute of Technology, China
Heming Huang	Qinghai Normal University, China
Tao Huang	Chinese Academy of Sciences, China
Jing Jiang	Jiangsu University, China
Zheng Kou	Guangzhou University, China
Fangjun Kuang	Wenzhou business College, China
Chun Li	Hannai Normal University, China
Dongdong Li	South China University of Technology, China
Yang Li	Shandong University, China
Yihao Li	Fordham University, USA
Jinxing Liu	Qufu Normal University, China
Nan Mi	Texas Tech University, USA
Baoli Peng	The Chinese University of Hong Kong, China
Xianya Qin	Fordham University, USA
Junliang Shang	Qufu Normal University, China
Xuequn Shang	Northwestern Polytechnical University, China
Yuan Shang	University of Arizona, USA
Xiaohong Shi	Xi'an Technological University, China
Yansen Su	Anhui University, China
Juan Wang	Guangxi University, China
Juan Wang	Qufu Normal University, China
Xinzeng Wang	Shandong University of Science and Technology, China
Zhuo Wang	Shanghai Jiaotong University, China
Yi Wu	Georgia Institute of Technology, USA
Juanying Xie	Shaanxi Normal University, China
Jie Xu	St John's University, USA
Peng Xu	Guangzhou University, China
Tingyun Yan	Fordham University, USA
Xuemei Yang	Xianyang Normal University, China
Lianping Yang	Northeastern University, China
Shengli Zhang	Xidian University, China
Liang Zhao	University of Michigan, USA

Contents

Novel Data Science Theory and Applications

Business Data Science: Fintech, Management, and Analytics

Locally Linear Embedding for High-Frequency Trading Marker Discovery

Henry Han[1,4(✉)], Jie Teng[2], Junruo Xia[2], Yunhan Wang[2], Zihao Guo[1], and Deqing Li[3]

[1] Computer and Information Science, Fordham University, Lincoln Center,
New York, NY 10023, USA
{xhan9,zguo50}@fordham.edu

[2] The Gabelli School of Business, Fordham University, Lincoln Center,
New York, NY 10023, USA
{jteng8,jxia18,ywang920}@fordham.edu

[3] Department of Business, Management, and Accounting, University of Maryland,
Eastern Shore, MD 21853, USA
dli@umes.edu

[4] Department of Innovation and Entrepreneurship, Long Island University,
Greenville, NY 10458, USA

Abstract. High-frequency trading (HFT) has been challenging fintech and data science. HFT trading marker prediction is a rarely investigated but important problem in finance and data science. In this study, we first propose locally linear embedding based HFT marker prediction algorithm to tackle this problem and HFT trading marker evaluation algorithms for validation. Our results demonstrate locally linear embedding (LLE) outperform its peers in capturing trading markers in terms of accuracy and complexity for its local data structure keeping mechanism in embedding.

Keywords: Locally linear embedding · High-frequency trading · Manifold learning

1 Introduction

In the last few decades, high-frequency trading (HFT) has progressively dominated the financial market [1]. Around 55% of trading volumes in the U.S. equity market and 80% of foreign-exchange (FX) future volumes attributes to HFT in 2016 [2]. Unlike traditional trading, HFT can finish the equities or futures transaction execution tasks in a few milliseconds by computer algorithms and machines – highly-sophisticated, automated and lightning-fast. Therefore, HFT provides the opportunity to exploit profit from a marginal or a small price change in seconds with large trading volumes. Since the ultra-fast trading mode generates a large volume of financial data, HFT data poses challenges in both the financial market and data science fields. HFT makes the market more insecure somewhat by bringing 'flash crash', in which large and abrupt decline happens in security prices in the market. This is probably because almost all trading systems may employ similar trading strategies when price drops below a threshold.

© Springer Nature Singapore Pte Ltd. 2020
H. Han et al. (Eds.): IDMB 2019, CCIS 1099, pp. 3–17, 2020.
https://doi.org/10.1007/978-981-15-8760-3_1

HFT data is characterized by a large or huge number of observations (transactions) but very few variables. A billion of transactions can only share several variables (e.g. price, volume). The variables are not completely independent with each other and many observations can be quite similar or even same periodically [3, 4]. Different stocks under HFT can have different trading frequencies and even demonstrate periodic patterns or diurnal patterns [3]. Technically, HFT data is structured big data for its huge volumes and structured transaction format. For example, five stocks under the HFT trading mode can have more than 200 million transaction records in a month and require more than 80 Gigabytes storage.

HFT data has been drawing attention in fintech and data science. Some pioneering investigations were conducted from different standing points. For example, Brogaard *et al.* examined the role of high-frequency traders (HFTs) in price discovery and price efficiency [4]. Conrad *et al.* studied the relationship between high-frequency quotations and stock price behavior [5]. Kirilenko *et al.* studied the 2010 flash crash brought by HFT [6]. Son *et al.* employed machine learning to forecast the trends of high-frequency KOSPI200 index data [7]. Nevmyvaka *et al.* studied the reinforce learning to optimize trade execution [8].

Despite previous endeavors, almost no work exists to identify market markers, i.e. trading markers in the HFT market. We interchangeably use trading markers, market markers or markers in our context. The trading marker, which can be categorized as global and general markers according to their price change ratios, is a meaningful buying or selling point for profitability in trading. The global market markers, which can be viewed as more profitable buying/selling points, refer to the transactions with large price change ratios. For example, those transactions with large abrupt price changes could be typical global markers. On the other hand, the general market markers consist of all transactions with the ratios in a relatively small range. They are less profitable buying/selling points but have more likelihood to appear in trading.

HFT market marker prediction aims to answer a core query in HFT, given a set of historical transactions, how can we determine whether the stock price of the existing time is a trading marker, i.e., traders can buy or sell stocks at the corresponding period to make a profit? Besides the profit gain from trading, detecting the real trading marker is critical to understand the dynamics of different stock under HFT. Detecting buying or selling points contribute to a different aspect compared with predicting stock movement. Previous literature has tried to answer questions in predicting the future up-down stock price tendency via all kinds of machine learning and deep learning algorithms [9–15]. Up-down trend prediction studies aim to predict whether the stock price will go up or down in the future given the transaction history in specified previous time intervals rather than searching the market signal to execute worthwhile trading. But trading marker identification or prediction study has no previous literature available, compared with the abundant and consistent publications on stock movement [9–15].

The unique characteristics of HFT data put a challenge to seek meaningful trading markers. Each HFT dataset has relatively few variables but a huge amount of observations even within a short period interval. Furthermore, given a time interval, the large number of transactions could contain redundant information since HFT stock price may barely change or change on a small or even tiny scale [16]. One possible solution is to go through observation sampling to change high-frequency trading data into 'low-frequency trading' data to fit the traditional data analysis schemes [17]. But it can be risky because sampling can remove trading markers because an abrupt price change can start from anywhere due to a stochastic event (e.g. breaking news). Since different stocks under HFT can have different trading frequencies and even demonstrate periodic patterns, different sampling techniques have to be invented to accommodate different stocks.

Dimension reduction techniques (e.g. PCA) offers another promising possibility to discover trading markers. It has a potential to unveil the underlying HFT data structures in an alternate subspace spanned by important information-carrying bases (e.g., PC). The trading markers can sometimes distinguish themselves from the rest of common transactions through subspace visualization or clustering. Previous studies applied dimension reduction techniques as the intermediate steps for stock price prediction, up-down trend forecasting, or financial market early warning signal retrieval [15, 18–21]. However, it remains unknown which dimension reduction techniques fit HFT data better and how to capture markers in the subspace. Recent work shows that HFT data demonstrate quite different characteristics under principal component analysis. The first principal component of HFT data can even reach more than 60% variance explained ratio, which is much higher than traditional finance data [18].

In this paper, we propose a novel approach to detect and predict HFT trading markers by employing locally linear embedding (LLE) and a state-of-the-art clustering technique: density-based spatial clustering of applications with noise (DBSCAN) by viewing HFT data a manifold [22–24]. A manifold is a set of data points that locally behave like Euclidean space. Unlike PCA conducting a linear dimension reduction by mapping data into a subspace spanned by maximum variance directions, LLE seeks a low-dimensional manifold embedding, i.e., a subspace, to represent the original manifold in the high-dimensional space by preserving distance metrics locally and neighborhood relationships [24, 25]. It is also a feature extraction procedure so that data with exceptional behaviors will stand out in the embedding for its 'special coordinates'. Our study exploits the special characteristics of HFT data to construct two-dimensional embeddings of HFT data via LLE. It aims to unveil the latent underlying structure of HFT data and distinguishes potential trading markers as outliers in embedding. The trading markers are searched among the outliers in density-based clustering by exploiting its extreme deviation nature from general transactions.

We further present HFT trading marker evaluation algorithms to validate marker prediction on behalf of general and global markers. Our results demonstrate locally linear embedding (LLE) capture trading markers better than its peers (e.g., t-SNE, ISOMAP, and KPCA) on behalf of time complexity and accuracy [25–28]. The underlying neighbor structure keeping mechanism in LLE seems to unveil the dynamics of HFT better and contributes to identifying potential markers well. Almost all peer algorithms demonstrate an equivalent level of performance to capture the general markers, but LLE shows advantages in capturing the global markers that record large price changes.

2 Methods

This section introduces data acquisition, preprocess, and proposed trading marker discovery model, trading marker evaluation, and related LLE and DBSCAN clustering.

2.1 Data and Preprocessing

The study is based on the HFT data acquired from the" intraday prices" API provided by IEX cloud [29]. We develop our own intraday data retrieval software to acquire the trading data given any period with an interval of 1 min. In our experiment, we acquire 20 HFT large-cap stocks from four representative markets: IT, Bank, Retail, and Fashion, starting from 2/1/2019 through 2/22/2019 during 14 trading days. Unlike raw HFT data, each observation in our data represents the trading transaction in a one-minute time interval. Each market consists of five representative stocks. For example, the IT market contains MSFT, AMZN, ITEL, GOOG, and APPL. The Bank market includes JPM, BAC, HSBC, C (Citigroup), and GS. The retail market consists of WMT, TGT, JCP, KSS, and HD. The fashion market consists of TPR, AEO, GES, CPRI, and TIF.

Each HFT dataset consists of 5850 observations across 9 features. Besides standard variables such as close, high, low, open, and volume, we also include *ChangeOverTime*, *MarketAverage*(price), *NotionalValue*, and *NumberOfTrades* for their importance in trading. The *ChangeOverTime* refers to the stock price change ratio in a minute and is generally a small ratio (e.g. 0.002816). The *numberOfTrades* indicates the frequency of trading of the stock in the one-minute time interval. The *NotionalValue* refers to the total value of the position in trading.

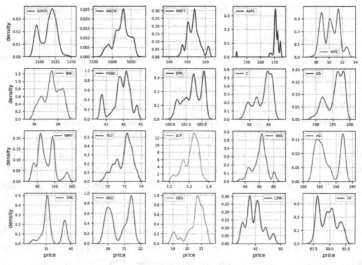

Fig. 1. The price density function estimations of all 20 HFT stocks from IT, bank, retail, and fashion industries.

Figure 1 summarizes the close price probability density functions of the 20 HFT stocks using Gaussian kernel density estimation, where each stock demonstrates different price distributions [30]. Almost no distributions are subject to a normal or log-normal distribution. Most stocks follow platykurtic distributions for their negative kurtosis values except AAPL, MSFT, KSS, and GES. AAPL could be riskier and have more outliers

than the others for its very large kurtosis value: 22.2974. At the same time, the probability density functions also reflect the trading patterns. Those most frequently traded
stocks (e.g. GOOG & AAPL) usually have much lower density values than the least
frequently traded ones (e.g. JCP). The retail stock JCP presents itself as an outlier for
its very high-density value (e.g. >12).

2.2 Trading Marker Discovery Model Design

We aim to capture trading markers from input HFT data $X = \{x_i\}_{i=1}^{n}$ with n transactions
in a time interval [t1, t2] across p features $x_i \in \Re^p$, from a manifold learning approach
by viewing HFT data as a manifold. Since trading markers are a set of transactions
with remarkable price changes, their exceptional data behavior can be detected well by
examining the intrinsic structure of HFT data. However, it is not easy to directly exam-
ine the intrinsic structure of multi-dimensional HFT data. Manifold learning like LLE
calculates a low-dimensional manifold embedding to preserve the intrinsic geometric
structure of the original data [24]. LLE generates nonlinear low-dimensional embeddings for high-dimensional data while keeping the underlying neighborhood structure
[24, 25]. It models local data behavior well and forces those outliers standing out in the
embedding compared to its peers such as ISOMAP that are good at modeling holistic data
behavior. LLE maps a high-dimensional manifold locally isometric to a low-dimensional
embedding. It keeps local data geometric structures by maintaining the local linear relations in each neighbor. It assumes each point can be represented by its neighbors in a
linear combination and the weights that represents the intrinsic neighborhood geometric
structures are used to reconstruct corresponding points in the embedding. Since those
markers with abrupt price change may not be represented by its neighbor points well, its
corresponding points in the embedding will easier to stand out in the embedding from its
neighbors. Thus, LLE has a built-in advantage to locate those markers with large price
oscillations.

Although direct neighbor price comparisons or more complicate price function optimization might grasp some potential markers for individual stocks, it may lack generalized prediction capabilities for generic data, especially because different stocks can have
different frequencies. Moreover, the proposed manifold learning approach can analyze
all variables' contributions in trading marker identification rather than only focus on
a specific one. It can be more meaningful to predict market markers by exploiting all
involved variables via a comprehensive data-driven approach.

The Embedding Space Dimensionality. he embedding should contain the original data
variance as much as in selecting its dimensionality. Since the first PC explained variance
ratio of almost all HFT data exceeds 60%, the data variances would concentrate on the
projections on the two major bases in the embedding space $\Re^l$. Thus, it is reasonable to
consider the 2D manifold embedding of HFT data: $Y \in \Re^{n \times 2}$ in LLE to seek trading
markers. The 2D manifold embedding provides geometric support for trading marker
discovery under LLE. It retrieves the intrinsic structure of HFT data in a 2D subspace
by mirroring original data behaviors in the original high-dimensional space. The trading
markers are the transactions with exceptional behaviors and will be mapped as outliers
in the 2D-embedding geometrically. Those common transactions will stay together in
the embedding for their similar data behaviors to form clusters.

Manifold Embedding Clustering. It is desirable to employ a clustering algorithm to validate and collect all the outliers as potential markers from the manifold embedding. The manifold embedding of HFT data can have arbitrary-shape clusters with noise because of different trading frequencies and scenarios. Classic center-based K-means clustering can't handle such data well because it requires that clusters are convex shaped [22]. We employ a density-based clustering algorithm: density-based spatial clustering of applications with noise (DBSCAN) to cluster the arbitrary-shaped manifold embedding [22, 23]. It determines whether a sample belongs to a specific cluster by a concept of reachability that is measured by the minimum number of data points within a radius from that single point and an outlier is an unreachable point. Given a manifold embedding DBSCAN clustering is more likely to detect more outliers than other clustering algorithms.

Algorithm 1 describes the HFT trading marker prediction. The DBSCAN outputs an outlier set including the transactions labeled as noise in clustering as the trading marker candidate set. We tune the neighborhood size: $min\_samples = 10$ and the maximum distance $eps = 0.5$ in DBSCAN in our implementation. We set a marker filtering threshold $\varepsilon = 0.01\%$, which means the price change ratio should reach at least 0.01% compared with its neighbor transactions for all potential markers collected from clustering.

Algorithm 1: HFT trading marker prediction

// Goal: identify HFT trading markers
Input:
      HFT dataset X
      Trading marker filtering threshold ε(0.01%)
      The maximum distance eps between two points in a DBSCAN neighbor
      The neighborhood size in DBSCAN: min_samples
Output:
      The predicted trading marker set: M

1.   // Do locally linear embedding to get the embedding of input data
2.   Embedding ← LLE(X)
3.   // Use DBSCAN clustering to capture clustering an outlier set O.
4.   O ← DEBSCAN(Embedding, eps, min_samples)
5.   // Outlier analysis to find a trading marker set M
        for each transaction point o in O do:
6.   if (o.price-o.leftNeighbor().price)× (o.price-o.RightNeighbor().price)>=0 then
7.   // Do nearest neighborhood search in left and right time interval
8.   if |o.price-o.leftNeighbor().price|/o.left > ε
    i.   or |o.price-o.RightNeighbor().price|/ o.price > ε then
9.   M.add(o)
10.   end
11.   return M

2.3   True Marker Identification

To validate the predicted trading markers, we need to identify the 'true markers' of HFT data. It is done by collecting all extrema by viewing whole trading price signals as a piecewise price function $p(t)$ with time t. Given an HFT sequence $X = \{x_i\}_{i=1}^{n}$ in a given time period [t1, t2] with corresponding trading price curve $p(t)$, we collect a set Δ that includes all the maxima and minima of $p(t)$ according to a price change ratio threshold η:

$$\Delta = \left\{ x : \frac{|x.p(t) - x.p(t_0)|}{|x.p(t_0)|} \geq \eta \right\} \tag{1}$$

where $x.p(t)$ and $x.p(t_0)$ are prices of transaction x at time t and t_0 respectively. The markers in the set Δ are obtained through derivative-based extrema collection methods. However, the extrema collection alone may fail to capture some abrupt price changes (e.g. sharp peaks) that do not appear on extremes because they do not have derivatives mathematically. But the sharp peaks can be critical markers trigged by breaking news or unexpected events in the market. To identify the sharp peaks, we exploit continuous wavelets to transform HFT price data into a wavelet space and collect the transactions with abrupt price changes into a set S_p by using the method developed by Du et al. [30].

The final true marker set is a union of the two sets: $S = \Delta \cup S_p$.

Partitioning True Markers as General and Global Markers. We partition true markers as general and global markers for the sake of marker evaluation. General makers have a small price-change ratio during a time period of interest. It models those small perturbations in HFT trading. On the other hand, global markers model those large or exceptional perturbations that happen in HFT trading. Traditional extrema -finding ways may not be able to capture all of them. But they are more likely to generate a large profit or loss in trading. The following equations define the general and global markers in [t0, t], which is implemented as a two-minute interval in our study, respectively. We set $\gamma = 0.01\%$ and $\zeta = 0.1\%$ in this study.

$$T = \left\{ x : \gamma \leq \frac{|x.p(t) - x.p(t_0)|}{|x.p(t_0)|} < \zeta \wedge x \in S \right\} \tag{2}$$

$$G = \left\{ x : \frac{|x.p(t) - x.p(t_0)|}{|x.p(t_0)|} \geq \zeta \wedge x \in S \right\} \tag{3}$$

2.4 Predicted Trading Marker Evaluation

It is worthwhile to point out that the size of true markers can be much larger than the number of predicted markers. The predicted marker evaluation is done via a neighbor search approach to find the best match between true markers and predicted markers. To evaluate the effectiveness of trading marker prediction, we calculate the mean square error (MSE) for all entries in the predicted marker set M as follows:

$$MSE = \frac{1}{|M|} \sum_{\hat{p}_j \in M, p_j \in S} \left(p_j - \hat{p}_j \right)^2 \tag{4}$$

where $\hat{p}_j$ is the price of the predicted marker $\hat{x}_j \in M$ and p_j is the price of the corresponding true marker $x_j \in S$ identified through a neighborhood search. The MSE evaluates whether the predicted marker is a true marker by quantifying how close between the predicted markers and correct answers.

Given a predicted marker x' and the true marker set $S = G \cup T$, it needs to know whether it is a predicted global marker or general marker. We propose a nearest neighborhood search approach to solve the problem. We use true global markers to seek their corresponding best-matched predicted global markers. If the time index of a predicted marker m falls in a tolerance interval of a true global marker g: $m_t \in [t_g - \delta, t_g + \delta]$ (e.g. $\delta = 5$ min), it is believed to hit the true global marker. Otherwise, we search the nearest neighbor of g in M: $\mathcal{N}(g) = \{m \in M : ||m - g|| < \varepsilon'\}$ to identify a predicted global marker b closest to g. If more candidates are available in $\mathcal{N}(g)$, it is picked as the one with closest price and time index to closest to those of the global marker g. We then collect the predicted global markers in a set M_G and get the predicted general marker set M_T besides calculating the global marker MSE: $GlobalMSE = \frac{1}{|M_G|} \sum_{\hat{p}_j \in M_G, p_j \in G} (p_j - \hat{p}_j)^2$.

As to general marker evaluation, we seek the closest true general marker in a neighborhood of each predicted general marker. Given a predicted general marker $\hat{x}$, we search its nearest neighbor to identify the corresponding true general marker x with the closest time index as that of $\hat{x}$. In implementation, such a search is done in an interval $\mathcal{N}(\hat{x}, \alpha) = (t_{\hat{x}} - \alpha, t_{\hat{x}} + \alpha)$ based on its time index $t_{\hat{x}}$. The neighborhood radius α is selected as a certain number of minimum time units in transactions (e.g. $\alpha = 2$ min). Finally, the MSE of the predicted general markers: $GeneralMSE = \frac{1}{|M_T|} \sum_{\hat{p}_j \in M_T, p_j \in T} (p_j - \hat{p}_j)^2$ is calculated to evaluate the effectiveness of general marker prediction. The final MSEs of all predicted markers is the sum of $GeneralMSE$ and $GlobalMSE$. The smaller the MSE indicates the better trading marker discovery.

2.5 Locally Linear Embedding and Density-Based Clustering

Locally linear embedding (LLE) generates nonlinear low-dimensional embeddings for high-dimensional data while keeping the underlying neighborhood structure in embedding [24]. It assumes that a point in the high dimensional manifold can be represented linearly by its adjacent points and reconstructs each point using its nearest neighbors by solving an optimization problem. The solution to the problem is a weight matrix $W = (w_{ij})$ that represents the intrinsic neighborhood geometric structures.

Given a dataset $X = \{x_1, x_2, \dots x_n\}$ in a high-dimensional manifold: $x_i \in \Re^p$, the k nearest neighbors of x_i are determined first by minimizing the cost function: $\varphi(w) = \sum_{i=1}^{n} x_i - \sum_{j=q(i)} w_{ij} x_j^2$, where $q(i)$ is an index set of the nearest k neighbors of x_i, $|q(i)| = k$, and x_j is a nearest neighbor of x_i, and w_{ij} is a linear weight assigned to x_i and x_j. $w_{ij} = 0$, if x_j is not in the neighborhood of x_i. The row sum of the weight matrix is required to be 1: $\sum_j w_{ij} = 1$ to impose translational invariance in embedding.

The weight matrix $W = [w_1, w_2, \dots w_n]$ can be obtained by solving the Langangian: $\mathcal{L}(Y, \lambda) = w_i^T G_i w_i - \lambda (\vec{1}^T w_i - 1)$, i.e., $w_i = G_i^{-1} \vec{1}$, where $G_i = (x_i - x_j)(x_i - x_j)^T, j \in q(i)$, and $\vec{1} \in \Re^k$ is a vector with all entries being "1"s. When $k = |q(i)| > p$, penalty terms need to add to G_i to get its inverse: $w_i = (\alpha I + G_i)^{-1} \vec{1}$.

LLE then seeks a corresponding low-dimensional embedding $Y = \{y_1, y_2, \dots, y_n\}, y_i \in \Re^l, l \ll p$, to maintain the 'local linear relations' by minimizing another loss function: $\varphi(y) = \sum_{i=1}^{n} y_i - \sum_{j=p(i)} w_{ij} y_j^2$, where $p(i)$ is the index set of the nearest k neighbors of y_i. The embedding Y is required to be zero-mean data:

$\frac{1}{n} \sum_{i=1}^{n} y_i = 0$ and the covariance matrix of Y is the identity matrix: $\frac{1}{n} Y Y^T Y = I$. The embedding finding is equivalent to solving a sparse eigenvalue problem. $\varphi(Y) = Y^T M Y$, $M = (I - W)^T (I - W)$. The embedding Y can be obtained by doing eigenvalue decomposition for matrix M.

2.6 Density-Based Clustering

DBSCAN is a density-based clustering algorithm that handles arbitrary-shape clusters with noise. It groups the points with many close neighbors in the same clusters for their high-density. It also marks the points with far-away neighbors as outliers for their low-density. DBSCAN classifies points as core, reachable, and outliers [22]. A core point simply refers to a point whose neighborhood (a neighborhood with a radius ε) has an enough number of points. A reachable point is a point that can be reached by one or a sequence of core points. An outlier is an unreachable point. In our context, it will be a transaction with exceptional trading behaviors that are potentially to be trading markers.

Given a point to be clustered, DBSCAN retrieves its ε-neighborhood. If the neighborhood size is >= the minimum number of points (minpts) required to form a 'dense region', i.e., a region with the enough number of close points, the neighbor will be initially as a cluster and the point is marked as a core point. Otherwise, the point is marked as noise. If the point is a reachable point for a cluster, its ε-neighborhood will be marked as a part of that cluster. All points in the ε-neighborhood will be added to the cluster until the density condition is satisfied. This procedure continues until all clusters and outliers are identified [23].

The quality of DBSCAN clustering depends on the parameters setting of ε and minpts. In our context, the input of DBSCAN is the 2D-embedding of HFT data. Since we believe the outliers will be the potential market marker in HFT for their 'unreachable' properties compared to general transaction points. To make sure effective marker identification, we set the neighbor radius ε and Minpts so that only 2.5% outliers, which are trading marker candidates, are identified among all transactions.

3 Results

We first demonstrate how our proposed trading marker discovery method works through a visualization approach by employing the peer manifold learning methods of LLE: t-SNE, and KPCA. Figure 2 illustrates the embedding clustering of MSFT and BAC stocks under t-SNE, LLE, and KPCA, where different colors differentiate different clusters [24, 25, 28, 29]. The (a) and (b) subplots show results on the t-SNE embeddings, (c) and (d) illustrate results on the KPCA embeddings, and (e) and (f) demonstrate those on the LLE's embeddings, separately. Each subplot pictures the latent data structure through similar points and outliers' location under DBSCAN. t-SNE on (a) and (b) subplots show the preponderant advantage on aggregating data points with similar information into one group. But DESCAN identifies relatively few outliers. KPCA encircles all outliers inside with a smooth oval shape by demonstrating an ambiguous clustering structure. The clusters in KPCA feature space are not as compact as those of t-SNE and LLE. KPCA fails to discover the relationship between outliers and normal transactions because too many outliers are generated in visualization and clusters do not show a good coherence besides increasing interpretive difficulty.

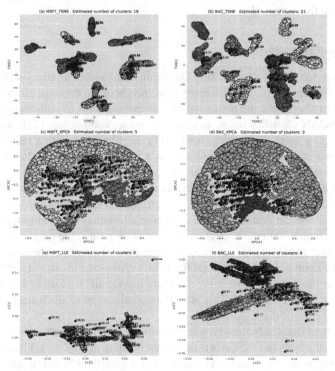

Fig. 2. The embedding clustering visualization of HFT stock data (MSFT and BAC) under t-SNE, kernel KPCA, and LLE, where each color differentiates each cluster identified by DBSCAN. (Color figure online)

LLE on (e) and (f) subplots achieve an alternative visualization and clustering for its good local neighborship maintenance in embedding. It illustrates more outliers with respect to different clusters than t-SNE and provides more trading marker candidates. In particular, it demonstrates a much faster computing compared to t-SNE for its low complexity [25]. The resulting embedding of LLE depicts more local but less global hidden trading patterns for its neighborhood-preserving mechanism [24–26]. It suggests locally isometric keeping can isolate trading markers better than t-SNE and KPCA.

3.1 LLE Achieves Leading Performance in Trading Marker Discovery

We have found that LLE and its peers demonstrate an almost equivalent or close performance in capturing general markers, but quite different in global marker discovery. It is probably because general markers recording relatively small price changes can be easily captured compared to global markers. Figure 3 compares the global MSEs of all 20 stocks from the four industries across LLE and its comparison methods: t-SNE,

ISOMAP, KPCA, PCA and SPCA (sparse PCA) [31]. LLE achieves the lowest individual global MSEs and t-SNE actually achieves the second-lowest ones among all the methods. It is probably because LLE's local linear relationship maintenance can 'force' global markers to stand out in embedding that makes them easily caught as outliers in clustering. Similarly, t-SNE locality keeping contributes to outlier capturing in DBSCAN. KPCA has the worst performance probably because KPCA may bring noise in the feature space that blurs the difference between trading markers and common transactions. PCA, SPCA, and ISOMAP seem to have almost equivalent performance though SPCA seems to achieve better for the fashion stocks (e.g. AEO). The good performance of SPCA on the fashion stocks may indicate different data demonstrates different model preference in marker discovery besides the sparseness constraints can enhance data locality in dimension reduction [31].

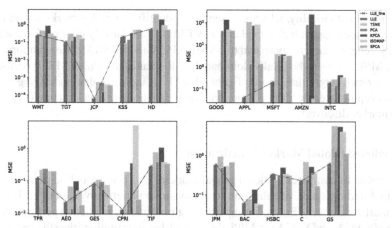

Fig. 3. The comparison of global MSEs of six methods on 20 HFT stocks from four industries: Retail (NW subplot) IT (NE subplot), Fashion (SW subplot), and Bank (SE subplot). The LLE performance is heightened by a line plot.

Figure 4 presents an overview of different MSE performance on behalf of six dimension-reduction methods for all 20 HFT datasets. Besides LLE, almost all methods demonstrate an equivalent level general MSE performance. LLE's local linearly relationship maintenance may not contribute to capturing general markers but global markers. It pushes global markers out of the clusters in embedding, caught by DBSCAN and labeled as outliers in clustering.

In contrast, three manifold learning methods t-SNE, KPCA, and ISOMAP display nearly equally lowest general MSEs comparing with the rest. t-SNE, KPCA, and ISOMAP all maintain the holistic-level fundamental data structure in embedding to a different degree. The holistic structure keeping mechanism contributes to unveil the big picture of data and those general markers are more likely to be components in a holistic embedding to depict the trading tendency in a more continuous manner. Therefore, it is more likely to be labeled as unreachable outliers in DBSCAN. In other words, holistic data structure keeping in embedding can benefit general marker retrieval but is usually

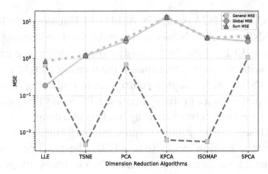

Fig. 4. The comparisons of different MSEs on behalf of 6 dimension-reduction methods in trading marker discovery.

at the price of data locality, which is essential to identify the global markers with large and abrupt price change. It also explains why LLE has a relatively poor performance in capturing general markers compared to t-SNE, KPCA and ISOMAP. More importantly, the sum MSE seems to be totally dominated by the values of global MSE because its values are much larger than those of general MSE. It indicates that underlying neighborhood keeping in embedding is more important than holistic data structure keeping in trading marker discovery.

3.2 Predicted Global Marker Visualization

It is desirable to compare predicted global markers and true global markers through an original trading space visualization to demonstrate the effectiveness of LLE-based trading marker discovery. Figure 5 compares the global marker prediction performance of LLE, t-SNE, PCA, and KPCA for APPL data, which has more outliers than the rest of data during the trading period. We drop SPCA and ISOMAP for their equivalent performance with PCA. It suggests that the global markers predicted by LLE match the original ones best among the four methods. Both LLE and PCA capture the marker with the lowest price in the two-week transaction period, but t-SNE miss it. However, LLE captures the first true marker but PCA misses it. It indicates LLE's local neighborhood configuration preservation can distinguish global markers more efficient in embedding than its peers. Similar results can be found for the rest of data also. Data locality keeping can be an essential in high-frequency trading marker discovery because it will be more likely to make the global markers in embedding to be unreachable outliers in the density-based clustering [23].

Fig. 5. The comparisons of general and global MSEs for IT, bank, retail, and fashion industries.

4 Conclusion and Discussions

We investigate unexplored but important HFT market marker prediction problem by proposing an LLE-based marker discovery approach in this study. Such a novel approach can learn and predict potential trading markers by seeking outliers in the two-dimensional embedding that keeps neighborhood relationships. LLE demonstrates a good time cost compared to most peers except PCA that has the least complexity. The expected computation time of 20 datasets for LLE, PCA, and t-SNE, KPCA, ISOMAP, SPCA are 2.4192, 0.0025, 441.5876, 3.1771, 64.6532, and 8.1302 s respectively. This is because the complexity of PCA is decreased to $O(n)$ when p is a constant and relatively small (<10) for given input data $X \in \Re^{n \times p}$. The time complexity of LLE is $O(n^2)$ given a constant small number of features p, compared to t-SNE's $O(n^4)$ and ISOMAP's $O(n^3)$.

Such a study can argument with HFT trading price discovery techniques to contribute to building powerful trading system to recommend whether the trader should buy/hold/sell, at what price and share numbers, before execution. It overcomes the weakness of traditional price-based marker discovery by taking advantages of all features in an automatic machine learning approach in a more data-driven approach. Unlike existing concept that HFT should do trading under a small price change, our global marker discovery shows that high frequency trading can be more profitable if trading is conducted under the global markers. Thus, investigating global marker capturing can be more important than general markers on in trading marker discovery.

It is possible to employ more HFT datasets to investigate generalization of the proposed marker discovery method besides identifying models for HFT data from different markets or groups. This is because we find that different data may show their own model preference in trading marker discovery. We doubt different trading frequencies for different group of HFT data will lead to some prototype models for efficient trading marker discovery.

Furthermore, the existing parameters setting is a fixed way. For example, we apply a fixed perplexity in t-SNE and the neighborhood size in LLE for all HFT data. It can be desirable to look for a more adaptive approach so that different HFT data can determine optimal parameter selection for each dataset for the sake of HFT trading marker discovery. Furthermore, we are interested in comparing new state-of-the-art manifold learning methods such as using universal manifold approximation and projection (UMAP) in the proposed trading marker discovery method by extending our existing discovery mechanism [32].

References

1. Cespa, G., Vives, X.: High frequency trading and fragility. Working Papers Series. European Central Bank (2020) (2017)
2. Hendershott, T., et al.: Does algorithmic trading improve liquidity? J. Finance **LXVI**, 1–33 (2011)
3. Brownlees, T., Cipollini, F., Gallo, M.: Intra-daily volume modelling and prediction for algorithmic trading. J. Financ. Econ. **9**(3), 489–518 (2011)
4. Brogaard, J., Hendershott, T., Riordan, R.: High-frequency trading and price discovery. Rev. Financ. Stud. **27**, 2267–2306 (2014)
5. Conrad, J., Wahal, S., Xiang, J.: High-frequency quoting, trading, and the efficiency of prices. J. Financ. Econ. **116**, 271–291 (2015)
6. Kirilenko, A., et al.: The flash crash: high-frequency trading in an electronic market. J. Finance **72**, 967–998 (2017)
7. Son, Y.: Noh, S, Lee, J, Forecasting trends of high-frequency KOSPI200 index data using learning classifiers. Expert Syst. Appl. **39**(14), 11607–11615 (2012)
8. Nevmyvaka, Y., Feng, Y., Kearns, M.: Reinforcement learning for optimized trade execution. In: Proceedings of the 23rd International Conference on Machine Learning, pp. 673–680. ACM (2006)
9. Fischer, T., Krauss, C.: Deep learning with long short-term memory networks for financial market prediction. European J. Oper. Res. **270**, 654–669 (2018)
10. Dixon, M., Klabjan, D., Bang, J.H.: Implementing deep neural networks for financial market prediction on the Intel Xeon Phi. In: Proceedings of the Eighth Workshop on High Performance Computational Finance, pp. 1–6 (2015)
11. Huang, C.-L., Tsai, C.-Y.: A hybrid SOFM–SVR with a filter-based feature selection for stock market forecasting. Expert Syst. Appl. **36**, 1529–1539 (2009)
12. Huanga, S.-C., Wub, T.-K.: Integrating ga-based time-scale feature extractions with SVMS for stock index forecasting. Expert Syst. Appl. **35**, 2080–2088 (2008)
13. Kazem, E., Sharifi, F.K., Hussain, M., Saberi, O.K.: Hussain Support vector regression with chaos-based firefly algorithm for stock market price forecasting. Appl. Soft Comput. **13**, 947–958 (2013)
14. Patel, J., et al.: Predicting stock and stock price index movement using Trend Deterministic Data Preparation and machine learning techniques. Expert Syst. Appl. **35**, 259–268 (2015)

15. Berradi, Z., Lazaar, M.: Integration of principal component analysis and recurrent neural network to forecast the stock price of casablanca stock exchange. Procedia Comput. Sci. **148**, 55 (2019)

16. Han, H., Li, M.: Big data analytics for high frequency trading volatility estimation. In: Tavana, M., Patnaik, S. (eds.) Recent Developments in Data Science and Business Analytics. SPBE, pp. 351–359. Springer, Cham (2018). https://doi.org/10.1007/978-3-319-72745-5_39

17. Andersen, T.G., et al.: The distribution of realized stock return volatility. J. Financ. Econ. **61**(1), 43–76 (2001)

18. Aït-Sahalia, Y., Xiu, D.: Principal component analysis of high-frequency data. J. Am. Stat. Assoc. **114**(525), 287–303 (2019)

19. Anagnostidis, P., Emmanouilides, C.J.: Nonlinearity in high-frequency stock returns: evidence from the athens stock exchange. Phys. A **421**, 473–487 (2015)

20. Juárez, C.A.Q., Escobedo, A.V.: Analysis of stock market behavior of the major financial exchanges worldwide using multivariate analysis (principal component analysis PCA) for the period 2011 to 2014. Revista CEA **2**, 25 (2015)

21. Huang, Y., Kou, G., Peng, Y.: Nonlinear manifold learning for early warnings in financial markets. Eur. J. Oper. Res. **258**(2), 692–702 (2017)

22. Ester, M., et al.: A density-based algorithm for discovering clusters in large spatial databases with noise. In: KDD 1996 Proceedings of the Second International Conference on Knowledge Discovery and Data Mining, pp. 226–231 (1996)

23. Schubert, E., Sander, J., Ester, M., Kriegel, H.P., Xu, X.: DBSCAN revisited, revisited: why and how you should (still) use DBSCAN. ACM Trans. Database Syst. (TODS) **42**(3), 19 (2017)

24. Roweis, S.T., Saul, L.K.J.S.: Nonlinear dimensionality reduction by locally linear embedding. Science **290**(5500), 2323–2326 (2000)

25. Donoho, D., Grimes, C.: Hessian eigenmaps: locally linear embedding techniques for high-dimensional data. Proc. Natl. Acad. Sci. U.S.A. **100**, 5591 (2003)

26. Der Maaten, L.V., Hinton, G.E.: Visualizing high-dimensional data using t-SNE. J. Mach. Learn. Res. **9**, 2579–2605 (2008)

27. Tenenbaum, J.B., De Silva, V., Langford, J.C.: A global geometric framework for nonlinear dimensionality reduction. Science **290**, 2319–2323 (2000)

28. Schoelkopf, B., Smola, A.J., Mueller, K.-R.: Kernel principal component analysis. In: Advances in Kernel Methods, pp. 327–352. MIT Press, Cambridge (1999)

29. IEX-API. https://iexcloud.io/docs/api/#intraday-prices. Accessed 2019

30. Du, P., Kibbe, W.A., Lin, S.M.: Improved peak detection in mass spectrum by incorporating continuous wavelet transform-based pattern matching. Bioinformatics **22**(17), 2059–2065 (2006)

31. Jenatton, R., Obozinski, G., Bach, F.: Structured sparse principal component analysis. In: ICML (2009)

32. McInnes, L., Healy, J., Melville, J.: UMAP: uniform manifold approximation and projection for dimension reduction. ArXiv arXiv:1802.03426 (2018)

Implied Volatility Pricing with Selective Learning

Henry Han[1,4(✉)], Haofeng Huang[2], Jiayin Hu[2], and Fangjun Kuang[3]

[1] Computer Information Science, Fordham University, Lincoln Center,
New York, NY 10023, USA
xhan9@fordham.edu
[2] The Gabelli School of Business, Fordham University, Lincoln Center,
New York, NY 10023, USA
{hhuang96,jhu101}@fordham.edu
[3] Wenzhou Business College, Wenzhou 32500, Zhejiang, China
[4] Department of Innovation and Entrepreneurship, Long Island University,
Greenville, NY 10458, USA

Abstract. Machine learning presents itself as an alternative data-driven approach to predict implied volatilities in Fintech. However, such an approach suffers from a relatively low prediction accuracy besides a model selection issue. In this study, we propose a novel selective learning approach to enhance machine learning implied volatility pricing. It boosts different machine learning models' performance on different option data on behalf of moneyness, besides identifying optimal machine learning models in implied volatility prediction. In particular, selective learning can be an excellent way to enhance implied volatility pricing for the option datasets with more noise. In addition, we find out-of-the-money (OTM) and in-the-money (ITM) options fit machine learning prediction better than near-the-money (NTM) options. This pioneering work first provides a robust way to enhance implied volatility pricing via machine learning and will inspire similar studies in the future.

Keywords: Option · Implied volatility · Machine learning

1 Introduction

Since Black, Scholes and Merton introduced the ground-breaking option pricing model, the Black-Scholes-Merton (BSM) model has been ubiquitous in financial research and practice areas [1, 2]. Notwithstanding critiques for its not all empirically valid assumptions such as risk-free interest rates, efficient markets, and constant volatility, the model not only inspires following explorations in option pricing, but also continues to be a practical guide in trading for its useful approximation to reality [3]. Since volatility in the BSM model cannot be observed directly in the market, implied volatility, which measures investor's confidence about the future risk of the stock, needs to be priced under the model. It indicates the current market expectation of future stock volatility and impacts market option price seriously. In the financial industry, traders are used to quoting options by implied volatility rather than the price [4].

© Springer Nature Singapore Pte Ltd. 2020
H. Han et al. (Eds.): IDMB 2019, CCIS 1099, pp. 18–34, 2020.
https://doi.org/10.1007/978-981-15-8760-3_2

There are quite a lot of implied volatility pricing methods derived from the BSM model. They can be classified as iteration and closed-form methods. The former aims to solve a corresponding nonlinear equation numerically and the latter seeks to find a closed-formula to model implied volatility by using at-the-money (ATM) option price as an initial point [5].

Manaster and Koehler first applied Newton-Raphson pricing method to calculate the implied volatility [4]. However, this approach is highly dependent on the starting value of the iteration, because the Newton-Raphson method itself only converges towards a local optimal solution. Especially, the existence of deep in-the-money (ITM) and deep out-of-the-money (OTM) options may lead to unsatisfied results or failure in pricing under the approach.

ITM refers to an option has a positive intrinsic value. A call (put) option is ITM if the stock's market price is above (below) the strike price. Similarly, OTM means a call (put) with a strike price higher (lower) than the market price of the stock. An OTM option has no intrinsic value. An ATM option means its strike price is identical to the market price of the stock. Though an ATM option has no intrinsic value, it can have a time value before expiration.

The deep in-the-money call/put options have strike prices much lower/higher than the stock price (e.g., $10). Their option prices are highly sensitive to the change of their stock prices instead of implied volatilities. On the other hand, the deep out-of-the-money (OTM) options are more affected by volatilities for its low intrinsic values. The Newton-Raphson pricing cannot handle such options well unless delicate initial values are selected for them adaptively [6].

On the other hand, those iterative methods with a global convergence such as bisection, Brent-Dekker method or their variants can guarantee to find implied volatilities [7]. But they may suffer from high time complexities especially when dealing with the large option datasets. Jäckel improves the iteration methods by providing a better initial guess that reduces the steps of iterations. However, it also leads to complicated implementations [8, 9].

The closed-form approximation methods are non-iteration methods and aims to find an approximated formula expression for implied volatility calculation. The main idea of the methods is to employ Taylor expansions on an at-the-money (ATM) option price point to obtain an implied volatility formula [10]. These methods can be computed extremely fast if the formula is obtained. Brenner and Subrahmanyam, Corrado and Miller, Chambers and Nawala make several attempts to optimize the approximation form of implied volatility [10–13]. However, the methods may perform poorly on out-of-the-money (OTM) options, which need some time to be profitable for option buyers [12].

The methods mentioned above only can cover European options because of its model-driven nature. To explore implied volatilities of other types options (e.g. Asian options), different implied volatility pricing models have to be built accordingly. However, it may not only face the challenge from a modeling complexity itself but also the gap between the real market and theoretical market because theoretical or even unrealistic assumptions are usually made for the sake of modeling.

During the recent two decades, the application of machine learning has been developing rapidly and actively in predicting implied volatilities [14]. Machine learning implied volatility pricing is different from the previous model-driven methods for its data-driven. Almost all machine learning models do not rely on theoretical assumptions about markets and options. Instead, they dig knowledge and learn implied volatilities from input option data and construct an implied volatility prediction function.

Malliaris and Salchenberger apply neural networks to predict S&P 100 implied volatility by using past volatilities and other options market factors [14]. Gavrishchaka and Banerjee forecast stock market volatility using support vector machines (SVM), suggesting the efficiencies of working with high-dimensional inputs [15]. Yang and Lee predict the implied volatility distribution by using Bayesian kernel machines [16]. Zeng and Klabjan design an online adaptive a primal support vector regression model to explore the implied volatility surface and realize dynamically updating support vectors to improve efficiency [17].

The machine learning approaches present unique advantages over traditional model-driven ways. First, machine learning methods can utilize more option-related features (e.g., ask-bid difference) in pricing instead of only using those few included in a pricing model. It can fully exploit the impacts of all possible variables on implied volatilities in a data-driven learning procedure. Second, the volume of data and the velocity of generating data in the financial market significantly increases with the surge of fintech. It calls for a data-driven way to exploit a large amount of data in implied volatility pricing. The machine learning approaches meet such an urgent demand. It makes it possible to derive an appropriate model by listening to data talks' and achieve 'more data, better prediction'. On the other hand, it enables implied volatilities more interpretable for the real market. It is easy interpreted as a function of a set of parameters such as option price, stock price, strike price, and other related ones in machine learning. Third, machine learning approaches can do implied volatility pricing for almost any type of options given enough available data. It brings a good generalization in pricing for different types of options and avoids the modeling complexities in building different prediction models.

However, different challenges remain for the existing machine learning approaches. Implied volatility prediction can be viewed as a 'learning-hard' problem because the nonlinearity between implied volatility and its parameters and noise from the stochastic nature of the market. How to enhance its prediction accuracy that generally suffers from a low pricing accuracy compared to the traditional model-driven ways remains unclear. However, since implied volatility is quoted often as option price in trading, the low accuracy is becoming an urgent problem, especially because more and more machine learning methods are employed in implied volatility pricing [14–16]. Besides, it remains unknown which machine learning models are 'good-fit' ones though different ones have been employed [17]. It is unknown how options in different moneyness status behave under the machine learning approach. How OTM, ITM, and ATM options react differently in machine learning pricing? Which one is more suitable for the data-driven approach? The answer to the query will unveil latent knowledge in implied volatility pricing on behalf of moneyness.

In this study, we propose novel selective learning to tackle the challenges. Such a generic learning algorithm boosts each machine learning model's prediction by providing a selective mechanism to pick high-quality data and seeking a partial but better learning result. It is particularly suitable for solving the learning-hard problem such as implied volatility prediction. Technically, it enhances implied volatility pricing by removing problematic points and potentially problematic points, namely, 'bad guys', in both training and test data, by probing learning and nearest neighborhood search. The 'bad guys' are those options that cause or potentially cause inaccurate predictions for a given machine learning model. The proposed algorithm no longer seeks the whole solutions with a mediocre performance, but a partial solution with a better performance

We apply the selective learning method to three large option datasets under six state-of-the-art machine learning models. The datasets are collected from our option acquisition software: *OptionGlean* developed for this study. The proposed selective learning boosts each model's prediction at least 33.68% averagely on behalf of moneyness. Furthermore, we identify gradient boosting (GB) as the best model for pricing among the six machine learnings models. It has leading performance over its peers in the original learning and selective learning.

We also present a new moneyness classification by grouping options as OTM, ITM, and NTM (near-the-money) for the sake of the robustness of machine learning pricing. The OTM and ITM data are the remaining data of their original data by removing shallow OTM and ITM options. The NTM data includes the shallow OTM and ITM besides ATM data. Our study demonstrates that OTM and ITM data are more suitable for machine learning pricing than NTM data. It suggests that NTM implied volatility can be more unpredictable though options trading activity tends to be high for them. Furthermore, we propose the second stage selective learning algorithm to handle the potentially problematic points dropped in selective learning. The second stage selective learning works well for OTM data under extremely randomized forests (ET). To the best of our knowledge, it is the first generic learning acceleration algorithm in pricing and will inspire more similar methods in the topic.

2   Selective Learning for Implied Volatility Pricing

Selective learning is a novel generic acceleration algorithm for any supervised machine learning. Its basic idea is to eliminate 'bad guys', the problematic points or potentially problematic points with or potentially with high prediction errors under a given machine learning model. Then the following learning is based on good-quality training and test data. In our context, the proposed selected learning gets rid of the options that may negatively affect the prediction.

Machine Learning Pricing Evaluation

Machine learning pricing is essentially a regression procedure, where implied volatility is a response variable. We use mean square error (MSE) and prediction error to evaluate a learning model's performance. The prediction error (*Err*) evaluates the performance of the model on an individual data point (option).

$$Err = \left| \widehat{IV} - IV \right| \tag{1}$$

where IV is the true implied volatility, and $\widehat{IV}$ is the prediction of implied volatility. The mean square error (MSE) represents the average performance of the model for n options. The less MSE value signifies better performance of the model.

$$MSE = \frac{1}{n} \sum_{i=1}^{n} \left(IV_i - \widehat{IV}_i \right)^2 \tag{2}$$

2.1 Selective Learning

Selective learning assumes we have enough training and test data in learning. It has three basic inputs: a machine learning model θ, training data $X = \{u_i, v_i\}_i^N$, and test data $X' = \{x_i, y_i\}_i^n$, where u_i and x_i are a training and test option, and v_i and y_i are their corresponding implied volatilities. But the implied volatilities in the test data are supposed unknown.

Unlike traditional machine learning using the whole training and test data unselectively, the proposed selective learning eliminates those problematic points and potentially problematic points in both training and test data by probing learning and nearest neighbor search. It aims to obtain good-quality training and test data that fits the model θ better. Selective learning consists of probe learning, nearest neighbor search, and data clean three main components.

Probing Learning

Probing learning is a warm-up learning procedure on the 'whole training data' to find initial problematic points under the machine learning model θ. It views training data as an entirety to perform machine learning. To conduct probing learning, we first randomly split training data into two parts: train-train X_t^{train} and train-test X_t^{test} according to the threshold ratio between the training and test data size: $\tau = \frac{|X|}{|X'|+|X|} : \frac{|X'|}{|X'|+|X|}$. For example, if $\tau = 80\% : 20\%$, then X_t^{train} and X_t^{test} will count 80% of training and 20% test samples respectively.

Problematic Point Sets. The machine learning model θ is fitted by X_t^{train} and predict implied volatility for each entry of X_t^{test}. Since all true implied volatilities are known for X_t^{test}, we can identify a set of problematic points, whose absolute errors are top-ranked (e.g., 90th percentile) entries among the absolute errors of all entries in X_t^{test}, i.e., $S_p = \{x : Err(x) \geq Err_\delta\}$, where Err_δ is δ^{th} percentile of all train-test absolute errors (generally $\delta = 90$). In our context, each point is an option, and the problematic point set includes the options with poor prediction errors under the model θ.

Nearest Neighbor Search (NNS)

The nearest neighbor search finds the potentially problematic points, which are defined as the union of the points close to each option $x \in S_p$, in the whole training data X and test data X'. For any point $x \in S_p$, the nearest neighbor search is employed to find its k nearest neighbors in the training data X according to a distance measure,

$$S_{x,k}^{train} = \{a : D(x, a) < \tau_k\} \tag{3}$$

It answers the query: 'which ones are closest neighbors to the problematic point x,' where τ_k is the distance to find the k nearest neighbors. Similarly, its potentially problematic point set in the test data can be obtained by finding k' neighbors of x,

$$S_{x,k'}^{test} = \{a : D(x, a) < \tau_{k'}\} \tag{4}$$

where k' is usually set as k. Thus, the sets of potentially problematic points in the training and test are $X_k = \bigcup_{x \in S_p} S_{x,k}^{train}$ and $X'_{k'} = \bigcup_{x \in S_p} S_{x,k'}^{test}$, respectively. We employ the Manhattan distance in the nearest neighbor search for its advantage over other distance measures in identifying effective potentially problematic points for option data.

Data Clean

This step clean and update training and test data by removing the potentially problematic points X_k, $X_{k'}$ and the problematic point set S_p $X_{clean} = X - X_k - S_p$ and $X'_{clean} = X' - X_{k'}$. The cleaned training data X_{clean} will be used to predict implied volatilities for options in the cleaned test data X'_{clean}. Since the problematic points perform poorly under the machine learning model θ and its associated points potentially to have a similar behavior under the model, it is reasonable to believe that removing them and their nearest neighbors in training and test data will enhance the following machine learning pricing. The algorithm 1 describes the details of the proposed selective learning.

Algorithm 1 Selective learning

Input:
        Training data $X = \{u_i, v_i\}_i^N$
        Test data $X' = \{x_i, y_i\}_i^n$
        Machine learning model θ
        Training data partition threshold τ
        The percentile cutoff δ
        The number of nearest neighbors k, k'
Output:
        Cleaned training data X_{clean}
        Cleaned test data X'_{clean}

// Probing learning

1. //Partition training data as train-train and train-test under a threshold τ
2. $X_t^{test} \leftarrow partitionTrainingData(X, \tau)$
3. //training machine learning model θ by train-train data
4. $\theta \leftarrow fit(\theta, X_t^{train})$
5. //Predict implied volatilities for train-test data
6. $\hat{v} \leftarrow \theta.predict(X_t^{test})$
7. // Calculate absolute error for all entries in train-test data
8. $Err = |\hat{v} - v|$
9. // Identify problematic points
10. $S_p \leftarrow FindProblematicPoints(Err, \delta)$

// Nearest neighbor search

11. // Find potentially problematic points in training and test data
12. $X_k = X_{k\prime} = \{\}$
13.   For $x \in S_p$
14. $X_k.add(NearestNeighborSearch(x, k, X))$
15. $X'_{k\prime}.add(NearestNeighborSearch(x, k', X'))$

// Data clean

16. $X_{clean} \leftarrow X - X_k - S_p$
17. $X'_{clean} \leftarrow X' - X_{k\prime}$

2.2 Machine Learning Models for Selective Learning

As mentioned before, selective learning is a generic method applied to any machine learning models. We employ six state-of-the-art models to evaluate its performance. They include k-nearest neighbor (k-NN), support vector machines (SVM), random forests (RF), gradient boosting (GB), extra tree (ET) and deep neural networks (DNN) [18–23]. The representative models range from instance learning (k-NN), kernel-based learning (SVM), bagging and boosting ensemble learning (RF, GB), and deep learning (DNN). We describe all the models except k-NN briefly as follows.

Given training option data $x_1, x_2 \cdots x_m$, and their implied volatilities $y_i \in R^+, i = 1, 2 \cdots m$, different machine learning models build their own implied volatility prediction function differently.

Support vector machines (SVM) seek to find the optimal hyperplane by solving the optimization problem about normal vector the hyperplane w:

$$min_w \frac{1}{2} w^T w + C \sum_{i=1}^{m} \max\left(1 - w^T \varphi(x_i) y_i, 0\right)^2 \tag{4}$$

The implicit feature map $\varphi(\cdot)$ is implemented by a radial basis function ('*rbf*') kernels, and the complexity tradeoff parameter C is set as 1.0 in this study. Given an option x, the prediction function is $f(x) = w^T x + b$, where b is the offset item for the hyperplane.

Random forests (RF) computes an implied volatility prediction function by aggregating the prediction functions of $|B|$ independent decision trees: $f(x) = \frac{1}{|B|} \sum_{k=1}^{|B|} h_k(x)$, where the kth decision tree, whose decision function is $h_k(x)$, is built by a set of randomly selected training samples. RF decreases the risk of overfitting of the original decision tree learning by combining the prediction functions from uncorrelated learners.

Gradient Boosting (GB) seeks an implied volatility prediction function by optimizing the prediction functions of weak learners along the negative gradient directions of the loss function. Unlike bagging ensemble learning that averages the prediction functions from the independent weak learners, the weak learns of GB are no longer independent. On the other hand, each weak learner is added sequentially to the procedure to 'boost' learning results. GB learns its prediction function in an iteration model,

$$\hat{f}_k = \hat{f}_{k-1} + r_k \frac{\partial L}{\partial f_k} \tag{5}$$

where $L(.)$ is the loss function in learning. The GB prediction function is initially formulated as the form of a weighted sum of decision functions $h_i(x)$ of the weak learners $\hat{f}(x) = \sum_{i=1}^{K} \gamma_i h_i(x)$, where the weights γ_i grow in each step when a new weak learner is introduced. It is further optimized in gradient learning. The samples are no longer equally likely to be selected to train a weak learner in GB. Instead, those with larger predation errors are more likely to be selected for training, because GB learns from mistakes committed by the weak learners in the previous iterations. As a result, GB does not demonstrate overfitting robustness as RF. There are different loss functions, but we choose the least square for its mathematical efficiency.

Extra trees standing for extremely randomized trees (ET) are an optimized RF. ET eliminates bootstrap in RF. Each subtree is built by using all samples instead of bootstrapped local samples [21]. Moreover, it splits nodes for each subtree by choosing cut-points, which are randomly specified thresholds, instead of seeking the best cut-points according to randomly selected local samples. It focuses on building an RF without bagging processing by emphasizing 'randomness'. As a consequence, ET no longer relies on the optimization procedure to identify the best cut-points in RF. Such techniques enhance each decision tree's generalization and accuracy and further contribute to the final prediction function.

Deep neural networks (DNN) DNN, an MLP with more than 2 hidden layers, is a representative of deep learning and a foundation of other deep learning models such as

DBN and CNN [23]. Its learning topology is a set of layered fully connected neurons where each learning unit is a neuron that simulates human being neuron's information process. Its learning mechanism is based on a back propagation (BP) algorithm or state-of-the-art variants (e.g. stochastic gradient descent (SGD)). It is more likely to encounter overfitting for its complicate learning topologies and gradient-based learning mechanism. In our study, in order to avoid overfitting, although we establish 8 layers and 200 neurons each layer, the dropout rate is set to 30%, which means 30% of neurons will be omitted randomly when assigning weights.

3 Results

3.1 Data Acquisition and Preprocessing

We develop option acquisition software: *OptionGlean* in python for this study. It is designed specifically to retrieve option data from Yahoo Finance (https://finance.yahoo.com) and Nasadq (https://www.nasdaq.com)). *OptionGlean* requires ticker name as input besides accepting user-specified input about options such as type, expiration, moneyness, and exchanges, *etc.* Three option datasets acquired by using *OptionGlean* include options in 2015, 2017 and 2018. We name corresponding datasets as option data 2015, 2017 and 2018, which contain 25701, 14251, and 36646 options respectively. The options can be any type of options (e.g. American options) in the market.

We separate each option dataset into three groups according to 'updated' moneyness: in-the-money (ITM), near-the-money (NTM), and out-of-the-money (OTM). Unlike traditional ways, we define OTM as call options with strike prices S > the sum of the strike price K and a threshold δ, which is selected as 50 cents in this study. i.e. $K > S + \delta$ or puts satisfying $K < S - \delta$. Similarly, ITM options are those calls and puts satisfying $S - \delta \leq K \leq S + \delta$. The near-the-money (NTM) includes at-the-money (ATM) options and the shallow out-of-the-money (OTM) and in-the-money (ITM) options.

The reason to introduce a threshold in moneyness classification is to enhance machine learning pricing generalization. Traditional classification may not match the market reality very well. A very shallow OTM or ITM option is equivalent to an ATM option because of a trade commission. Adding a threshold in option moneyness classification can make machine learning results in pricing close to the market reality. Furthermore, using NTM data rather than only ATM data can prevent the failure of machine learning because the small size of ATM data can lead to a learning failure.

Figure 1 illustrates ITM, NTM, and OTM of the three datasets on behalf of their own features via Radar plots, where calls and puts are visualized in different colors. All datasets have the key features: option price (last price), strike price, time to maturity, and stock price. The 2018 and 2017 datasets share almost the same features except for volatility. Figure 1 shows that the distributions of 2015 option data are 'more regular' than those of the 2017 and 2018 option data. It seems that that the two datasets may have more noise or outlier options from their distributions.

The call and put distributions generally are balanced for 2015 and 2017 option data. But they are imbalanced for 2018 OTM and ITM data. The OTM data has 13849 and 6501 calls and puts, but the ITM data has 4443 calls and 10553 puts. Compared to OTM and ITM data, NTM data has the least amount in each dataset. For example, 2015 NTM

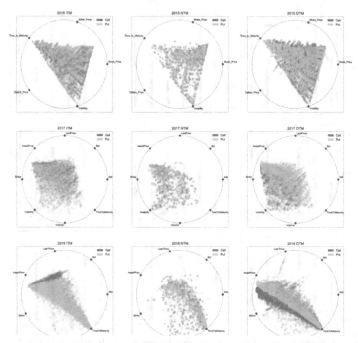

Fig. 1. The distributions of the three option datasets on behalf of OTM, ITM and NTM on a $[-1, 1] \times [-1, 1]$ plane. The 2015 option data has a fewer number of features than the 2017 and 2018 option data. (Color figure online)

data consists of 728 calls and 777 puts; 2017 NTM data only consists of 444 calls and 471 puts; 2018 NMT data has 684 calls and 618 puts.

3.2 Selective Learning in Implied Volatility Pricing

We employ the three benchmark datasets to evaluate the effectiveness of the proposed selective learning in implied volatility prediction under the six machine learning models in terms of MSE. Each dataset is partitioned as 80% training and 20% test in learning. We choose the percentile cutoff $\delta = 90$ in selective learning to identify potential 'trouble makers'. The neighborhood size in the nearest neighbor search (*NNS*) are set as $k = k' = 10$. We use the '*manhattan*' distance to search the neighbors for each problematic point for the sake of its effectiveness than other distance metrics (e.g., Euclidean distance).

Figure 2 compares the MSE values of selective learning under the six models for the OTM, ITM and NTM datasets. The MSE values of all the models under selective learning are lower or even much lower than those of original ones. This suggests the effectiveness of the proposed selective learning in implied volatility pricing. In particular, selective learning achieves impressive advantages on the 2015 OTM, ITM and NTM datasets, but almost all original models have a poor performance on them.

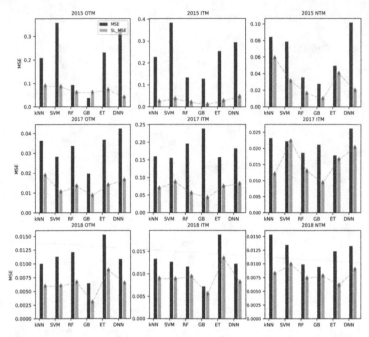

Fig. 2. The MSE comparisons of the six machine learning models using selective learning and its original models on different ITM, OTM and NTM options.

Given a dataset, we define an enhancement ratio η to evaluate the impact of selective learning under a machine learning model θ :

$$\eta = \frac{mse - mse_{sl}}{mse} \tag{6}$$

Where mse and mse_{sl} represent the MSEs before and after selective learning. Except for only one negative enhancement ratio, the other ratios on the 2015 option data are all between 18.84% and 92.24%. DNN achieves 87.74%, 84.47%, and 80.56% enhancement ratios for 2017 OTM, ITM, and NTM data respectively

The ITM and OTM options have more favorable enhancement ratios than the NTM options. We validate it by calculating the harmonic mean of the enhancement ratios as follows.

An enhance ratio matrix (η_{ij}) is generated by the six machine learning models on OTM, ITM, and NTM datasets. For example, 2015 option data will generate a 3×6 enhancement ratio matrix. We calculate the median of each row $\eta_{m,i} = median(\eta_{i.})$ and group them as a median enhancement ratio row vector $\eta_m = [0.6317, 0.8950, 0.5768]$. By doing the same calculations for 2017 and 2018 option data, we have a 3×3 matrix by grouping all median enhancement ratio vectors.

$$\eta_m = \begin{pmatrix} 0.6317 \ 0.8950 \ 0.5768 \\ 0.6077 \ 0.5585 \ 0.2644 \\ 0.4378 \ 0.2914 \ 0.2949 \end{pmatrix} \tag{7}$$

Each column in the matrix represents the median enhancement ratio for OTM, ITM, and NTM data respectively. We calculate the harmonic mean $\eta_h = \sum_{i=1}^{n} \frac{n}{1/\eta_{ij}}$ for each column to get the expected enhancement ratios on behalf of moneyness: $\eta_h = [0.5442, 0.4732, 0.3368]$. It indicates that OTM, ITM, and NTM data achieve 54.42%, 47.32%, and 33.68% average enhancement ratios in selective learning. Therefore, OTM and ITM options may be more suitable for machine learning implied volatility pricing than the NTM data. Such a result is also verified by each model's median enhance ratio, i.e., OTM and NTM datasets always have a better median enhancement ratio for each model. For example, DNN has 84.47% and 54.98% median ratios for the OTM and ITM data, but only 33.30% for the NTM data.

Besides, GB outperforms its peers in pricing no matter in selective learning or original learning for its almost the lowest MSE, though it does have a poor case for 2015 OTM data. It is not the machine learning model with the highest enhancement ratio in selective learning. But it always attains lower MSEs than others no matter in the original learning or selective learning.

In summary, selective learning can improve the performance of the machine learning models in pricing for different option data at different moneyness It enhances pricing via machine learning more efficient by dropping possible bad guys in training and test by providing a better pricing by screening a portion of high-quality options.

Checking Error Distribution in Pricing

It is desirable to check the changes in prediction error $Err = \left| \widehat{IV} - IV \right|$ distribution from the models under selective learning compared to original models. We compare the Q_1, Q_2, and Q_3 values of the prediction errors to evaluate the impacts of selective learning on each model. All Q_1, Q_2, and Q_3 values decrease after selective learning for each machine learning model. The only slight difference is probably that the decrease of Q_3 can be more obvious than those of Q_1 and Q_2. It suggests that selective learning can help to enhance the dataset with large prediction errors for its 'bad guys removing mechanism'.

Figure 3 compares median prediction errors (Q_2) of the original learning models with those of selective learning (SL) on behalf of OTM, ITM and NTM data. The selective learning Q_2 values, denoted by 'SL_error' in Fig. 3, decrease for almost all models compared to the original errors no matter the statuses of moneyness. Only DNN and RF have non-decreasing median prediction errors on the NTM 2017 dataset. Furthermore, it echoes that GB outperforms all its peers in pricing for its smallest Q_2 values under original or selective learning for all data.

Similarly, Fig. 4 compares Q_3 values of the prediction errors on behalf of OTM, ITM, and NTM datasets. All lines have an increasing slope values than their corresponding ones in Fig. 3. It means each model has a 'better' enhancement in terms of the Q_3 values than the Q_2 ones. GB still overperforms other peers in pricing though the others have quite different patterns than those in Fig. 3. It indicates that different models may be suitable for different options at different statues of moneyness. For example, the 8-layer DNN model may not work well under selective learning for some 2017 NTM options with small or median prediction errors, it has a good improvement for the options with large prediction errors in the dataset.

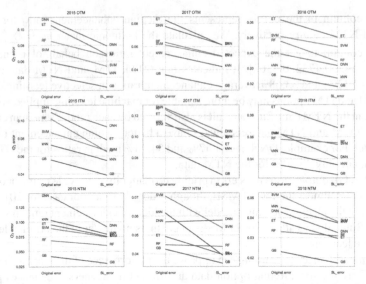

Fig. 3. The comparisons of the median prediction errors (Q_2) of the six machine learning models under selective learning and its original models on different option datasets.

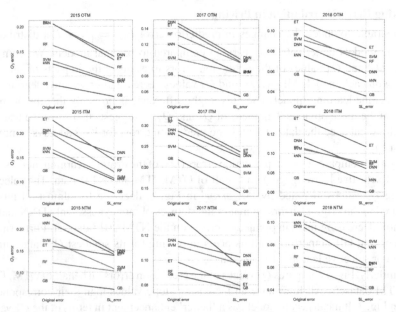

Fig. 4. The Q_3 comparisons of the prediction errors of selective learning and original learning on different ITM, OTM, and NTM data.

3.3   The Second Stage of Selective Learning

The proposed selective learning identifies and gets rid of potentially problematic points $X_{k'}$ in test data X' besides the problematic points and its nearest neighbors in training data. It brings a new learning mechanism by fitting a learning machine with high-quality data to find a partial but better solution instead of a whole solution. But sometimes we may not be able to simply disregard the 'bad-guys' in learning. They can be urgent options in pricing.

We propose the secondary stage of selective learning to handle this situation via a local training dataset construction technique. The basic idea is to generate a local training dataset S_x for each $x \in X_{k'}$ by searching its m nearest neighbors in the training data X based on a distance (e.g. 'manhattan' distance). The local training dataset of x consists of its nearest neighbors, i.e. proximity points, which will have more advantages to predict its implied volatility than a genetic training dataset. The local training dataset S_x is employed to fit a machine learning model before predicting the implied volatility of x.

In our experiment, we set $m = 20$ to build the local training dataset for each potentially problematic point in the test data. We drop DNN in the secondary stage of selective learning (2nd SL) because it is not suitable for small datasets. Figure 5 compares the MSEs of original learning and selective learning for all potentially problematic points of each test dataset. Each model is colored by the color of the OTM, ITM and NTM data they are applied to. The points above the dashed-curve in each scatter plot indicate their original MSEs are lower than the selective learning MSEs (SL-MSEs). It means the 2nd stage selective learning fails on them. The points on the curve indicate the 2nd stage selective learning has the same-level performance as the original one or only slight improvements. The points below the curve are those whose 2nd stage selective learning results are better than their ones and save the 'bad guys' that are dropped in the first stage of selective learning.

The OTM data can be more favorable than the others in the 2nd stage selective learning (2nd SL), though different data demonstrate quite different patterns. The 2nd SL still slightly improves performance of the potentially problematic points. For example, ET and SVM both achieve decent improvements for the 'bad guys' of OTM and ITM in the 2015 option datasets. However, only ET attains a good performance on the 'bad guys' of 2017 OTM data. Alternatively, ET and RF both have good pricing performance for the 'bad guys' of 2018 OTM, NTM and ITM data.

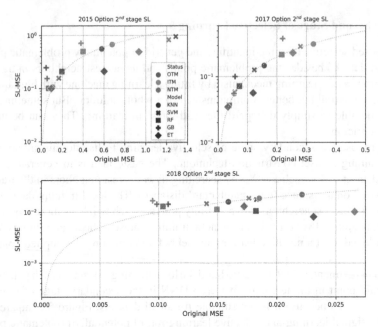

Fig. 5. The comparisons of different MSEs of the potentially problematic points in the test data before and after the secondary state of selective learning (2nd SL) on behalf of OTM, ITM, and NTM.

4 Conclusion

We propose a novel generic selective learning approach to enhance machine learning implied volatility pricing. It overcomes the weakness of implied volatility pricing via machine learning by boosting each model's prediction accuracy on different option data on behalf of moneyness. It somewhat challenges the existing learning philosophy by proposing a new learning concept, that is, fit a learning machine with the most suitable high-quality data to seek an approximate partial but better learning results. It is particularly good for solving the difficult learning problems such as implied volatility pricing by machine learning.

This study finds that OTM and ITM data have a favorable status in machine learning implied volatility pricing than NTM data besides identifying GB as a robust pricing model. It is possible to take advantage of the findings to apply our approach to price deep OTM and ITM options that are challenging the traditional model-driven approaches [6, 7]. Furthermore, we find OTM data can be more suitable for the 2nd stage selective learning under ET.

An interesting concern can be about the possible overfitting risk of the proposed selective learning. The generalization of the learning machine under selective learning will increase or decrease? The answer is that selective learning will contribute to decreasing overfitting risk because of its approximate learning scheme. The implied volatility prediction function generalization is optimized actually because of high-quality training and test data.

Our on-going work is to optimize selective learning pricing by identifying possible prototypes of OTM, ITM, and NTM through traditional model-driven approaches so that the potentially problematic point search can be augmented with prior knowledge. It is likely to conduct selective learning for some specific types of options such as deep OTM/ITM to further polish it and observe their behaviors. At the same time, we plan to integrate selective learning in genetic algorithm oriented implied volatility pricing besides other specifically designed deep learning models for implied volatility prediction [24–26].

References

1. Black, F., Scholes, M.: The pricing of options and corporate liabilities. J. Polit. Econ. **81**(3), 637–654 (1973)
2. Merton, R.: Theory of rational option pricing. Bell J. Econ. Manag. Sci. **4**(1), 141–183 (1973). https://doi.org/10.1142/9789812701022_0008
3. Bianconia, M., MacLachlan, S., Sammon, M.: Implied volatility and the risk-free rate of return in options markets. North Am. J. Econ. Finance **31**, 1–26 (2015)
4. Manaster, S., Koehler, G.: The calculation of implied variances from the Black-Scholes model: a note. J. Finance **37**(1), 227–230 (1982)
5. Jiang, G.J., Tian, Y.: The model-free implied volatility and its information content. Rev. Financ. Stud. **18**(4), 1305–1342 (2005)
6. Andersen, L., Andreasen, J.: Jump-diffusion processes: volatility smile fitting and numerical methods for option pricing. Rev. Derivat. **4**(3), 231–262 (2000)
7. Benninga, S.: Financial Modeling. MIT Press, Cambridge (2000)
8. Jäckel, P.: By implication. Wilmott **26**, 60–66 (2006)
9. Jäckel, P.: Let's be rational. Wilmott **2015**(75), 40–53 (2015)
10. Li, M.: Approximate inversion of the Black-Scholes formula using rational functions. Eur. J. Oper. Res. **185**(2), 743–759 (2008)
11. Brenner, M., Subrahmanyan, M.: A simple formula to compute the implied standard deviation. Financ. Anal. J. **44**(5), 80–83 (1988)
12. Corrado, C., Miller, T.: A note on a simple, accurate formula to compute implied standard deviations. J. Bank. Finance **20**(3), 595–603 (1996)
13. Chambers, D., Nawala, S.: An improved approach to computing implied volatility. Financ. Rev. **36**(3), 89–100 (2001)
14. Malliaris, M., Salchenberger, L.: Using neural networks to forecast the S&P 100 implied volatility. Neurocomputing **10**(2), 183–195 (1996)
15. Gavrishchaka, V., Banerjee, S.: Support vector machine as an efficient framework for stock market volatility forecasting. CMS **3**(2), 147–160 (2006)
16. Yang, H., Lee, J.: Predicting a distribution of implied volatilities for option pricing. Expert Syst. Appl. **38**(3), 1702–1708 (2011)
17. Zeng, Y., Klabjan, D.: Online adaptive machine learning based algorithm for implied volatility surface modeling. Knowl.-Based Syst. **163**, 376–391 (2019)
18. Friedman, J.H.: Greedy function approximation: a gradient boosting machine. Ann. Stat. **29**, 1189–1232 (2001)
19. Breiman, L.: Random forests. Mach. Learn. **45**(1), 5–32 (2001)
20. Shawe-Taylor, J., Cristianini, N.: Support Vector Machines and Other Kernel-Based Learning Methods. Cambridge University Press, Cambridge (2000)
21. Geurts, P., Ernst, D., Wehenkel, L.: Mach. Learn. **63**(1), 3–42 (2006)

22. Hornik, K.: Approximation capabilities of multilayer feedforward networks. Neural Networks **4**(2), 251–257 (1991)
23. Diederik, K, Ma, J.: Adam: A method for stochastic optimization. arXiv preprint arXiv:1412. 6980 (2014)
24. Abdelmalek, W., et al.: Selecting the best forecasting-implied volatility model using genetic programming. Special issue in Intelligent Computational Methods for Financial Engineering. J. Appl. Math. Decis. Sci. (2009). Intelligent Computational Methods for Financial Engineering
25. LeCun, Y. Bengio, Y., Hinton, G.: Deep Learning. Nature **521**, 436–444 (2015). https://doi. org/10.1038/nature14539
26. Sezer, O., Ozbayoglu, A.: Algorithmic financial trading with deep convolutional neural networks: time series to image conversion approach. Appl. Soft Comput. **70**, 525–538 (2019)

A High-Performance Basketball Game Forecast Using Magic Feature Extraction

Tiange Li[1] and Henry Han[2,3,4](✉)

[1] The Gabelli School of Business, Fordham University, Lincoln Center,
New York, NY 10023, USA
tli115@fordham.edu

[2] Computer and Information Science, Fordham University, Lincoln Center,
New York, NY 10023, USA
xhan9@fordham.edu

[3] Department of Innovation and Entrepreneurship, Long Island University,
Greenville, NY 10458, USA

[4] Academy of Plateau Science and Sustainability, School of Computer Science and Technology,
Qinghai Normal University, Xining 810008, China

Abstract. It remains a challenge to predict real basketball game in a high-performance manner. In this study, we develop a magic feature extraction algorithm and its supporting techniques to tackle this problem. Our model achieves 79% stable accuracy for 2019's NCAA and successfully predicts the black horses in the early rounds. To the best of our knowledge, this is the first systematic work to address NCAA basketball game prediction and will inspire the future work in this area.

Keywords: Sports analytics · Feature selection · Genetic programming (GP)

1 Introduction

With the surge of all kinds of sports data, sports analytics is becoming a rising field in business analytics [1]. The goal of sports analytics is to predict correctly the results of games or to evaluate the relative advantage among teams and therefore facilitate relevant decision-making processes. Generally, there are two aspects of sports analytics: on-field analytics and off-field analytics. Off-field analytics is a 'global-level' analytics that focuses on the global business sides of sports. It aims at enhancing a sports organization's ticket sale, fan engagement, team winning ratios or other profitable goals in the long run. It digs the business value for a specific game and builds decision-making strategies to increase a sports business' profitability. In contrast, on-field analytics is 'local-level' analytics that aims to improve the accuracy of prediction of games using relevant team or coach data. The data it depends on is primarily historical statistics, showing the past performance of players and teams and also indicating possible trends.

Basketball game prediction is a typical and challenging area in on-field analytics. While models based on off-field data may achieve some satisfying result, for example,

© Springer Nature Singapore Pte Ltd. 2020
H. Han et al. (Eds.): IDMB 2019, CCIS 1099, pp. 35–50, 2020.
https://doi.org/10.1007/978-981-15-8760-3_3

models from sports gambling or those models considering stochastic processes, there are few good models and practical methods to predict results accurately based on mere on-field data [2]. On-field data plays an essential role to determine the outcome of a game than off-field data because of uncertainty in basketball competitions.

Quite a few research efforts have been made recently in on-field analytics. Goldsberry proposed a new way to quantify the shooting range of NBA players as well as defensive skills [3, 4]; Cervone *et al.* proposed a framework using player-tracking data to assign a point value to each moment of possession according to a novel expected possession value (EPV) [5]; Cervone *et al.* also proposed a method that combines various techniques and models to enable a richer quantitative characterization of defensive performance [6]; Sailofsky *et al.* examined the NCAA statistics and pre-draft player factors to identify the best players [7]; Skinner and Guy applied a network-style model to learn player skills and predict team performance [8]; Zou produced a metric called regularized adjusted plus-minus (RAPM) to measure attributes of basketball previously thought unmeasurable such as defensive contribution [9]; Stančin and Jović applied statistical methods to analyze the most important factors to a winning team in a whole season [10]; Hamdad *et al.* aim to focus on weaknesses of a team using data mining methods [11]. Xin *et al.* proposed a model to cluster similar team members and capture their interactive network [12].

Almost all the studies deal with evaluating a team's or a player's performance from different aspects and only one addressed the winning or losing result and its relation to regular on-field factors. This is probably because basketball games constitute of many contingencies: at each time point, players can decide what movement he would do, also, the interactions between players are full of uncertainties. Thus, compared to winning or losing results, evaluating a team's or a player's performance is more stable and meaningful. However, the most important question can be 'how to predict the game outcome from on-field data describing the uncertainties in competitions?'.

It is a rarely investigated, important, and challenging topic in on-field analytics or even machine learning because no dominating informative features available in on-field data can predict game results well, even though attributes are measured correctly (e.g., shooting percentage and defensive efficiency). The on-field features may be important to evaluate a general team or players' performance statistically. But none of them may matter a lot to predict the outcome of a specific game. Moreover, redundant features present a hurdle in prediction because it is still unclear how to extract useful features to capture essential data characteristics and evaluate their performance in prediction. On the other hand, the size of the on-field data is usually not big enough to build an effective and reliable prediction model. The sample size might also make the final predictions highly contingent and bring a generalization issue besides result interpretation. However, a good solution to the problem will not only advance our understanding of the impacts of different features on the game outcome but also, unveil the latent special characteristics of the on-field data from a machine-learning perspective.

In this paper, we investigate this problem by predicting the results of 63 games in 2019 NCAA Men's Basketball Tournament using on-field data via a novel magic feature generation approach. Compared to NBA game prediction, the NCAA competition that is held among universities adds extra difficulty in prediction. This is mainly because team players in each university may keep changing every year, resulting in unstable team performance and challenges in prediction. We have the following major contributions in this study.

We build a robust prediction model that is based on a novel magic feature extraction algorithm developed in this study. Our model, which stacks several base-classifiers together, achieves 79% stable accuracy for 2019's NCAA and successfully predicts the black horses in the early rounds. The proposed algorithm enlarges the number of features, transforms the whole dataset using genetic algorithms (GA), and then reduces the uncorrelated features in the pool with supervised feature selection. Results show that most generated 'genetic features' are given large weights when predicting games, which means they have possibly dig into the dataset and reveal hidden correlations between features. Besides, we append the previous features in past seasons to the following several years of data to capture more essential time-series features and potential trends. Furthermore, we make several single-game analyses, to reveal how features contribute to the result and evaluate the relative performance level of two teams. To interpret prediction results better, we employ *Shap* values to explain model outputs besides visualizing features interactions. To the best of our knowledge, this is the first systematic work to address NCAA basketball game prediction.

2 Methods

We introduce data preprocess, proposed magic feature extraction, and its related supporting techniques in this section.

2.1 Data and Preprocessing

The dataset is released by March Madness Competition held by Deloitte, containing 1112 historical games of NCAA Division I Men's Basketball Tournament from 2012 to 2018 [13]. For each game, there are 108 columns in the dataset, 54 same features for team 1 and team 2 respectively, where team 1 stands for the winning team. The original data does have missing items because poll rankings record only top 25 teams and coaches. To handle data missing, we filled the blank for other teams with 100 to calculate the difference. At the same time, we added a binary feature to indicate whether the ranking previously existed or not.

The original features in the dataset include basic competition information, team performance technical statistics, coach experience, and poll ranking information. The basic competition information refers to competition seed, season, score, location, and hostname; The team performance technical statistics mainly consist of shooting percentage on two points field goals, averaged opponent's steals, offensive efficiency, and estimated possessions; The poll ranking information includes ranks for teams and coaches. The coach experience is represented by the number of NCAA Tournament appearances at current school and career overall number of losses. We drop 8 unrelated features such as location, hostname, season, and similar ones for the sake of preprocessing.

To effectively analyze the dataset based on games, we preprocessed data by generating meta-features from the original features to represent the relative performance of the two teams in each game. Namely, all the meta-features are represented by the difference or ratio of the original features of two teams. The original number of records are kept for reference. We calculated the difference between a team and its averaged opponents for box score features; and the win-loss percentages for the coach experience feature. Besides, we computed a general box score for each team, which is the sum of the score in different point fields as well as opponent's blocked shots. After generating 22 meta-features and dropping the original 8 features, we totally have 68 features for the preprocessed dataset.

2.2 Magic Feature Extraction

Pretraining. We first applied four state-of-the-art classification models, namely, extra trees, *catBoost*, neural networks, and bagging logistic regression to the preprocessed dataset [14]. However, all the accuracy scores were not satisfying, and they fluctuated to a large degree instead of converging to a specific value. In addition, according to the permutation of feature importance, classifiers demonstrate a strong dependence on largely different features, which offers little insight into the result explanation and further processing. The poor classification performance is probably because few informative features are extracted for learning. Lots of features might be ignored or given little weight by the classifiers even though the meta-features are added to the data. It is desirable to have some genetic features that capture the essential characteristics of data that contribute to learning and prediction.

We propose a novel magic feature exaction approach to generate genetic features to enhance baseball game prediction. The magic features are defined as generic features constructed from genetic programming (GP) rather than the original features or meta-features [15]. The key idea of magic feature extraction is to generate a set of genetic features via symbolic classifications and regressions driven by genetic programming (GP).

The genetic programming (GP) algorithm receives multiple variables and a target and it tries to summarize the relationship using detected underlying mathematical expression. It begins by building a bunch of naïve random formulas and initializing them with a stochastic optimization process. In each generation, the fittest individuals are selected and help to further evolve the population. We generate one genetic feature each time in the evolution procedure. After a new genetic feature is generated, the dominating original features it relies on will be removed, while the supplementary original features that appear in the expression would still be kept. The reason for the removal is to prevent feature similarity or even the same features generated over time. The same process would be applied again after removing, the next feature would be generated to generalize the remaining dataset.

The generating process will be repeated several times until one of the two stop conditions are satisfied. The first is the percentage of original features that new genetic feature relied on is larger than a cutoff $\alpha = 0.3$, which indicates there are few overwhelming features and many remaining features contribute altogether to the result; The second is more-than one "fake features", i.e., the original features, are selected into the new genetic features. In our magic feature selection process for the preprocessed dataset, we found that 3 original features were identified as "fake features" in the 11th feature generating process, and the percentage of relied features α is 25.6%, so we chose to stop the process.

To extract more informative features, we create an extra genetic feature using regression: the target feature for fitting was changed to a continuous score difference from the previous binary win/loss classification. Besides, the linear discriminant analysis (LDA) feature of the remaining dataset was also generated in case the extra genetic feature is useless [14].

2.2.1 Magic Feature Extraction Insights

Several insights can be grasped from the magic feature extraction process. First, the most important features are the team's winning probability and coach's ranking, which is consistent with the results from extra tree classification. Second, the ranking of feature importance is similar to the human experience. For example, offensive efficiency is more important than defensive efficiency; All the generated percentage fields are more important than their corresponding absolute number. However, some 'rules' obviously counter common sense and may need to be further explored. For example, a team's pre-season ranking is more important than its final ranking. Furthermore, the generating process largely depends on the coach's data rather than the on-field performance of the team. It seems the averaged opponents' performance level seems to be even more important than that of the team itself.

2.2.2 Outstanding Original Feature Identification

Although genetic features were already generated to transform the whole dataset, we still want outstanding original features to be kept. Thus, the Boruta algorithm was chosen to be applied to serve for this purpose [16]. The core of Boruta is to select all relevant features instead of only the non-redundant ones. Boruta determines relevance by comparing the relevance of the real features to that of those random probes. It creates corresponding 'shadow' features, whose values are obtained by shuffling values of the original ones across different samples. Then it performs classification using all features and computes the importance of all features [16].

The Extra Tree classifier was selected to be the base classifier of the Boruta algorithm for its best performance in the pre-training phase. We used the maximum shadow features scores (e.g. 100) as the threshold in deciding which real features are doing better than the shadow ones. With this harsh feature selection process under 200 iterations, 26 features were selected from the 68 original features. Among the 26 selected original features, most of them are related to coaches. "Fake features" kept by Boruta algorithm are exactly the same as the "fake features" kept by genetic algorithm, justifying the previous selection in the magic features generating process. The final dataset consists of the 12 genetic features, LDA major component feature, and the 26 original features, 39 features in total. Figure 1 justifies the genetic features by feature importance, where the best accuracy on validation set is around 74%.

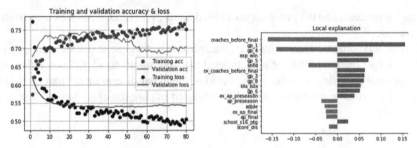

Fig. 1. The left plot shows baseline model performance on training and validation dataset and the right plot shows feature importance.

2.3 Outlier Detection Before Learning

In basketball games, the most exciting part is that there will always be unexpected black horses. These black horses would produce an extra challenge for reliable machine learning models. To improve the performance of the model, we did an outlier detection in advance applied to both training and test dataset and mark data points that are "few and different".

We employed Isolation Forest to do the outlier detection before learning [17]. This algorithm aims to detect isolated anomalies instead of profile normal points. an isolation tree is constructed to isolate every single instance. According to isolation forest, anomalies are isolated closer to the root of the tree; whereas normal points are distributed at the deeper end of the tree. Only points with shorter leaf paths are considered since they are more likely to be anomaly.

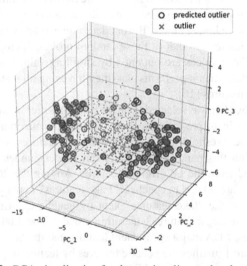

Fig. 2. PCA visualization for detected outliers and real outliers

Figure 2 illustrates the detected outliers and real outliers, where most of the real outliers were not detected considering game results. While most of the detected outliers distributed at the margin of the cluster, real outliers reside in the center. One of the explanations is that the anomalies being isolated is probably because of its extreme value, however, considering that the game is represented by the difference of two teams, the more extreme the value is, the more predictable the game would be. The outliers would more possibly appear where most of the features (difference) are at the median. The good news is that, the model detected correctly data points at the right bottom side of the cluster, and that the detected anomalies distributed more densely at this area in testing data compared to training data. Thus, we would keep the anomalies predicted by the model and add it as a new feature.

2.4 Historical Data Handling

Season constitutes a very important part of predicting results. A team's level is not only based on the current season's performance, but it is also based on its previous performance. This is because (1) team members change every year; (2) it is important to judge whether the team advance or retrograde; (3) data in the earliest year may not be informative to predict the result of the latest years. Therefore, we implement the sliding windows method to make use of the previous data record. Figure 3 illustrates the structure of the sliding window method.

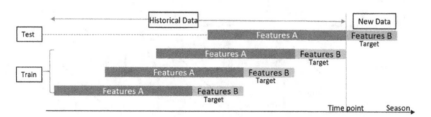

Fig. 3. Detailed structure for the sliding window method

The features used for predicting consist of two parts: all features for the past and features for this season. Features A in Fig. 4 is constructed using historical data and target and Features B is constructed using "new data" for this time point. In this case, we used data in the last three years to predict this year's target, together with this year's data. The past features consist of four parts, seed, ranking, offensive/defensive efficiency and PCA of on-field performance. The resulting probability of prediction using these features is concatenated to the transformed dataset.

2.5 Learning Model Selection and Ensemble

Base Learning Models. We selected six state-of-the-art learning models as base classifiers to evaluate the effectiveness of our magic feature selection in game prediction. They include Logistic Regression (LR), Extremely Randomized Trees (EX), and three boosting methods: Xgboost (XGB), LightGBM (LGBM), and Catboost (CATB), and neural networks (MLP) [14, 18–21]. Table 1 shows their performance in terms of AUC values under training, validation, and testing and Fig. 4 shows the ROC plots of the six base learning models.

Table 1. Training, validation, testing accuracy and testing log loss

	LR	EX	XGB	LGBM	CATB	MLP
Training AUC	0.744	0.796	0.825	0.839	0.794	0.759
Validation AUC	0.731	0.717	0.732	0.725	0.731	0.724
Testing AUC	0.716	0.742	0.729	0.727	0.765	0.728
Testing log loss	0.519	0.530	0.541	0.532	0.542	0.535

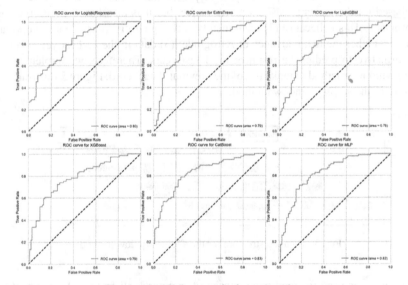

Fig. 4. The ROC curves for base classifiers

Model Ensemble. We used voting stacking methods to ensemble the base models. The voting method is set to "soft", which means predicting the class label based on the argmax of the sums of the predicted probabilities. The weights of voting are set to be equal to their accuracy score of cross validation. After that, we used an extra stacking transformation to generate meta features for the meta classifier. We made a change to stacking, namely, changing the input of meta classifier to a probability instead of binary classification predicted by the base models. This is because we want the meta classifier to learn from the probability distribution for each base classifier and make proper decisions, not to just sum them up as in voting classifier. The base models used for stacking are Logistic Regression, *Catboost* classifier and the previous voting classifier. Table 2 illustrates the ensemble model performance in terms of learning accuracy and Fig. 5 shows the ROC curves of the voting and stacking methods.

Table 2. Learning accuracy and testing log loss for ensemble models

Ensemble	Training AUC	Val AUC	Testing AUC	Testing log loss
Voting	0.808	0.742	0.739	0.515
Stacking	0.815	0.756	0.772	0.479

From the result, we can see that both voting and stacking classifiers achieve good improvements compared to previous base classifiers. This might result from the features base classifiers depend on vary a lot. Classifiers make use of different features. Of these two ensemble models, the final stacking performs a better log loss and testing accuracy. Considering the number of samples in this dataset is relatively small, the metrics log loss is more reliable than accuracy, since there may be some contingency in accuracy prediction. Stacking can positively make use of previous probability prediction made by base classifiers and therefore reduce the punishment of log loss to the maximum degree.

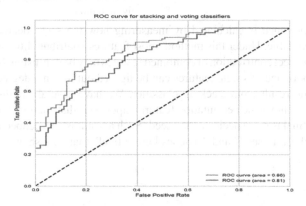

Fig. 5. ROC curves for voting and stacking classifiers

3 Results

To understand the overall data structure better, we plotted the clusters for features and data distribution considering different game results. Figure 6 shows feature clusters after PCA 2D visualization. Actually, the most important features, ranking, and primary genetic features point to the lower-left direction. Other genetic features stand in the direction of the composition of team and coach performances in vector space. Besides, previous features are really important since they point to a unique upper left direction. Figure 7 shows all features in a circle and plot the distribution of data according to these feature positions. Green points represent winning and blue points represent losing. From the result, the winning games are distributed closer to the generated genetic features and previous season features seem to play important roles to determinate win or lose.

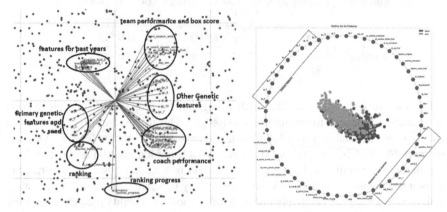

Fig. 6. The PCA visualization for feature clusters.   **Fig. 7.** Data overall distribution labeled with winning and losing categories (Color figure online)

Feature importance is calculated by measuring how the score decreases when a feature is not available. Under this method, most features contribute little to the overall score and the undirect contribution of combined features may be confusing. Nevertheless, the relative importance of single features can be figured out to some degree. The figure below shows the result from stacking classifier. We have several findings about the features. Figure 8 shows the permutation feature importance distribution. It shows coach ranking before final is more decisive than seed. Such a result is also echoed in our Target plot and PDP plot for "seed" and "coaches before final" (data not shown for the space limitation).

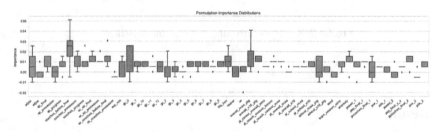

Fig. 8. Permutation feature importance distribution: coach ranking before final is more decisive than seed

3.1 SHAP Value Justification

A more indicative method to analyze features and predictions is to use Shap (Shapley Additive exPlanations) values. However, we only reaffirm some of our findings rather than drawing conclusions since the calibration may be different. Taken from game theory, Shap value of a feature interprets its influence on the model by calculating all combinations with and without this feature. A much simpler model is trained on the top of the original model so that it approximates the original predictions. Instead of explaining the whole complex model, Shap values try to explain how the model behaved for each data point [18]. Besides, since the order of features being seen is important to how model assigns their values, this process is repeated in every possible order, to make sure that features are nicely compared. Shap value is given by the expression: $\emptyset_i(p) = \sum \frac{|S|!(n-|S|-1)!}{n!}(p(S \cup i) - p(S))$ where p means the prediction, i is the feature being evaluated and S means the set of all possible combinations except for feature i. Of a high level, this equation is basically rooted in calculating the importance of feature i by the difference of the prediction with and without it.

Figure 9 left-plot shows feature contribution to the output using Shap value. This figure is more reasonable, if compared to the previous feature importance Figure, because it considered feature interactions. Shap dependence plot show how the model output changes by feature value. Each dot is a game, and the vertical dispersion at a specific value results from interaction inside. The feature for coloring (on the right bar) explains the main drive for this dispersion. From these graphs, we can reaffirm our process of feature generating and have clearer intuitions about features interaction.

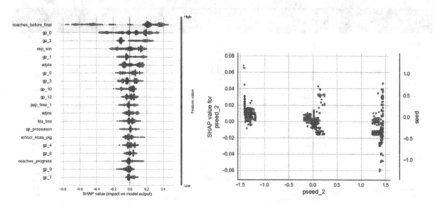

Fig. 9. Shap summary plot and dependency plot (Color figure online)

For example, Fig. 9 right-plot shows that the main drive for the dispersion of previous year seed's impact on model output is this year's seed, which is very reasonable. The points on the upper left corner says, team1's seed in the last year was largely smaller than team2's, however, this year's team1's seed is largely bigger than team2's. Points on the lower right corner is on the contrary. These are two group of points having the largest interaction. From another angle, when this year's seed difference is relatively large (pink), the impact of previous seed difference is majorly around 0. In other words, our model successfully captured the trend of a team's ranking.

3.2   Game Result Analysis

Figure 10 shows our final game prediction and the probability of predicting correctly for each game. Games with red box are final wrong predictions. Games with crosses following are wrong predictions without using previous seasons features. For final prediction, of the 63 real games in 2019, we have 51 correct prediction, with overall accuracy 81.0%. Note that we have no errors in predicting "sweet sixteen", which is a robust evaluation checkpoint for different models. For prediction without previous seasons features, the accuracy is 70.0%, with 19 wrong predictions. On one hand, this difference shows the importance of the sliding window method; on the other hand, this difference is resulted from the "black horses" in the games. They may not be real black horses, but they are definite black horses for the model. To be more specific, Michigan St, Purdue, and Auburn, they caused continuously wrong predictions if no previous seasons features used.

By adding them, the model can either evaluate its stability or capture time trends. In the final prediction, the model corrected several errors and successfully predicted these "black horses" in their early rounds, even though the predicted probability seems quite uncertain. For example, it says Texas Tech would win Gonzaga at probability 0.51 and Auburn would win Kansas at 0.52. In the next step, we would take a simple look at whether the model predicted so coincidentally and how features contributes to these predictions.

Fig. 10. The final prediction for NCAA 2019 with and without previous seasons' data (Color figure online)

Figure 11 shows single game analysis for four correct predictions, Auburn vs Kansas, Virginia vs Texas Tech, Michigan vs Texas Tech and Virginia vs Oregon. The first game Auburn vs Kansas turned out to be a Blackhorse game for Auburn, we can see that "anomaly" plays an important role in our prediction. "Anomaly" contributes the most to push up the predicted value, while all main features, e.g. primary genetic features, coach rankings, are pushing down the base value. Without "anomaly", the winner would be predicted otherwise. The second game in the Figure is the champion battle. We can get insights from the analysis and evaluate team levels. For Virginia, features that tend to increase its winning probability almost cover all of the dimensions, including coach, team ranking, offensive efficiency, estimation efficiency and genetic features. Compared to Virginia, room for improvement for Texas Tech is larger. The main drive to predict Texas Tech winning is "coach progress", which suggests that Texas Tech performed satisfactorily or even amazingly in this season (compared to Virginia). The third game Michigan vs Texas Tech shows a battle between teams of similar level. The seed for Michigan is 2 and the seed for Texas Tech is 3 this year. With all team statistics being similar, coach ranking would become an important reflection for the result. Also, previous season features cannot be ignored in reducing the winning probability for Michigan. The fourth game is for contrast, which shows a game between teams with huge level discrepancy. We can see that nearly all features are pushing up Virginia's winning probability (Fig. 11).

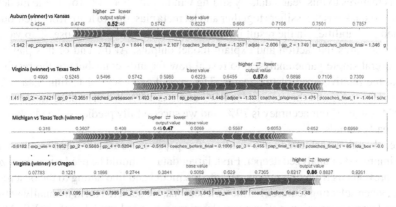

Fig. 11. Single games analysis for Auburn vs Kansas and Virginia vs Texas Tech. The number means the probability of the first team defeating the second game

We also checked the optimal discrimination threshold for the stacking (Fig. 12). Discrimination threshold is the threshold value at which the 'wins' class is chosen over the 'losses' class when predicting, and it is set to 0.5 generally. But this probability can be adjusted to adapt to the sensitivity to false positives or false negatives. From the graph, we see the visualizer is tuned to find the optimal F1 score with higher recall rate, which is annotated as a threshold of 0.35. Therefore, the predicted probabilities that just over 0.5 were not coincidentally to be "win". If we move this threshold, we would get an extra credit for game North Carolina vs Auburn.

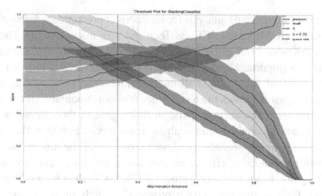

Fig. 12. Threshold plot for stacking classifier, showing precision, recall, f1 and queue rate at different threshold.

4   Conclusion and Discussions

In this paper, we propose a novel magic feature extraction algorithms and support techniques to predict basketball game results. In the feature generating section, we apply genetic algorithms (GA) to transform original dataset, Boruta algorithm to select outstanding original features, isolation forest to do anomaly detection and concatenate previous features to this year's data by sliding window method [14–16]. In the modeling section, we stack base classifiers together and let meta classifier to learn from their output predicted probabilities. In the result interpretation section, we mainly use Shap value to visualize features interaction and explain model outputs. At the end of this paper, we make several single game analyses, to reveal how features contribute to the result and evaluate the relative level of two teams. The overall accuracy for testing data is 77% with log loss 0.47, and the area under ROC curve is 0.89. When applied to this year's real NCAA games, our accuracy is 79%, and we successfully predicted the black horses in their early rounds.

Even though we achieved good results in this paper, there are still some problems that can be improved or analyzed deeper. First, larger dataset should be used since the size of dataset prevents us from delving deeper into the model and fine tuning to the best result. Second, when selecting features for the past seasons, we selected them manually without machine learning techniques. For example, more advanced model such as LSTM can be applied to extract time sequence impact along with entropy analysis [22–25]. Third, even though we find that the optimal threshold should not be 0.5, we made no changes but some qualitative evaluations. Fourth, we expect the predicted winning probability should have a linear relationship with real score difference, however, it is not the case in our essay. Further research can be done to fill this gap.

References

1. Sha, L., et al.: Interactive sports analytics: an intelligent interface for utilizing trajectories for interactive sports play retrieval and analytics. ACM Trans. Comput. Hum. Interact. **25**, 1–32 (2018). https://doi.org/10.1145/3185596

2. Percy, D.F.: Strategy selection and outcome prediction in sport using dynamic learning for stochastic processes. J. Oper. Res. Soc. **66**(11), 1840–1849 (2015)

3. Goldsberry, K.: CourtVision: new visual and spatial analytics for the NBA. In: MIT Sloan 6th Annual Sports Analytics Conference (SSAC 2012), pp. 1–7, 2–3 March 2012

4. Franks, A., Miller, A., Bornn, L., Goldsberry, K.: Characterizing the spatial structure of defensive skill in professional basketball. Ann. Appl. Stat. **9**(1), 94–121 (2015)

5. Cervone, D., D'Amour, A., Bornn, L., Goldsberry, K.: A multiresolution stochastic process model for predicting basketball possession outcomes. J. Am. Stat. Assoc. **111**, 585–599 (2014). https://doi.org/10.1080/01621459.2016.1141685

6. Cervone, D., D'Amour, A., Bornn, L., Goldsberry, K.: POINTWISE: predicting points and valuing decisions in real time with NBA optical tracking data. In: 8th Sloan Sports Analytics Conference (2014)

7. Sailofsky, D.: Drafting errors and decision making theory in the NBA Draft. Master of Arts Thesis, Brock University (2018)

8. Skinner, B., Guy, S.J.: A Method for using player tracking data in basketball to learn player skills and predict team performance. PLoS ONE **10**(9), e0136393 (2015)

9. Zou, S.: Open source data science pipeline for developing "Moneyball" statistics in NBA Basketball (2020). https://basketball-analytics.gitlab.io/rapm-data/open-source-data-nba.pdf

10. Stančin, I., Jović, A.: Analyzing the influence of player tracking statistics on winning basketball teams. In: 41st International Convention on Information and Communication Technology, Electronics and Microelectronics (MIPRO) (2018)

11. Hamdad, L., Benatchba, K., Belkham, F., Cherairi, N.: Basketball analytics. Data mining for acquiring performances. In: Amine, A., Mouhoub, M., Ait Mohamed, O., Djebbar, B. (eds.) CIIA 2018. IAICT, vol. 522, pp. 13–24. Springer, Cham (2018). https://doi.org/10.1007/978-3-319-89743-1_2

12. Xin, L., Zhu, M., Chipman, H.: A continuous-time stochastic block model for basketball networks. Ann. Appl. Stat. **11**(2), 553–597 (2017)

13. March madness competition 2019. https://www.gabelliconnect.com/announcements/basket ball-data-analysis/

14. Hastie, T., Tibshirani, R., Friedman, J.: The Elements of Statistical Learning: Data Mining, Inference and Prediction. Springer, Cham (2009). https://doi.org/10.1007/978-0-387-84858-7

15. Poli, R., et al.: A Field Guide to Genetic Programming (2008). ISBN 978-1-4092-0073-4

16. Kursa, M.B., Rudnicki, W.R.: Feature selection with the Boruta package. J. Stat. Softw. **36**(11), 1–13 (2010)

17. Liu, F.T., et al.: Isolation forest. In: Eighth IEEE International Conference on Data Mining (2008)

18. Chen, T., Guestrin, C.: XGBoost: a scalable tree boosting system, pp. 785–794 (2016). https://doi.org/10.1145/2939672.2939785

19. Minastireanu, E., Mesnita, G.: Light GBM machine learning algorithm to online click fraud detection. J. Inf. Assur. Cybersecur., 12 (2019). https://doi.org/10.5171/2019.263928

20. Dorogush, A.V., Ershov, V., Gulin, A.: CatBoost: gradient boosting with categorical features support (2018)

21. Lundberg, S.M., Lee, S.I.: A unified approach to interpreting model predictions. In: Advances in Neural Information Processing Systems 30 (NIPS 2017) (2017)

22. Sampaio, J., McGarry, T., Calleja-Gonzalez, J., Jimenez Saiz, S., del Alcazar, X.S.I., Balciunas, M.: Exploring game performance in the national basketball association using player tracking data. PLOS ONE **10**(7), e0132894 (2015)

23. Goldsberry, K., Weiss, E.: The dwight effect: a new ensemble of interior defense analytics for the NBA. In: MIT Sloan 7th Annual Sports Analytics Conference (SSAC13), pp. 1–11, 1–2 March 2013

24. Cheng, G., et al.: Predicting the outcome of NBA playoffs based on the maximum entropy principle. Entropy **18**(12), 450 (2016)
25. Hochreiter, S., Schmidhuber, J.: Long short-term memory. Neural Comput. **9**(8), 1735–1780 (1997)

A Study on the Effect of Flexible Human Resource Practice on Group Citizenship Behavior

Ying Huang[1(✉)], Ziyao Li[2], and Jing Wang[1]

[1] Guangxi University, Nanning 53007, Guangxi, China
sunnyg1@gxu.edu.cn

[2] China Mobile Group Guangxi Co., Ltd, No. 2 Guichun Road, Nanning 53007, Guangxi, China

Abstract. Based on the mediating effect of perceived organizational support and group cohesion, this paper explored the effect of flexible human resource practice on group citizen behavior in Chinese context. With a questionnaire survey among 443 employees of enterprises in Beibu Gulf, we construct and validate the relationship model of flexible human resource practice, organizational support, group cohesion and group citizenship behavior by statistical analysis and SEM. The results showed that perceived organizational support and group cohesion played a partially mediating role between flexible human resource practice and group citizenship behavior. It is recommended to foster and motivate group citizenship behavior by implementing flexible human resource practice

Keywords: Flexible human resource practice · Group citizenship behavior · Perceived organizational support

1 Background

Competition for talent has become the core part of enterprises activity under the background of globalization and information tide. In order to maintain their sustainable competitive advantages, work groups are required to have better extra-role performance, such as consciousness of the overall situation, excelsior attitude, team spirit and so on. Thus, group citizenship behavior which is beyond duty and not for the purpose of commendation has become research focus. Organ (1983) believes it is voluntary, not for the purpose of obtaining formal reward, and conducive to the operation and development of an enterprise [1]. Afterwards, it has been proved that Citizen behavior is beneficial to reducing the organization's internal friction and it urges employees to make the greatest efforts to achieve organizational goals (Huang Ying, 2012) [2]. In addition, Researchers defined citizen behavior at the group level as group citizenship behavior, which can effectively coordinate the internal and external relations of the working group, and promote the realization of enterprise goals.

At present, employees who were born after 1980 with strong self-consciousness and prominent personality have become the main force in the workplace. In order to make these young employees and groups show more team spirit and cooperate spontaneously,

© Springer Nature Singapore Pte Ltd. 2020
H. Han et al. (Eds.): IDMB 2019, CCIS 1099, pp. 51–64, 2020.
https://doi.org/10.1007/978-981-15-8760-3_4

enterprises need to actively implement effective human resources practices to shape and cultivate employees' initiative behavior. The practice of flexible human resources is a kind of flexible management thought which fully embodies people-oriented. Through flexible incentives, employees can cultivate their sense of identity with enterprises, achieving "harmony but difference" among employees spontaneously for the enterprise goals. Based on this, this paper mainly discusses how flexible human resources practice stimulates and influences group citizenship behavior.

Human resource practice focused on motivating employees spontaneously is helpful to guide the group citizen behavior. What is the essence of flexible human resource practice? How it impacts group citizen behavior? What measures we can take with low cost to promote flexible human resource practices? The related problems, however, have not been well revealed. Therefore, this paper explores the impact of flexible human resource practices on group citizenship behavior, does research in enterprises in China, establishes a model of the relationship between flexible human resource practices and group citizenship behavior, takes organizational support and group cohesion as intermediary variables, and broadens and enriches the study of flexible human resource practices and group citizenship behavior.

1.1 Variable Definition

Flexible Human Resource Practices

Scholars believe that flexible human resource practice abandons the shackles of rigid rules to employees, guides the employees to obey spontaneously, and dealing with the relationship between work and non-work flexibly as well (Li Jing, 2014; Matthijs et al. 2015) [3, 4]. Based on the definition proposed by Nue Huiping et al. (2007), Ketkar and Sett (2009), this paper define flexible human resource practice as a series of activities used to manage human resource with respect for human nature, which pay attention to psychological rules and diversified needs of employees [5, 6]. Flexible human resource practice is divided into two dimensions here named functional flexible human resource practice and incentive flexible human resource practice as well. The former of the dimension is mainly aimed at improving employees' ability, enabling employees to obtain diversified skills, and helping employees to adapt to diverse tasks. The latter develops the content of time flexibility and compensation flexibility in previous studies. It mainly refers to the human resource practice fits with the psychological behavior while facilitating the coordination of work and family relationships, and improves employee job satisfaction as well.

Perceived Organizational Support

Eisenberger (1986) proposed that perceived organizational support is an integrated perception of employees about whether organization pays attention to their effort and happiness. It reflects an emotional expectation [7]. He confirmed that perceived organizational support is closely related to variables such as leadership, member switching, procedural justice, job satisfaction, and the perceived organizational support is one-dimensional. This paper uses his definition and dimension division of perceived organizational.

Group Cohesion

This study argues that group cohesion is a kind of attraction and centripetal force of group members' solidarity, cooperation and interdependence. This study use the idea of Zaccaro et al. (1988) and Mumin et al. (2008) that Group cohesion should be divided into two dimensions: task cohesion and interpersonal cohesion.

Group Citizenship Behavior

Under the collective culture background, Lu Zhengbao (2010) named citizen behavior at group level group citizen behavior. He defined group citizen behavior as behavior which can effectively coordinate the internal and external relations of the working group, and spontaneously promote the realization of enterprise goals. This study follows his definition. Group citizenship behavior in this study is divided into three dimensions: organizational loyalty, team spirit and helping behavior

1.2 Research Hypothesis

Flexible Human Resource Practice and Group Citizenship Behavior

When organizations implement flexible human resource practice, employees tend to be more responsive to the organization and perform extra-role behaviors spontaneously which is beneficial to the organization motivated by an exchange of reciprocity. Human resource practice such as flexible selection, creation of a fair atmosphere, flexible work design, and diversified incentives can effectively guide organizational citizenship behavior (Zhang Jundong, 2008; Chen Zhixia, 2010) [8, 9]. At group level, researches also confirmed that the human resource practice based on flexibility positively influence group citizenship behavior (Huang Ying, 2012; Zhao Haixia, 2013) [2, 10]. Thus, this study argues that flexible human resource practices play an active role in guiding group citizenship behavior, and makes the following hypotheses:

H1: Flexible human resource practices have a positive influence on group citizenship behavior

H1-1: All dimensions of flexible human resource practice have a positive influence on organizational loyalty;

H1-2: All dimensions of flexible human resource practice have a positive influence on team spirit;

H1-3: All dimensions of flexible human resource practice have a positive influence on helping behavior.

The Mediating Role of Organizational Support and Group Cohesion
The Mediating Role of Perceived Organizational Support

Flexible human resource practices emphasize the inner feelings of employees who feel support and care from organizations easily, so that employees are more willing to take the responsibility to create wealth and success, and prefer more organizational citizen behaviors as well (Moorman, 1998; Kurtessis et al. 2015) [11]. It is the same situation for working groups. Based on the previous research, the following hypotheses are proposed:

H2: Flexible human resource practice and its dimensions have a positive influence on organizational support;

H3: Perceived organizational support has a positive influence on group citizenship behavior and its dimensions;

H4: Perceived organizational support plays an intermediary role between flexible human resource practices and group citizenship behavior.

The Mediating Role of Group Cohesion

Group cohesion is a kind of attraction and centripetal force. The influencing factors of group cohesion include organizational relationship, group size, leadership style and communication effectiveness as well. Flexible human resource practice is more popular among employees, and it can improve employees' cohesion in a deep level (Zhi Yucheng, 2010) [12]. Relevant research results show that employees in cohesive groups are more willing to unify their pace, adopt group behavior and be more willing to show citizen behavior (Zeng Shengjun, 2010; Ni Changhong et al. 2013) [13, 14]. Based on these studies, we try to analyze group cohesion as a mediating variable between flexible human resource practices and GCB, hypotheses is as following:

H5: Flexible human resource practice and its dimensions have positive influence on group cohesion;

H6: Group cohesion has positive influence on group citizenship behavior and its dimensions;

H7: Group cohesion plays intermediary role between flexible human resource practice and group citizenship behavior.

1.3 The Model of Relationship Between Flexible Human Resource Practice and Group Citizenship Behavior

According to the relationship we mentioned before, a theoretical model is showed in Fig. 1.

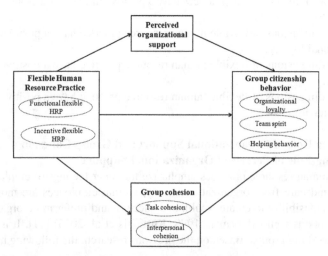

Fig. 1. Relationship model of flexible human resource practice, perceived organizational support, group cohesion and group citizenship behavior

2 Empirical Research

2.1 Sample Selection and Data Collection

Based on the preliminary investigation, a total of 520 questionnaires were issued in formal investigation, and 469 questionnaires were collected. After screening and sorting, the number of effective questionnaires was 443 and the effective recovery rate was 85.19%. The proportion of people under 35 years of age was as high as 87.4%. The education level of the subjects in the survey was generally high, and most of the subjects had the background of undergraduate and graduate students. The numbers of the members of most of the surveyed departments or teams are large, 56.4% of the participants said that there more than 10 people in their working group. And most of the surveyed departments or teams set up more than two years. Moreover, the participants mainly come from large and medium-sized enterprises, most of the subjects worked in state-owned enterprises, and 105 subjects worked in private enterprises (accounting for 23.7%).

2.2 Variable Measurement

We use object transfer consensus model to consider the design of questionnaire questions. All questions about employees in the scale are asked by "my working group". And then, calculate the average scores of individual members and aggregate them into group level for statistical analysis.

Measurement of Flexible Human Resource Practice

Flexible human resource practice scale consists of two main parts. Functional flexible human resource practices include 5 items, evaluating work design, empowerment, employee training and career development. The incentive flexible human resource practice dimension mainly measures the flexibility of work hours, performance management and salary management, and has 13 items.

Measurement of Perceived Organizational Support

Eisenberger (1986) developed a perceived organizational support scale containing 36 items, which has been confirmed applicable in different cultural backgrounds. However, a large number of measurement items are not conducive to our investigation. We try to use 6 items and believe perceived organizational support is one-dimensional.

Measurement of Group Cohesion

Mumin et al. (2008) divided group cohesion into two dimensions: task cohesion and interpersonal cohesion, and formed a measurement scale of group cohesion, which was proved to be highly reliable. This study will follow this scale, and summarize and optimize part of the measurement items combining the study of Sally, Caroline (2000) and Ye Qianheng (2003). Our scale measure the group cohesion by 8 items.

Measurement of Group Citizenship Behavior

In previous studies, Wu Xin (2005), Chen Xiaoping (2005), Lv Zhengbao (2010) and other scholars have compiled a measurement scale for citizen behavior at the group level. Lu Zhengbao (2010) developed a four dimension scale under the Chinese cultural background, and the scale showed good results. On the basis of his study, and combining the scale of Huang Ying (2012), we divided group citizenship behavior into three dimensions: organizational loyalty, team spirit and helping behavior, and measured by 12 items.

2.3 Research Method

This study uses SPSS22.0 and AMOS21.0 to analyze and verify the relationship between flexible human resource practice and group citizenship behavior.

Reliability and Validity of Variables

The reliability of the four scales were shown in Table 1 with high reliability whose Cronbach α are greater than 0.9.

Table 1. Reliability flexible human resource practice, perceived organizational support, group cohesion, and group citizenship behavior scale

Scale	Items	Cronbach α
Flexible human resource practice scale	18	0.949
Perceived organizational support scale	6	0.96
Group cohesion scale	7	0.938
Group citizenship behavior scale	10	0.933

We use Amos23.0 statistical software to do confirmatory factor analysis for group citizenship behavior, knowledge sharing, organizational commitment of the three theoretical measurement model, and judge the rationality of the model by goodness-of-fit. Goodness-of-fit of the models are shown in Table 2.

Table 2. Model suitability test of flexible human resource practice, perceived organizational support, group cohesion, and group citizenship behavior

Model	χ^2	df	$\chi^{2/df}$	GFI	CFI	IFI	NFI	RMSEA
Evaluation criterion	–	–	>0.9	<0.05	>0.9	>0.9	>0.9	<0.08
Flexible human resource practice	491.071	131	3.749	0.884	0.932	0.932	0.91	0.079
Perceived organizational support	22.228	8	2.778	0.984	0.994	0.994	0.991	0.063
Group Cohesion	43.325	12	3.61	0.974	0.988	0.988	0.984	0.077
Group citizenship behavior	105.382	31	3.399	0.956	0.98	0.98	0.973	0.074

It can be seen that the fitness index of two-dimensional model of flexible human resource practice, one-dimensional model of perceived organizational support, two-dimensional model of group cohesion and three- dimensional model of group organizational behaviorall meet the basic requirements of model fitting. Generally speaking, the four variable models are ideal.

Hypothesis Verification

This study uses SPSS22.0 to analyze the correlation of the variables, and lay the foundation for the subsequent of regression analysis and relation model verification in Tables 3 and 4.

Table 3. Correlation of variables at the overall level

Variable	1	2	3
1. Flexible human resource practice	1		
2. Perceived organizational support	0.846^{**}	1	
3. Group cohesion	0.696^{**}	0.724^{**}	1
4. Group citizenship behavior	0.726^{**}	0.745^{**}	0.831^{**}

Table 4. Correlation of variables at the dimension level

Variable	1	2	3	4	5	6	7
1. Functional flexible HRP	1						
2. Incentive flexible HRP	0.742^{**}	1					
3. Perceived organizational support	0.709^{**}	0.837^{**}	1				
4. Task cohesion	0.590^{**}	0.667^{**}	0.688^{**}	1			
5. Interpersonal cohesion	0.543^{**}	0.618^{**}	0.671^{**}	0.751^{**}	1		
6. Organizational loyalty	0.560^{**}	0.628^{**}	0.668^{**}	0.692^{**}	0.728^{**}	1	
7. Team spirit	0.624^{**}	0.686^{**}	0.699^{**}	0.701^{**}	0.746^{**}	0.788^{**}	1
8. Helping behavior	0.547^{**}	0.613^{**}	0.677^{**}	0.662^{**}	0.714^{**}	0.690^{**}	0.748^{**}

The correlation research results basically verify the H1, H2, H3, H5, H6, both from the overall level or from the dimension, there is a significant positive correlation between the flexible human resource practices, perceived organizational support, group cohesion and group citizenship behavior.In order to verify the mediating effect of perceived organizational support and group cohesion, this study adopts the method of testing intermediary function. The results of regression analysis are shown in Tables 5 and 6.

According to the data in the table, when the flexible oriented human resource practice is used as independent variable and the group citizenship behavior as the dependent

Table 5. Mediating effects of perceived organizational support on flexible human resource practice and group citizenship behavior

Independent variable	Dependent variable		
	Model 1	Model 2	Model 3
	Group citizen behavior	Perceived organizational support	Group citizen behavior
Flexible human resource practice	0.726^{***}	0.846^{***}	0.337^{***}
Perceived organizational support			0.46^{***}
R^2	0.527	0.715	0.587
Adjusted R^2	0.526	0.714	0.585
F	490.982^{***}	1170.117^{***}	312.753^{***}

Table 6. Mediating effects of group cohesion on flexible human resource practice and group citizenship behavior

Independent variable	Dependent variable		
	Model 1	Model 2	Model 3
	Group citizen behavior	Group cohesion	Group citizen behavior
Flexible Human Resource Practice	0.726^{***}	0.696^{***}	0.285^{***}
Group cohesion			0.633^{***}
R^2	0.527	0.485	0.733
Adjusted R^2	0.526	0.484	0.732
F	490.982^{***}	415.31^{***}	603.896^{***}

variable, the regression coefficient is 0.726, and the result is significant. When flexible human resource practice is used as independent variable and perceived organizational support as dependent variable, the regression coefficient was 0.846, and the result was significant. When perceived organizational support is used as an intermediary variable, the regression coefficient C' is 0.337, which is 0.389 lower than the coefficient C. It shows that the perceived organizational support plays a partial mediating role. In the same way, group cohesion also plays a partial mediating role between flexible human resource practices and group citizenship behavior.

We use bootstrap to test the mediating effect further. The results of the mediating effect of perceived organizational support did not include 0 (LLCI = 0.2745, ULCI = 0.4812), which shows the mediating effect of perceived organizational support was significant, and the mediating effect was 0.3781. the results of the mediating effect of group cohesion did not include 0 (LLCI = 0.3579, ULCI = 0.5032), which shows the mediating effect of group cohesion was significant, and the mediating effect was 0.4285. These fully verify H4 and H7.

Analysis of the Relation Model

The standardized results of the model are shown as Fig. 2.

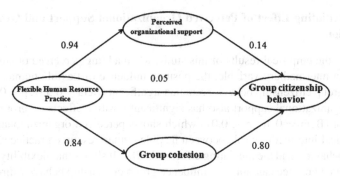

Fig. 2. The model of flexible human resource practice, organizational support, group cohesion and group citizenship behavior

The fitting indexes of the model are shown in Table 5. The absolute fitness index is 2.421, and the RMSEA is 0.57, both of which have reached a good standard. The IFI and CFI values were 0.934, which are also reached the standard. These show that the model fits well and can validate the relationship model Table 7.

Table 7. Test of model fitness

Model	χ^2	df	χ^2/df	GFI	NFI	IFI	CFI	RMSEA
Evaluation criterion	The smaller the better	–	<3	>0.9	>0.9	>0.9	>0.9	<0.08
Test model	1844.635	762	2.421	0.820	0.893	0.934	0.934	0.057

3 Conclusions and Discussions

3.1 Flexible Human Resource Practices Have a Positive Impact on Group Citizenship Behavior

Through correlation analysis and regression analysis, this study confirms that flexible human resource practice has a significant positive impact on group citizenship behavior. According to the results of the study, it is concluded that respecting employees and paying attention to their inner needs can effectively stimulate their work enthusiasm and organizational loyalty as well. When organization provides adaptable and diversified jobs to meet the needs of team growth, it can inspire employee in a deep level and work team as well, so that work team coordinate efficientlly. Flexible human resource practices based on this kind will help to promote more out-of-role behavior of employees, strengthen communication and collaboration with other groups, participate in team building and enterprise building, and actively maintain the honor and image of enterprises.

3.2 The Mediating Effect of Perceived Organizational Support and Group Cohesion

According to the empirical results of this study, when adding perceived organizational support as an intermediary variable, the positive influence of flexible human resource practice on group citizenship behavior remains significant (Beta = 0.337, p < 0.01), and perceived organizational support also has significant positive effects on group citizenship behavior (Beta = 0.46, p < 0.01) which shows perceived organizational support plays a piratical intermediary role between flexible human resource practice and group citizenship behavior, and the mediating effect is 0.3781. It shows that flexibility oriented human resource practice can not only influence group citizenship behavior directly, but also influence group citizenship behavior through perceived organizational support. This result is consistent with our hypothesis. It means that employees and their groups can feel the concern which can meet their needs, and is conducive to their growth. When adding group cohesion as an intermediary variable, the positive influence of flexible human resource practice on group citizenship behavior remains significant (Beta = 0.285, p < 0.01), and group cohesion also has significant positive effects on group citizenship behavior (Beta = 0.633, p < 0.01) which shows group cohesion plays a piratical intermediary role between flexible human resource practice and group citizenship behavior, and the mediating effect is 0.4285. It shows that flexibility oriented human resource practice can not only influence group citizenship behavior directly, but also influence group citizenship behavior through group cohesion. This result is consistent with the hypothesis of this research. Flexible human resource practice not only focuses on diversified training and incentive for employees, bur also pays more attention to the comprehensive evaluation of employees and their working groups. It has a positive impact on both task cohesion and interpersonal cohesion of working groups. And cohesive groups tend to spontaneously protect the image of the enterprise, actively provide advice for the organization, strengthen cooperation with other groups, help others share heavy tasks and solve difficulties.

4 Management Inspiration

4.1 Adopt Flexible Work Design and Give Employees More Rights

The lack of flexibility and flexibility in work design seriously affects the enthusiasm and long-term development of employees. It is not conducive for enterprises to adapt to the increasingly complex competitive environment.

On the one hand, enterprises should adopt flexible work design, respect the individual needs of employees, and implement more flexible work methods. Employees are given broader responsibilities and work content, changing the traditional way of job design from horizontal and vertical aspects, and increasing the richness and challenge of work. At the same time, employees are encouraged to move among posts through job rotation, try more jobs, gain more work experience, tap their own potential and lay a good foundation for communication and communication between organizations. On the other hand, employees should be affirmed and respected, empowerment should be emphasized, control ability of employees should be improved, and work autonomy should be enhanced. In order to encourage employee to participation in management, enterprises should establish a good atmosphere to enhance confidence and motivation of employees while giving employees the full trust and providing comprehensive support for employees. Enterprises can set up workers' congress system, staff proposal construction system, staff supervision mechanism for managers and so on.

4.2 Implement Flexible Work Hours

Flexible human resource practices enable employees to balance work and family relationships, and then promote group citizen behavior. Therefore, enterprises should give employees flexible working hours, implement flexible working system, break through the limitations of traditional working hours and space. The working hours are divided into flexible working hours and core working hours. In addition to the required working hours, employees can flexibly choose the specific time and mode of their work. The enterprise can also allow employees to arrange and adjust their work time, work place and work manner under the premise of completing the work with guaranteed quality and quantity. Under this flexible working mechanism, employees can complete their tasks with high quality, enhance their sense of loyalty and trust to the organization, and trigger employee citizenship behavior.

4.3 Adopt Flexible Performance Appraisal System

Performance management is an essential part of human resource practice. It is the basis for staff training, job adjustment and salary determination. Scientific performance appraisal will improve employee skills and motivation, reduce turnover rate, improve job satisfaction and commitment to the organization. Therefore, the flexible performance evaluation system is conducive to guiding employees behavior, promoting members to perform better in their work attitude and behavior performance, and this practice will form a benign incentive cycle mechanism, so that employees are more willing to work in the organization and show more out-of-role behavior.

To establish a flexible evaluation system, first, enterprises should change the traditional concept of performance appraisal, use "comprehensive performance view" instead of "task performance point of view. When evaluating performance, enterprises should not only assess individual performance, but also assess team performance; not only to assess the results, but also assess employees' behavior, values, and competence. When assessing performance and achievements, enterprise should consider the objectives and contents of the work, take the quantity and quality of work into account. When assessing behavior and attitude, enterprises can consider loyalty, engagement, team spirit and other aspects of the employees (Specifically, pay attention to the behaviors like complying with norms, maintaining corporate image, respect others and so on). Then, enterprises should make a comprehensive judgment according to identification of employees, the observation of their direct leadership, and the assessment by assessment team. Second, enterprises should pay attention to the practicability and operability of performance appraisal methods, and ensure that the performance evaluation standards match the nature of post categories. Assessment methods should also be chose and adjusted according to the environment, business strategy, and management practices. At the same time, enterprises should also quantize assessment index basic on the content and essence of work, and combine the quantitative index and qualitative index. Under the conditions permit, enterprises can allow staff representatives to participate in the evaluation of the formulation to ensure that the performance appraisal system is fair, public and objective. Third, Performance communication and feedback should be timely. Flexible performance appraisal should be a process of study, improvement and control. Besides fair and impartial examination results, performance appraisal needs continuous and dynamic communication. Management and performance appraisal departments should be willing to communicate with employees for their performance, explore the existing problems in the past and the future improvement with employees through formal and informal communication, help staff to solve the existing problems, and guide them to perform better in their work. Fourth, enterprise should combine performance results with incentives.

4.4 Design Flexible Salary and Welfare System

Compared with the rigid salary system, flexible incentives can improve the subjective initiative of employees, improve job satisfaction, and thus enhance the core competitiveness of the company. Enterprises should make salary competitive first. Salary managers should combine the salary and welfare system with enterprise management strategy. Considering the value orientation of enterprises in the process of salary design and implementation, improve the assessment of the value of human capital, and enhance the competition advantage of the compensation system. In addition, Enterprise should link the salary of the employees with the growth of the enterprise, and provide them a competitive salary while fully considering the characteristics and ability of the enterprise. Moreover, enterprises should ensure that the salary system can be adjusted flexibly with the changes of market environment, enterprise growth, enterprise performance and staff performance. Secondly, in order to make employees show more organizational loyalty and helping behavior, enterprises should also pay attention to the diversity and selectivity of welfare. At present, personalization and diversification have become the main characteristics of employee demand. Therefore, welfare should be more diverse (enterprises could provide parking spaces, nutritious lunch, lactation lounge, study area, a longer paid annual leave for employees), meet the needs of all aspects of staff, so that employees get more satisfaction. On the other hand, enterprises should also consider optional welfare, and launch a "buffet" welfare (Some employees may not need the lunch provided by the enterprise, then the enterprise may consider to supplement other welfare for them), to satisfy different needs of the employees, motivate employees better, and improve the performance of employees. Establishing such a flexible welfare system can better motivate employees, improve and enhance employees 'behavior, enhance their sense of identity and belonging to the organization, and self-publishing shows out-of-role behavior conducive to the organization.

4.5 Establish an Effective Communication System

First of all, it is important to create a good communication atmosphere in the enterprise, creating the prerequisite for communication. Communication should be comprehensive, so either leaders or grass-roots employees, working groups should be aware of the importance of communication. Leaderships and managers should create opportunities to communicate with employees, encourage the emotional recognition of employees through communication, so as to promote the smooth development of the work. Second, communication should be consistent and involved in all aspects of human resource practice. Through communication, managers could know the views of the employees while helping them to understand the goals and systems of the enterprise. When employees encounter difficulties in their work, enterprises should communicate and guide with them timely helping them to complete the task smoothly. Enterprises should also regularly fed back work performance to employees, and help them to improve. Third, enterprises should broaden the channels of communication, and communicate in different ways such as staff meetings, department or team meetings, one-on-one and so on. Open office is also one of the ways to promote effective communication between employees and managers.

4.6 Building Flexible Team Culture

Firstly, the organization can adopt "team" instead of "individual" work mode, establish cross-departmental work teams or project teams, etc. to promote employees to increase their understanding of team members in the process of completing tasks, adapt to each other, and form a good communication atmosphere. Secondly, group building activities can be organized regularly. Relevant incentives, such as offering paid travel opportunities for team members and distributing souvenirs, are also provided. The organization of the activities has a positive role in promoting the communication of personnel within the department, as well as the communication of personnel across departments. It plays a more sustained incentive role in harmony working atmosphere with positive energy beneficial to enhance the execution of team.

References

1. Organ, D.W.: Organizational Citizenship Behavior: The Good Soldier Syndrome. Lexington Books, Lexington (1988)
2. Sanehez, R.: Strategic flexibility in product competiton. Strat. Manage. J. **16**(S1), 135–159 (1995)
3. Change, S.: Flexibility-oriented HRM systems, absorptive capacity, and market responsiveness and firm innovativeness. J. Manag. Official J. South. Manag. Assoc. **39**(7), 1924–1951 (2013)
4. Bal, P.M., De Lange, A.H.: From flexibility human resource management to employee engagement and perceived job performance across the lifespan: a multisample study. J. Occup. Organ. Psychol. **88**(1), 126–154 (2015)

5. Lai, J., Long, W., Lam, S.: Organization citizenship behavior in work groups: a team cultural perspective. J. Organ. Behav. **34**(7), 1039–1056 (2013)
6. Ketkar, S., Sett, P.K.: HR flexibility and firm performance: analysis of a multi-level causal model. Int. J. Hum. Resour. Manage. **20**(5), 1009–1038 (2009)
7. Eisenberger, R., Stinglhamber, F.: Perceived organizational support. J. Appl. Psychol. **71**(3), 500–507 (1986)
8. Smith, C., Organ, D., Near, J.: Organizational citizenship behavior: its nature and antecedents. J. Appl. Psychol. **68**(4), 653–663 (1988)
9. Lai, J., Long, W., Lam, S.: Organizational citizenship behavior in work groups: a team cultural perspective. J. Organ. Behav. **34**(7), 1039–1056 (2013)
10. Beeri, I.: Turnaround management strategies in public systems: the impact on group-level organizational citizenship behavior. Int. Rev. Adm. Sci. Int. J. Comp. Pub. Adm. **78**(1), 158–179 (2012)
11. Kurtessis, J., Eisenberger, R., Ford, M., et al: Perceived organizational support: a meta-analytic evaluation of organizational support theory. J. Manage. (2015)
12. Bhanthumnavin, D.: Perceived social support from supervisor and group members psychological and situation characteristics as predictors of subordinate performance in Thai work units. Hum. Resour. Dev. Q. **14**(1), 79–97 (2003)
13. Dayan, M., Benedetto, A.: Procedural and interactional justice perceptions and teamwork quality. J. Bus. Ind. Mark. **23**(8), 566–576 (2008)
14. Preacher, K., Hayes, A.: SPSS and SAS procedures for estimating indirect effects in simple mediation models. Behav. Res. Methods **36**(4), 717–731 (2004). https://doi.org/10.3758/BF0 3206553

Patent Text Mining on Technology Evolution Path in 5G Communication Field

Tingting Liu[1,3], Tie Wei[1,3(✉)], Henry Han[2,3], and Tiange Li[2,3]

[1] College of Business, Guangxi University, Guangxi 10593, China
weitie2000@sina.com
[2] The Gabelli School of Business, Fordham University, Lincoln Center, New York, NY 10023, USA
[3] Computer Information Science, Fordham University, Lincoln Center, New York, NY 10023, USA

Abstract. 5G communication is a pioneering topic in current technology development. Deciphering the evolution path of 5G technology is crucial for relevant companies to formulate technology strategies. This paper employs text mining to unveil the technological evolution path of 5G communication besides employing novel semantic classifications of technical content in the 5G field. It first discovers the path of technological evolution through text mining and visualization besides presenting a theoretical and empirical basis for the existing technological innovation management.

Keywords: Technological evolution · Patent · Text mining · 5G communication technology

1 Introduction

Technological innovation is increasingly becoming the key to gaining competitive advantage in today's world. In technological innovation, it is essential how to grasp the trend of technology development, adapt to the development trend of technology, and choose the right direction of technology development. Then the R&D investment can be carried out correctly, which is a crucial link in the formulation of the innovation strategy [1]. Technological development is a long-term and gradual process of constant change, which is often referred as evolution [2].

The evolution of technology has always been the focus of academics. Existing research generally believes that technological evolution is the physical performance of technology, changing or improving along a specific technology track over time [3]. The main research is to explore the evolutionary conditions and evolutionary trajectories of technology in the physical orbit from the perspectives of scientific paradigm [4], technological change [5, 6] and industrial development [7–9]. Its main purpose is to show the trend of various technologies in a certain field in time. However, most of the current research on technology evolution focus on changes in macro-technical paths, such as S-curves for technology development, technology jumps and so on [10, 11]. However,

© Springer Nature Singapore Pte Ltd. 2020
H. Han et al. (Eds.): IDMB 2019, CCIS 1099, pp. 65–81, 2020.
https://doi.org/10.1007/978-981-15-8760-3_5

there is still insufficient attention to the development and evolution of specific technical solutions at a more micro level. Enterprises are more concerned about specific technological developments. In particular, the degree of understanding of the evolutionary routes and development trends of specific new technologies is more valued by business managers when formulating technology selection strategies.

5G (fifth generation mobile communication communication) technology will bring revolutionary changes to the human being's life. Communication mobile communication has been developed for more than 40 years. With the popularity of smart devices around the world, people are increasingly dependent on cellular networks. To a large extent, the rapid development of 5G technology has been spawned. The emergence of 5G communication is called the fourth technological revolution. With the successful use of 5G and the deployment of operators this year that the 5G topic has been pushed to the forefront. At a deeper level, the successful development of 5G has brought about a new industrial revolution, which makes human being daily life more convenient. Thus, we use the relevant patents of 5G technology to verify the practicality of the evolutionary path.

However, it remains unknown for its possible evolution path from a technological innovation standing point. Understanding the 5G evolution path can decipher its key development process and predict its future trend, which will provide a valuable guide for enterprise technological innovation directions. The evolution path means that the activity trajectory of the technology cluster during a period of time. Different technical content will be classified into various evolution paths. We can unveil hidden but valuable technical content and its evolution status by establishing an evolutionary path.

How to decipher the evolution path of 5G communication? Patent data can be a useful resource because it is publicly available data that records almost all existing scientific and technological innovations. Its development often echoes the roadmap of a technology [12]. Patent information is often used to achieve different goals in the quantitative research of technological evolution. For example, patent citations, Number of family patents, and other relevant information [13, 14].

However, there is a large amount of unstructured information such as patent abstracts and claims, most of which do not follow any structured format. Patent texts are more effective than structured information (e.g. patent citations) in terms of quantitative research of technological evolution. On the other hand, unstructured data such as patent texts are so huge that they cannot be processed manually alone.

With the development of data mining, research based on text mining is becoming a research trend [15]. Previous patent analysis can only judge the patent information contained in title, author, and so on, but cannot handle the patent content well [16]. Text mining can process patent text through natural language processing (NLP) and extract key information. Thus, it can more accurately and comprehensively judge the trend of technological evolution. Text mining can extract patented technical subject terms, cluster similar technology subjects, and arrange the topic clusters by time [17, 18]. John and Lent first applied text mining to patent analysis in his 1997 study [19].

The existing research on technology evolution through patent text mining methods focuses on two aspects. The first is about patent technology opportunities. Sun used patent citation data to identify core patents based on three methods that are patent citation frequency, number of claims, and patent family size [20]. However, the method of patent measurement can only use structured data, but the technical information of unstructured data hiding in the patent text cannot be utilized. This reduces the accuracy of the judgment. Gong et al. established a three-dimensional patent map through text mining and kernel principal component analysis [21]. It analyzes the average cited frequency of patents around the blank spots of patent maps, the average number of patents of the same family, and the density of blank spots to explore potential technology opportunities. Han et al. judge the risk of patent infringement by calculating the Euclidean distance between the cited documents before and after the patent [22]. The survival time of patents is calculated based on survival analysis through Euclidean distance, the number of forwarded citations, patent nationality, and patent transfer. Zhang analyzed the influence of patent technology portfolio through text clustering [23]. They try to categorize patents in the technical field, but do not provide an effective life cycle of the patented technologies.

The second is on the changing trend of patent technology topics. Yoon et al. use text mining to analyze trends in high-tech technology [24]. Feldman et al. analyze the causes of technological mutations at different times by judging the development of technical topics in time [25]. Helen et al. establish a patent channel based on text mining, which is the deployment of patent clusters in time [26]. Its purpose is to help companies intuitively judge the choice of existing technology with development advantages or develop new technologies.

However, there is no research the evolution path of 5G technologies, which is an urgent topic both in technical innovation and wireless communication. So this paper intends to establish a path for technological evolution through patent text mining. And we bring in 5G technology to discover the evolution of the field.

The evolution analysis of 5G technology is even not investigated yet. In recent years, although there are a few articles that use patent information to analyze information on 5G and related technologies, the main research focuses on technical analysis and prediction in the field of communication. In 2016, Liu first made an analysis of the development trend and competitive situation of the 5G field in terms of the overall distribution and technology distribution of patents in the field of fifth-generation mobile communications [27]. Zhang et al. conducted a multi-country comparative analysis of the overall distribution, popular technology and competitive situation of polarization code technology in the 5G field [28]. Xiao et al. used the patent measurement method to analyze the competitive situation in the field of 5G communication based on patent data [29]. Some researchers have also analyzed the core technologies in the 5G field based on patented technology evaluation indicators [30].

However, the existing research does not have the study in evolution path of 5G technology. In particular, there is a lack of classification of 5G technologies. As a result, it remains unclear about the evolutionary distribution of various technologies in the 5G field. Thus, this paper intends to solve this problem by using text mining methods to establish the path of technological evolution. We have found that 5G technology can be classified into three categories via an in-depth analysis of the evolution path.

To the best of our knowledge, it is the first research to classify the 5G technical content and establish a meaningful evolution path.

2 Method

2.1 Data Sources

All patent data collected in this study are publicly available from Chinese Intellectual Property Office Patent Database (SIPO) and SooPAT database [31, 32]. We use the "fifth generation mobile communication", "5g" and 5G domain main specific block technologies "millimeter wave" and "MIMO" as the extended keywords to create a search term "theme = (fifth generation mobile communication OR 5G) AND summary = (millimeter wave OR MIMO)". The search date is December 2018, and the search scope is valid patents from 2015 to 2018 and is applied in China. Because of the high professional value of invention patents, this study only selected invention patents as raw data. The search yielded 175 invention patents, and obtained 160 invention patents after removing invalid patents and irrelevant patents. The main technical content of the patent is included in the patent abstract and sovereignty in the patent text. In order to avoid data redundancy, we select the abstract and sovereign items in the patent as the content of the patent text analysis. Finally, a document set is created from 160 patent abstracts and sovereign text for text analysis.

2.2 Research Method

2.2.1 Natural Language Processing

The original text obtained through the search is too large and the information is complicated, and there is a lot of unwanted noise. Before text mining, the data needs to be preprocessed. That is, cleaning data and converting unstructured data into structured data. The steps include word segmentation, removal of stop words and punctuation, and extraction of tags [33].

The preprocessing of data is called natural language processing. This study chose to use python's *jieba* package for natural language processing [34]. In this process, we mainly divide the text content into several words and remove meaningless words and punctuations. This process is to denoise the data for later processing. The preprocessed vocabulary includes a specialized dictionary. After getting the keywords of the document set and their word frequencies, we convert the document set into a preliminary sparse matrix through the *doc2bow* model to prepare for the subsequent calculation [35].

Based on the patented nature of this research document set, we add technically unrelated words to the stop words, such as "technology", "application" and so on. We use the word bag model to extract keywords and count the keyword frequency, and then filter the words with low technical relevance. Table 1 includes the keywords with frequency >80 in the text.

The key words are most informative words in 5G. For example, the core technologies in 5G communication are MIMO (Multiple-Input Multiple-Output) and millimeter waves, etc. The millimeter wave refers to an electromagnetic wave having a wavelength

Table 1. Key words frequency statistics table

Key word	Word frequency	Key word	Word frequency	Key word	Word frequency
Antenna	1075	Radiation	250	Power	115
MIMO	356	Channel	218	Substrate	114
Mrow	306	Array	217	Network	114
MSUB	300	Feed	213	Surface	109
Beam	296	Plate	158	Output	107
Millimeter wave	290	Patch	157	Information	106
Matrix	289	Terminal	128	Radio frequency	103
Signal communication	262	Code	121	Wave filter	94
Medium	252	Launch	120	MTD	84
Metal	251	Codebook	119	Crevice	84

of 1 to 10 mm, which is a core in 5G. 5G communications rely on the MTD (Microwave Traffic Detector) radar system. Alternatively, some key words indicate professional technical attributes. For example, Msub refers to the properties of the substrate dielectric material. MROW refers to Multiple readers, single writer locks, which means that access speed is increased by multiple read and single write threads in communication.

2.2.2 VSM Processing

After getting the initial sparse matrix, we use the Python *Gensim* package for lexical vectorization [36]. Since the extracted domain tags cannot be clustered, the tags need to be converted to be represented in digital form. We choose the Vector Space Model (VSM) to convert labels into digital representations [37]. The basic idea of the vector space model (VSM) is that the extracted words represent text features, regardless of the position and order in which the words appear, and only consider the frequency of occurrence of the words, i.e. the weight value. It uses the weight of each label as a dimension in the space vector to create a label matrix.

We choose Tf-idf to calculate the weight of feature items in our study, i.e. $D_{ij} = tf_{ij} * idf_j$. The D_{ij} is the j-th word on the document D_i, the tf_{ij} represents the frequency at which the word D_{ij} appears in the text, and idf_j (Inverse document frequency) indicates how important the word D_{ij} is in the entire text. The larger the value is, the lower the frequency of occurrence of the word. Finally, we build a document-vocabulary matrix through the VSM model.

2.2.3 Technical Path Subject Classification

We then convert the document-vocabulary matrix to a topic model to calculate its text similarity. We employ the LSI (latent semantic indexing) model which obtaining the latent semantics of the text to get the theme [38]. The LSI model classifies document sets by singular value decomposition. Singular value decomposition refers to the decomposition of sparse keyword matrix into several small matrices, so as to achieve the purpose of dimensionality reduction and noise reduction. It extracts the document by extracting the topic and its feature words, and establishing the topic model based on the frequency of occurrence of each topic and feature words [39]. The LSI model mainly classifies text topics based on SVDs values. Its formula is expressed as:

$$A_{m \times n} \approx U_m \times k \sum_k \times k V^T_{k \times n} \tag{1}$$

The meaning of this formula refers to the input of m texts, each of which has n words. A_{ij} corresponds to the eigenvalue of the jth word of the i-th text. The most commonly used here is based on the pre-processed standardized TF-IDF values. k is the number of topics we assume, generally less than the number of texts. After SVD is decomposed, U_{il} corresponds to the relevance of the i-th text and the first topic. V_{jm} corresponds to the relevance of the jth word and the mth word meaning. $\sum lm$ corresponds to the relevance of the first topic and the mth word meaning.

According on the LSI model, this article classifies all text into five categories. By calculating the maximum weighted value of the SVD of the keyword, five categories of topics and the main keywords within the theme are obtained. Figure 1 shows the classification results of SVD and their subject terms.

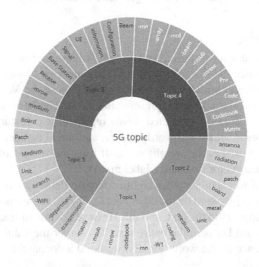

Fig. 1. Keywords for five themes.

2.2.4   Establishment of a Technical Path

By calculating the distance between all patents and various topics, the similarity between patents and topics is obtained. We then assign each patent to the technical topic with the highest similarity coefficient. The similarity matrix of all five topics is visualized as follows.

Figure 2 shows the classification results for all patents. Each patent is represented by a straight line, with five vertical axes representing five topics, and top-down numbers representing similarity coefficients. The degree of similarity between a patent and a subject is measured by the maximum absolute value of the value. The similarity coefficient of each line and each topic can be seen from the figure. The most similar to the topic is the technical subject, and finally the subject of the patent is represented in a different color. The distribution of each topic can be roughly seen from the figure.

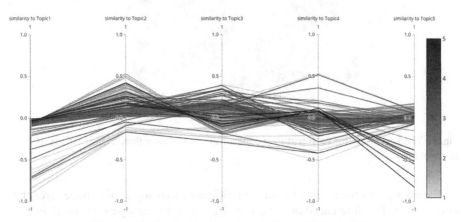

Fig. 2. The visualization of patent similar matrix by topics

2.2.5   Three Classifications of Technical Content

The biggest innovation of this paper is as follows. We first categorize the patent technical content into three levels according to its semantics. Such a semantics-based categorization can unveil the technical content in a more representative and focused manner. Then we dig technical evolution patterns for each level via analytics and visualization.

According the same categorization approach, this paper first divides all patents into five categories, which are called "technical topics." Each topic is classified again by using the patent text within each topic as a separate document set. According to the size of the number of patents within the subject, it is divided into two or three categories. Such a classification is called "technical branch".

It is observed that there are still a large number of patents in the evolutionary path of each technical branch not categorized. We further process the 'remaining data' as a separate document set to conduct the third categorization to find more clear evolutionary paths, i.e., "technical reclassification". Figure 3 illustrates the key words for each technical content after the three categorizations.

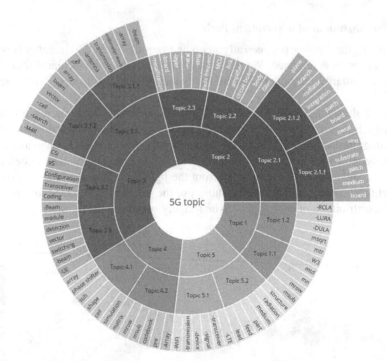

Fig. 3. The key words for each part of the technical content after three classifications. (Color figure online)

Figure 3 shows the keyword content of different technologies after three layers by visualization. The different colors represent different technical topics. Two to three circles from the inside to the outside represent different levels of technical content. Each small fan of the outermost circle represents each keyword. For example, theme 2 is represented by blue. The largest fan of the innermost area represents the first classification result, that is, the technical topic class. Secondly, the outer circle is the second classification result, that is, the technology branch class. In this category, the number of patents in topic 2.1, which is branch one, is still large, so the topic 2.1 is classified a third time. Expressed as two categories of outer ring again, that is, technical reclassification class. The small fan shape of the outermost circle represents the most representative keyword of this technical content.

The purpose of the three technical classifications is to carefully distinguish the technical content so that each technical content can avoid the influence of heterogeneous technologies as much as possible. This allows us to have a clearer judgment on the evolution of each technical content. More details can be found in the following Result selection.

3   Result

In this study, 5G communication patents are divided into three layers. The results of each stratification are as follows. Through simple and direct data visualization, we try to find some rules of technological evolution and future trends. The results of each stratification will be analyzed in turn below.

3.1 First Classification Result

The horizontal and vertical axes in Fig. 4 represent year and the number of patents respectively. The evolution trend of 5G technology with different themes and the comparison of each theme are represented by different colors.

(1) From the overall perspective of Fig. 4, all topics are showing growth momentum. By comparing different topics, we found that the number of patents in the paths of topics 2 and 3 is the largest and the growth momentum is the strongest. Therefore, themes 2 and 3 should be the main technical content.

(2) From different topics: Topic 1 shows a "U" development trend from 16 to 18 years. The study reached its best in 17 years and then weakened rapidly. But the overall research results are not much.

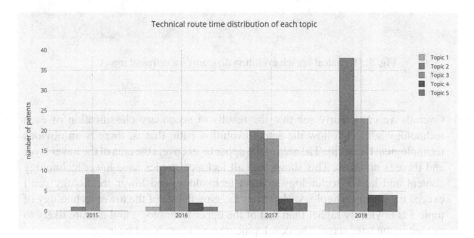

Fig. 4. Technical route time distribution of each topic. (Color figure online)

Topic 2 are the most common topic among the five technical topics. From 15 to 18 years, research results increased rapidly, showing a "blowout" growth. This shows that the subject technology has high research value and the existing research has not reached the bottleneck. There are still research directions to choose from.

Topic 3 are more in general from 15 to 18 years, and show a "smooth" growth. It shows that the research content of this technology topics is rich and it is in the core position of 5G field. It needs to be developed first and other technologies or research based on this technology. This technology topics has not been fully developed yet, which indirectly shows that there is still a large research space for 5G technology.

Topic 4 shows that the research results are generally less and the annual growth is not much, which indicates that the research value of this technology is not very high, and it may be in the edge of the technology field.

Topic 5 are similar to 4. But from 15 to 18 years there was a slight increase over topic 4. It because this research needs to be based on other research results. At this stage, the development is limited, and the follow-up may increase considerably.

After preliminary classification of technical content in 5G field, we have a general understanding of the distribution of technical topics and the evolution current of each topic. Following is a separate classification of each topic,we explore the evolution of technology branches within each topic.

3.2 Second Classification Result

The five maps in Fig. 5 categorize the five technical topics in turn. Different colors represent different technical branches.

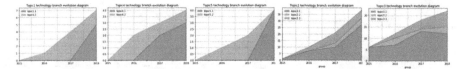

Fig. 5. Technical branch evolution diagrams for different topics.

(1) Overall, we can clearly see that the results of secondary classification of each technology subject follow the samel evolution path, that is, there is an obvious technological hierarchy. The area of the upper layer covers the area of the lower layer and there is no cross. This shows that all technical topics have lower technology content and higher technology content technology, and lower technology can't exceed the upper technology. However, the growth rate of the lower technology of topic 3 is obviously higher than that of the upper technology, and is more likely to exceed in the later stage. The development trend is analyzed later.

(2) From the different technology branches we can see that:

In topic 1, Each branch fits the trend of this topic. In 2017, the number of patents is large, which indicates that branch 1 and 2 are related to each other and grow synergistically. From different technical branches, the number of branch 1 is less than branch 2. Branch 2 is developing continuously from 2016 to 2018, which indicates that branch 2 or more supports the development of topic 1.

The technical branch 1 of topic 2 is basically the same as the evolutionary path of the topic, while branch 2 and branch 3 are different from the evolutionary path of the topic. It shows that the branch 1 is the core technology of the topic and supports the path development of topic 2. Branch 3 has reached the maximum number of patents on the subject before the maximum number of patents, indicating that it is an auxiliary technology for the subject and supports the development of core technology. Branch 2 has the least number of patents and its path is roughly the same as that of the subject, indicating that it is a non-essential ancillary technology of the subject.

The evolution path of Topic 3 is the result of the joint influence of each technology branch. From each technical branch, branch 1 has the largest number of patents, and the number of patents in each year is roughly the same, indicating that it should be the core technology of the subject, and has a large space for development. The number of branch 3 patents is next. The trend of change is roughly the same as that of the subject, which indicates that it develops with the development of the subject and depends on the development of the core technology in the subject. Branch 2 has poor regularity of technological change.

Each branch of the topic 4 has no exactly the same path as the topic evolution. From the each branch, the technological path of branch 1 is roughly the same as that of the subject. It should be the core technology in the subject. The technical path of branch 2 is contrary to the topic, indicating that it should be the assistant technology of the subject.It lays the foundation for the early development of core technologies.

The evolution path of topic 5 should be the result of the interaction of two branches. The technology of branch 1 should be the front-end technology of branch 2. When the technology of branch 1 reaches maturity, the technology of branch 2 has a sudden growth.

Through the analysis of Fig. 5, we have a more detailed understanding of 5G technology. However, at this stage, there are still many patents in some technical branches. We try to have a clear understanding of the evolution process. So we have a third classification of branches with a large number of patents.

3.3 Third Classification Results

The Fig. 6 reclassifies Branch 1 of Topic 2 and branch 1 of topic 3.

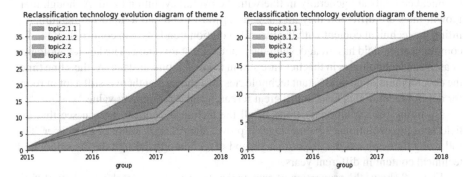

Fig. 6. Evolution diagram of technical reclassifications of different technology branches.

(1) We can see from Fig. 6: Firstly, there are upper technology and lower technology in the refined and differentiated technology content. Secondly, their evolutionary trends are roughly the same. Finally, the lower technology can not break through the upper technology.

(2) We can see from the left picture of Fig. 6, the two subclasses correspond to the evolutionary path of the branch and its technical topic. It shows that all the technologies in this branch are core technologies. From each technical reclassification, reclassification 1 is much larger than that reclassification 2. This indicates that the technology in reclassification 1 should be the key technology to support the development of this technology topic. Reclassification 2 is the key technology next to reclassification 1, and its technological content or less leads to less patent output.

On the right, the evolution path of reclassification 1 and 2 is roughly the same as that of the branch. However, the number of patents in reclassification 1 is much larger than that in reclassification 2, which indicates that it should be the key technology to support the development of this topic. However, the number of patents in reclassification 2 is small, indicating that its technology content or small.

By comparing two pictures, we can see that the key technologies can be roughly divided into two categories. One is the key technologies that support the development of the technology, the other is the auxiliary key technologies with less technology content.

3.4 Overall Trend Analysis

After three layers of technical content, we have evolved a relatively clean technical path. Next, we visualize the overall evolution trend of 5G. By fitting each technological path and the overall evolution trend, we can judge the valuable technological content and its future development trend. In this way, we can prove the value of our research. The following figure analyses the development characteristics and future trends of 5G from two perspectives.

From Fig. 7, we can find that the technological evolution in this field has increased greatly. Very few technical contents have a tendency to disappear. In addition, the other technical contents are generally in line with the overall evolution trend. Although their growth rates are different, they all show an upward trend and a rapid momentum. By fitting the technical content with the overall evolution trend, we can confirm that the 5G communication field has great development potential at this stage. The key technologies or assistant technologies in this field are still in the development stage. At the same time, the development of most assistant technologies also shows that there is still a lot of room for development in this field. It has great research and development value.

By fitting the evolutionary trend of all the technical content, we can see that the field is in the evolutionary rising stage. It proves the research value of this paper. The following figure shows the evolutionary trend of technology through the distribution of technical content in different years.

Figure 8 shows the proportion of each technology content in each year through four rings.

From Fig. 8 we can see that the color types of the four rings are gradually increasing. Combined with the previous data, we find that 5G market is in the stage of vigorous development. The technology content in this field is rapid and abundant. At the same time, all kinds of technologies have a common trend of development.

From the figure above, we can see the benefits of subdividing the patent content according to the number of patents. Decentralized patents include front-end technologies,

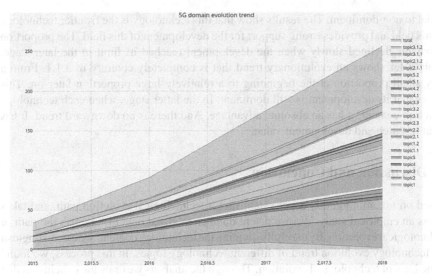

Fig. 7. Evolution of all technical content in the 5G field.

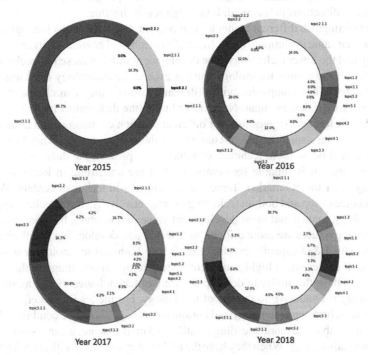

Fig. 8. Proportion of technical content in different years.

key technologies for subsequent development and ancillary technologies. For example, in topic 3.1.1, with the passage of time, the absolute proportion gradually decreases from initial to non-dominant. The results show that this technology is the frontier technology in this field and provides strong support for the development of this field. The proportion of patents declined slowly when the development reached its limit in the later stage. Topic 2.1.1 shows an evolutionary trend that is completely contrary to 3.1.1. From a very small proportion at the beginning to a relatively large proportion later on. This shows that its development is still dominant. In the latter stage, when each technology is rich in content, it has an absolute advantage. And there is no downward trend. It has great research and development value.

4   Discussion and Conclusion

Based on text mining, this paper establishes a technological evolution path, and takes 5G as an empirical study. We use patent number as visual data to establish the path of technological evolution. By carefully dividing the technology content, we can distinguish the technology evolution trend of different technology topics. In this process, we found some laws of technological evolution. Through the analysis we draw the conclusion that there is still a lot of room for development in 5G field on the whole. Secondly, the R&D of various technology content is mostly in a period of rapid growth. Finally, there are many research directions in this field. It has high research value.

We can distinguish different technological content, but different technological content follows the same evolutionary law. Most of the technical content exists in the upper technology and the lower technology. The development trend of these two technologies is the same, but the bottom technology can not exceed the technology thickness of the upper technology. For companies with different locations, they can choose the right technology for their development. Next, according to the different technology content, we provide development suggestions for different location communication companies.

Topic 1 has very small technical content, but the evolution is growing fast. In the future, there will be no more technical content, but at present, the development space is relatively large. It is suitable for companies that are not strong in R&D, but want to gradually enter the 5G market. Topic 2 and 3 are rich in technical content. We discusse the choice of top and bottom technologies respectively. Their top technology have high technology content and great development value in the future, but compared with lower technology, there are more difficult to research and develop. For small high-tech enterprises with R&D capabilities, they can choose high-end technologies with high R&D value. It can ensure a highly competitive technological advantage in the market, but it does not have the ability to develop a variety of technologies. Focusing on a single technology is conducive to the use of technology. For the leading enterprises with abundant funds, both top technology and bottom technology should be combined. While maintaining the absolute market leading position and market share, it can also ensure the technological advantages. When they have the ability, they can work with other high-tech companies to develop various technical themes and grasp market trends.

The trends in Topics 4 and 5 are weaker than other technologies, but still occupy a small share of the market. For companies whose main business is not communication technology and who wants to have sufficient financial support in the communication edge market through 5G communication, they can choose to develop topics 4 and 5. The current development rate of its technical content has slowed down, the main content is easy to understand, and there is still market value.

The trends in Topic 4 and 5 are weaker than other technologies, but still occupy a small share of the market. For companies whose main business is not communication technology and who wants to have sufficient financial support in 5G, they can choose to develop topic 4 and 5. The current development rate of its technical content has slowed down, the main content is easy to understand, and there is still market value.

This paper also provides the following countermeasures and suggestions for government departments in the formulation of 5G industrial policy: Governments at all levels should formulate policies to encourage scientific research institutions in colleges and universities to study related technologies, such as research subsidies, awards for scientific research achievements, establishment of laboratories, etc. At the same time, it should protect the interests of relevant communication companies, promote the marketization of relevant scientific research results, and play the role of supporter and protector in the overseas market of enterprises.

Of course, there are some problems in this paper. Firstly, because of the patent technology, the stop-use vocabulary can not filter the technology-independent vocabulary. Secondly, when establishing subject classification, excluding patents below threshold may omit technical topics. Finally, the same topic words will also appear in multiple paths, indicating that there may be interaction in the development of related technology topics. This paper only considers the development of technology theme of single-choice path.

In the follow-up study, we hope to use a large number of data to confirm the law of the technological evolution path we found. At the same time, we can consider the influence of path crossing.

References

1. Kelly, K.: What Technology Wants. Penguin, New York (2010)
2. Dong, J.R.: Technological innovation process management - theory, method and practice (2000)
3. Lu, W.M.Y.: A integrated of technology evolution and its new reseach focus—evolution of emerging technologies. Stud. Sci. Sci. **26**(3), 466–471 (2008)
4. Dosi, G.: Technological paradigms and technological trajectories : a suggested interpretation of the determinants and directions of technical change. Res. Policy **11**(3), 147–162 (1982)
5. Liu, X.: Technology track and independent innovation. Forum Sci. Technol. China **2**, 30–33 (1997)
6. Zhang, J., Liu, D.: Study of the interaction between innovation dynamics in local industrial clusters and industrial technology track. Stud. Sci. Sci. **25**(5), 858–863 (2007)
7. Xia, R., Xu, C., Huang, T.W.: Research on distribution characteristics on innovative opportunities based on difference of industrial technology trajectory. Sci. Technol. Prog. Policy **27**(24), 5–11 (2010)
8. Lei, L.S.L.P.X.: The change of emerging technology and the viewpoint of strategical resource. Chin. J. Manage. **2**(3), 304–306 (2005)
9. Sun, X.: Technological innovation and industrial evolution (2012)
10. Becker, R.H., Speltz, L.M.: Putting S-curve concept to work. Res. Manage. **26**, 31–33 (1983)

11. Green, S.G., Gavin, M., Aiman-Smith, L.: Assessing a multidimensional measure of radical technological innovation. IEEE Trans. Eng. Manage. **421**, 203–214 (1995)
12. Chen, Y., Huang, Y.Y., Fang, G.J.: Patent information collection and analysis (2006)
13. Bei, L., Xiangdong, C.: Comparative analysis of core and emerging technologies between Mainland China and Taiwan based on patent citation network. Sci. Res. Manage. **36**(2), 96–106 (2015)
14. Li, X.: Behind the recent surge of Chinese patenting: an institutional view. Res. Policy **41**, 236–249 (2012)
15. Tseng, Y.H., Lin, C.J., Lin, Y.I.: Text mining techniques for patent analysis. Inf. Process. Manage. **43**, 1216–1247 (2007)
16. Mogee, M.E.: Using patent data for technology analysis and planning. Res. Technol. Manage. **34**(4), 43–49 (2016)
17. Zhang, X., Fhang, S.: Review on Technology Evolution Research from Patent Citation Perspective. Sci. Sci. Manage. Sci. Technol. **37**(3), 58–67 (2016)
18. Liu, Y.: Research on evaluation of regional patent development based on principal component analysis and cluster analysis. Sci. Technol. Manage. Res. **28**(11), 80–82 (2008)
19. John, G.H., Lent, B.: SIPping from the data firehose. In: International Conference on Knowledge Discovery & Data Mining (1997)
20. Taotao, S.U.N., Xiaoli, T.A.N.G.: Research on evaluation of regional patent development based on principal component analysis and cluster analysis. Libr. Inf. Serv. **56**(4), 80–84 (2012)
21. Huiqun, G., Qiongze, L.: The research on discover robotics industry technology opportunities based on patent text mining. Sci. Technol. Prog. Policy **31**(5), 000070–000074 (2014)
22. Han, E.J., Sohn, S.Y.: Patent valuation based on text mining and survival analysis. J. Technol. Transfer **40**(5), 821–839 (2015). https://doi.org/10.1007/s10961-014-9367-6
23. Zhang, S., Wang, W., Yu, Y., et al.: Optimization of patent portfolio analysis method based on text mining technology. ITA **38**(10), 127–144 (2015). https://doi.org/10.1007/s10961-014-9367-6
24. Yoon, J., Choi, S., Kim, K.: Invention property-function network analysis of patents: a case of silicon-based thin film solar cells. Scientometrics **86**(3), 687–703 (2015)
25. Feldman, R., Dagan, I., Hirsh, H.: Mining text using keyword distributions. J. Intell. Inf. Syst. **10**(3), 281–300 (1998). https://doi.org/10.1023/A:1008623632443
26. Niemann, H., Moehrle, M.G., Frischkorn, J., et al.: Use of a new patent text-mining and visualization method for identifying patenting patterns over time: concept, method and test application. Technol. Forecast. Soc. Chang. **115**, 210–220 (2017)
27. Liu, Y., Li, Y.: The patent information analysis and strategic research of the fifth generation mobile communication technology. Sci. Technol. Manage. Res. **36**(9), 155–160 (2016)
28. Zhang, J., Pei, J., Yang, Z.: Research on competitive situation for the polar code patented technology in 5G era. Sci. Res. Manage. (S1).(2018)
29. Xiao, X., Zhao, H.: Analysis on competitive situation of 5G communication enterprises. Telecommun. Eng. Technics Stand. **30**(10), 47–53 (2017)
30. Wu, Y., Yang, D.: Research on identification of industrial core technology based on patent data—taking the field of fifth generation mobile communication industry as an example. J. Intell. **38**(3), 39–52 (2015)
31. http://www.sipo.gov.cn/
32. http://www.soopat.com/
33. Chen, L.Y.G., Zhang, J., et al.: Research on multiple main paths method oriented to analysis of technological evolution. Libr. Inf. Serv. **10**, 124–130 (2015)
34. https://pypi.org/project/jieba/
35. Wallach, H.M.: Topic modeling: beyond bag-of-words. Bibliometrics 977–984 (2006)

36. https://pypi.org/project/gensim/
37. Vector space model.https://en.wikipedia.org/wiki/Vector_space_model. Accessed July 2019
38. Zhang, W., Yoshida, T., Tang, X.: A comparative study of TF*IDF, LSI and multi-words for text classification. Expert Syst. Appl. **38**(3), 2758–2765 (2011)
39. Tao, L.: Application of machine learning methods in text classification. China Strateg. Emerg. Ind. **40**, 145–146 (2017)

Research on Advertising Click-Through Rate Prediction Model Based on Ensemble Learning

Xiaojuan He[1], Wenjie Pan[1(✉)], and Hong Cheng[2]

[1] School of Statistics and Information, Shanghai University of International Business and Economics, Shanghai 201620, China
hanspwj@126.com

[2] School of Statistics and Mathematics, Shanghai Lixin University of Accounting and Finance, Shanghai 201209, China

Abstract. The advertisement logs accumulated in the Internet have problems such as sparse data, large feature quantity, and extremely uneven distribution of positive and negative samples, which made it difficult to obtain interesting features and to improve precision for single prediction models. In response to these problems, this paper proposes a CTR prediction model based on GBDT-Stacking. GBDT-Stacking model uses the GBDT to automatically extract and transform features suitable for prediction and uses Stacking model to predict CTR of user, which improves the performance of baseline effectively. The experimental results in the real advertising dataset show that the GBDT-Stacking model of this paper uses increased by at least 4% compared to single model in AUC value.

Keywords: GBDT · Stacking · SMOTE · CTR

1 Introduction

Online Advertisement refers to advertising that uses the Internet as a carrier to spread through multimedia such as text and images, and sound (Gao et al. [1]). On October 14, 1994, Wired magazine opened the world's first website containing advertisements. About 12 sponsors paid for the banner ads placed on the entire website, including P&G, IBM and other world-famous big companies. This feat has become an important milestone in the history of online advertising. Since then, online advertising has begun to develop rapidly, and various online advertisements have emerged with a faster and newer appearance.

The original Internet advertising mostly uses a traditional media way to deliver advertising. By embedding fixed images and text in the page to display the advertising content, this advertising method is not flexible enough for the Internet platform, and it is difficult to match with the changing webpage, and the effect of the advertisement delivery is poor. With the rapid development of the mobile industry as a whole and the iteration of mobile terminals, the form of advertising has been fundamentally changed. The carrier of advertising has been gradually converted from online newspapers such as newspapers and magazines into online applications such as mobile applications. The content of the webpage and the characteristics of the visiting users can realize the targeted

© Springer Nature Singapore Pte Ltd. 2020
H. Han et al. (Eds.): IDMB 2019, CCIS 1099, pp. 82–93, 2020.
https://doi.org/10.1007/978-981-15-8760-3_6

delivery of advertisements, which is the development trend of the Internet company's advertising mechanism. Compared with traditional advertising, online advertising has a unique advantage in terms of coverage, flexibility, targeting, cost and effectiveness evaluation, and has developed into a multi-billion-dollar industry. Agents usually get revenue from advertisers according to the number of clicks on advertisements, so they can maximize their profits by predicting the probability of users clicking on advertisements accurately. And predict the probability of users clicking on advertisements is Click-Through Rate (CTR) prediction. CTR prediction is an advertising mechanism based on a given user and webpage content, which is calculated to obtain the best matching advertisement for the user (Zhou et al. [2]). This mechanism can greatly improve the ad Click-through rate, increase the number of visits to the websites served by the ads, help users to obtain high-quality information, and build a benign and harmonious advertising industry chain.

Accurately predict the behavior of users clicking on advertisements, recommend advertisements that users are interested in and increase users' click-to-advertising behavior, not only can obtain the best publicity results for advertisers, but also find ways for agents to realize large-scale liquidation means and avoid a waste of resources. However, most of the advertisement logs accumulated in the Internet is very sparse, the features are high dimension, and the proportion of samples is extremely uneven. How to use an efficient model to extract important information from the data to improve the ability of CTR prediction become a huge challenge. Only by solving these problems can agents provide personalized services to users, improve user experience satisfaction, and optimize the performance of advertising systems. Therefore, the prediction of advertising click-through rate is not only of great research significance, but also of high commercial value.

Online advertising click-through rate is an important research topic in computing advertising. Predictive models are the primary task of analyzing and predicting ad click-through rates. With the development of technologies such as data mining and natural language processing, there have been many mature researches on the prediction of ad click rate. Richardson et al. [3] uses Logistic Regression (LR) and Multiple Additive Regression Trees (MART) to predict the click-through rate. The results show that the LR is better than MART. Shen et al. [4] proposes an improvement of LR model based on the online optimization algorithm -Follow the Regularized Leader (FTRL) which uses mixed regularization to prevent training over-fitting. FTRL algorithm not only improves the calculation efficiency of the parameters, but also optimize the model parameters. Although the LR model has low complexity and strong interpretability, the performance of the model depends on artificial structural features. In order to learn the nonlinear relationship in the advertising data and improve the predictive ability of the model, Rendle [5] proposes the Factorization Machines (FM) model, and the FM model uses the idea of matrix decomposition, which can reduce the dimensions of the training parameters and can mine the relationship between the different feature components. However, when the FM model combines two features, it is assumed that the same feature produces the same influence. But in fact, when combined with features of different feature domains, the hidden vectors may exhibit different distributions. For that, Juan et al. [6] further develops the field-aware factorization machine (FFM), which is characterized by segmentation of features into several domains, each of which will be studied for implicit variables in different feature domains.

In recent years, deep learning technology has been continuously improved and the ability of data processing has been continuously improved. Deep learning has achieved great success in the fields of computer vision, speech recognition, and natural language processing, and researchers have also begun to explore deep learning on CTR prediction. Zhang et al. [7] consider the sequential behavior of users clicking on the ad, and construct Recurrent Neural Network (RNN) to predict click-through rate. However, when the RNN uses the gradient descent optimization algorithm, it will cause a gradient explosion problem. Zhang et al. [8] propose a feature dimension reduction method based on tensor decomposition, and utilize the deep learning technique to describe the nonlinear correlation in data to solve the feature learning problem of high-dimensional sparse data. In addition, the prediction models used to study the click-through rate of advertising include SVM (Dave et al. [9]), Probabilistic Graphical Model (Yue et al. [10]), hierarchical Bayesian models (Qin [11]), random forests (Xiao et al. [12]), and deep confidence networks (Yang et al. [13]), CNN-LSTM network (She et al. [14]) and so on.

In summary, most of the prediction models of online ad CTR use a single machine learning algorithm. However, the single classifier has poor ability in predicting online CTR due to the applicability and the constraints of the preconditions of the single classifier. With the emergence and rapid development of ensemble learning technology, the learner began to apply variety of machine learning algorithms to predict online ads CTR. Ensemble learning is effective way to improve the predictive performance of the learning system by training multiple predictive models and combining the results in a specific way, which can make up for the shortcoming of a single algorithm and get better performance. Therefore, combining the domestic and foreign scholars' research, this paper proposes to use the Stacking model to predict the online advertising click rate. In the process of feature construction, we draw on the thoughts of He et al. [15], use the Gradient Boosting Decision Tree (GBDT) to automatically construct features that make the click rate prediction performance more reliable.

2   The GBDT-Stacking Model

The process of GBDT-Stacking model proposed in this paper is as follows: firstly, a new feature X is constructed based on the original data using the GBDT; then choose the FM model, FFM model, random forest (RF), and eXtreme Gradient Boosting (XGBoost) as the primary predictive model of Stacking, their prediction probability of each data is $P\{P_{FM}, P_{FFM}, P_{RF}, P_{XGB}\}$; and finally use P as the input of the two-layer neural network model and get the result. The process is shown in Fig. 1.

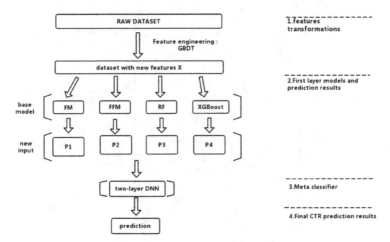

Fig. 1. CTR prediction process based on GBDT-Stacking model

2.1 Feature Engineering

The data used to predict click-through rates tends to be highly dimensional, and in reality, not the more features, the better the prediction will be. It is necessary to balance the relationship between effect and efficiency. Therefore, in the process of click rate prediction, it is critical to apply features that are highly correlated with the accuracy as much as possible to reduce the cost of time and material force. There are two common methods of feature transformation. One is a nonlinear transformation method for continuous features, which bin features and treat the bin as a new classification feature. The other is to construct the tuple input feature for categorical features, which uses Cartesian Product to obtain all possible values of the classification feature, and then deletes the combination that has no effect on predictions. The gradient boosting machines is a powerful method to achieve the above two feature transformation methods (He et al. [15]).

The idea of using GBDT for features transformations is to treat each individual tree as a classification feature, take the index of the leaf that each variable ends in as a result, and then use the One-Hot encoding to generate the new feature. The process is shown in Fig. 2:

The GBDT model with two subtrees is concerned, its output of each subtree is the new feature transformed. The left subtree contains three leaves, and the right contains two leaf. If an input variable falls on the third leaf of the left subtree and the second leaf of the right subtree, the new binary feature generated is [0,0,1,0,1]. The GBDT model is the classics L2-TreeBoost algorithm that Friedman [16] put forward.

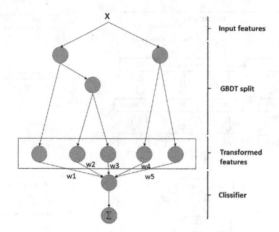

Fig. 2. GBDT feature transform process

2.2 Ensemble Learning Model Structure

Ensemble learning is an algorithm based on weak learners to generate strong learners. Because of its powerful generalization ability, it has become a research hotspot in the field of machine learning. At present, the common ensemble learning methods are mainly divided into two categories: one is to use the same learning algorithm in different data, and it improves the predictive performance of the final learner by constantly modifying the parameters of the model, such as Bagging and Boosting; the second is to use different types of algorithms in the same data set, and it improves the predictive performance by combining the advantages of different learners, such as Stacking. Ali Labs used the Stacking model to predict the probability of KKBOX's churn (Wang et al. [17]). They found that the prediction performance of the Stacking is better than that of the random forest, XGBoost and benchmark models.

The general steps of prediction using Stacking are following: First, using FM, FFM, RandomForest and XGBoost as the first layer of Stacking to train training data with the transformed features X respectively. Then using these four trained models to predict each test data and get corresponding prediction result. At last, using the result of previous layer as new input of neural network to get the final CTR probability. In order to avoid over-fitting caused by using the same data as training and test data, this paper uses the 5-fold cross-validation method to generate the secondary training set. The steps are as follows:

Step1, the training set is divided into 5 parts on average, one of which is used as a validation set, and the remaining are used as training sets, which are marked as {TR_1, TR_2, TR_3, TR_4, TR_5};

Step2, using FM model to fit training sets, then applying trained FM model to predict validation set and get its result P_1. The traversal is carried out 5 times in tune, and we can get all validation result P{P_1, P_2, P_3, P_4, P_5} used as new input to the second layer model; for test data, using each trained model to predict the test set, and each sample will get 5 output, which are averaged as input to the second layer model. As shown in Fig. 3.

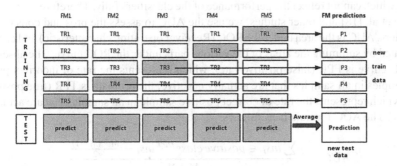

Fig. 3. FM model click rate prediction

Step3, repeating Step2 for FFM model, RF and XGBoost model, and finally get all predicted results of training set and test set in the first layer models. The results are used as input to train the next two-layer DNN model to obtain the final CTR probability y (Fig. 4).

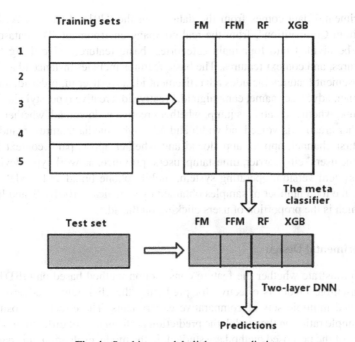

Fig. 4. Stacking model click rate prediction

2.3 Evaluation Metrics

Accuracy has certain disadvantage as measurement when the data is extremely unbalanced, which cannot reflect the performance of the classifiers fully. Therefore, according to Tian et al. [18] and other studies, we use the AUC to assess the pros and cons of the classifiers. AUC is the area under the ROC (Receiver operating characteristic) and can be calculated by summing the areas of the various parts under the ROC curve. The abscissa of ROC curve is FPR (false positive rate), which represents the rate of incorrect positive samples in all samples that are actually negative; the ordinate is TPR (true positive rate), which represents the rate of correct positive sample in all samples that are actually positive. The AUC formula is as follows:

$$AUC = \frac{\sum ins_i \in positive\ class\ ^{rank}ins_i - \frac{M * (M+1)}{2}}{M * N}.$$ (1)

Among them, $^{rank}ins_i$ is the serial number of samples, M is the number of positive examples and N is the number of negative samples. AUC has a good tolerance for sample proportion, and the larger the AUC value, the better the performance of the classifier.

3 Empirical Study

The experimental data comes from the data set in the "2018 iFLYTEX AI Marketing Algorithm Competition", from the real company environment. The dataset in the paper can be divided into four main categories: basic features, advertising features, media features, and context features. The basic features includes instance id and clicks; the advertisement features includes advertisement id, advertiser id, order id, advertiser industry inner, advertiser name, campaign id, creative id, creative type, style id, whether has deep link, whether creative is jump, whether creative is download, whether creative is js, whether creative is voiced, ad width and ad height; media features including app category, first channel, app id, app slot id and whether app is paid; context features including the users' city, carrier, timestamp, users' province, network type, device type, operating system version, operating system, mobile phone brand and model. For the given data, the total number of samples obtained by statistics is 1001650, and less than 20% of which is the proportion of users clicking on the ad.

3.1 Experimental Design

In order to illustrate whether the feature construction method based on GBDT model and the ensemble model is effective for predicting the click rate of advertisements, we conducted multiple sets of comparative experiments. The effects of positive and negative sample ratios on the model, the prediction performance of different models and the stability of the proposed method are studied. In the model evaluation, this paper uses the AUC value to evaluate the performance of the model prediction. In the process of constructing new features by GBDT, we chose 100 as the number of iteration trees, and finally constructed 800 new features.

In the process of preprocessing the data, in order to avoid the influence of sample imbalance on the prediction performance, the Synthetic Minority Oversampling Technique (SMOTE) is used to synthesize minority sample. We use the SMOTE to adjust the ratio of positive and negative samples to 1:1, and prove that the model can obtain the optimal prediction ability when the positive and negative samples are balances by experiment. Different from the random oversampling algorithm which based on the strategy of copying samples from minority samples, the SMOTE algorithm adds a new minority data by "interpolation" method to balance data. Therefore, the unbalanced data can be processed better, and the prediction performance and generalization ability of the machine learning algorithm are improved.

3.2 Analysis of Results

We first studied the influence of different sample ratios on the prediction performance of the model, and then measured the performance of different models. Finally, the GBDT-Stacking prediction model proposed in this paper is tested on different magnitudes to verify the stability of the model in the click rate prediction problem.

3.2.1 The Effect of Positive and Negative Sample Ratio on Model Prediction

The purpose of this set of experiments is to study the effect of the ratio of positive and negative samples on the predictive performance of the model. Training data were extracted from the original data in order of positive and negative samples ratio of 4:1 to 1:4, and the GBDT-Stacking model was experimented with these data respectively. Through experiments, we found a great proportion of positive and negative samples that can optimize the prediction performance, and chose the ratio to apply to the next experiment. The AUC values obtained from the seven sets of experiments are described

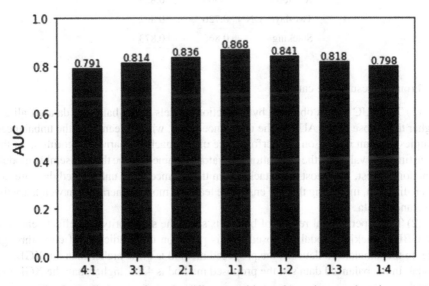

Fig. 5. Prediction results under different positive and negative sample ratios

in Fig. 5, where the horizontal axis is the ratio of samples and the vertical axis is the AUC value.

It can be seen from the figure that the AUC value obtained by the experiment is the highest when the ratio of samples is 1:1, which is 0.868. Other than this, the more unbalanced the ratio of samples, the lower the AUC value obtained by the model also can be observed. By analyzing the results of this set of experiments, we can find that the data sample imbalance will affect the click rate prediction performance, and the model on CTR prediction will get the best prediction performance when the positive and negative sample ratio is 1:1.

3.2.2 Comparison of Different Prediction Models

This set of experiments conducted experiments in the original data and the balanced data sets respectively, and compared the predictive performance of the single prediction method and the proposed method in the advertising click rate. Firstly, we use GBDT to construct feature for the two data sets. Then, using FM model, FFM model, Random Forest, XGBoost, the two-layer neural network model and Stacking for prediction and comparing their predictive ability. the AUC values of each mode are obtained as follows (Table 1).

Table 1. Comparison of different prediction model results

Model	Unbalanced	Balanced
FM model	0.691	0.798
FFM model	0.722	0.805
Random Forest	0.762	0.787
XGBoost	0.803	0.825
Two-layer DNN	0.726	0.802
Stacking	0.866	0.873

From the results, we can find:

(1) The AUC values obtained by prediction models in the balanced data set all are higher than those of the AUC in the unbalanced data, which means that the unbalanced samples have an impact on the performance of the machine learning algorithm. At the same time, the values of the evaluation indicators obtained by the three ensemble models (Random Forest, XGBoost and Stacking) in the balanced and unbalanced data are not much different, indicating that the ensemble learning model generally can well learn the unbalanced data.

(2) The experimental results of both sets show the superiority and effectiveness of the GBDT-Stacking model proposed in this paper on the prediction of click-through rate. In the unbalanced data set, the proposed method is 6.3% higher than the XGBoost model. In the balanced data set, the proposed method is 4.8% higher than the XGBoost model.

3.2.3 Stability Experiment

This set of experiments used random undersampling techniques to extract 100 thousand, 300 thousand, 500 thousand, 700 thousand, 1,100 thousand, 1,300 thousand, and 1,500 thousand sample sets from the balanced training data set for GBDT-Stacking experiments. We compared the prediction performance of the proposed model under different data volumes and evaluated the stability of the model.

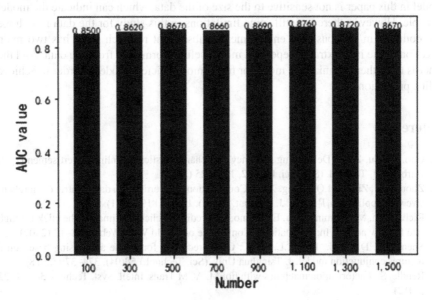

Fig. 6. AUC values of the model under different data volumes

Referring to Fig. 6, we can find that when the amount of data increases gradually, the prediction performance of the model shows an upward trend, which indicates that as the training data increases, the model is more fully trained to obtain more robust parameters. At the same time, the differences in AUC values obtained by the proposed model under the data of different magnitudes are not obvious, and the proposed model can also obtain relatively stable results in the case of sample imbalance as mentioned above. In summary, it can be considered that the proposed model is more stable than the single model in terms of click rate prediction.

4 Conclusion

In this study, we achieve the ad click rate prediction based on the ensemble learning model, and through a series of experiments study the effectiveness of the model.

First of all, due to the sparseness of the current advertising data, the non-linear relationship between features, and the time-consuming feature extraction, the paper uses the GBDT model to nonlinearly transform the original data, and generates new features suitable for prediction, which greatly reduced the cost of manual feature engineering.

Secondly, while using GBDT for feature engineering, the idea of ensemble learning was introduced, we use Stacking to predict the click-through rate on the constructed features. By comparing with the results of a single prediction model, the superiority and effectiveness of the GBDT-Stacking prediction model over the single prediction model are verified.

Finally, by experimenting in the order of magnitude of different magnitudes, the prediction results obtained have small fluctuations, so it can be shown that the proposed model in this paper is not sensitive to the size of the data, which can indicate the model is stable. However, there are still some limitations in this study for the data is so huge and complex in real business environment. Subsequent research work has two main directions. One is to extract deeper and more useful information from the data. And the other is to further optimize the model or find a more efficient model in order to achieve online prediction.

References

1. Gao, C., Lu, Z.M.: Developing tendency and characteristics of online advertisement[J]. J. Harbin Inst. Technol. (Soc. Sci. Ed.) **02**, 122–125 (2003)
2. Zhou, A.Y., Zhou, M.Q., Gong, X.Q.: Computational advertising: a data-centric comprehensive web application[J]. Chin. J. Comput. **34**(10), 1805–1819 (2011)
3. Richardson, M., Dominowska, E., Ragno, R.: Predicting clicks: estimating the click-through rate for new ads[C]. In: International Conference on World Wide Web, pp. 8–12 (2007)
4. Shen, F.Y., Dai, G.J., Dai, C.L., et al.: CTR prediction for online advertising based on a features conjunction model. J. Tsinghua Univ. (Sci. Technol.) **58**(04), 374–379 (2018)
5. Rendle, S.: Factorization machines with libFM. ACM Trans. Intell. Syst. Technol. **3**(3), 1–22 (2012)
6. Juan, Y., Zhuang, Y., Chin, W.S., et al.: Field-aware factorization machines for CTR Prediction. In: RecSys 2016 Proceedings of the 10th ACM Conference on Recommender Systems, 15. 9. 2016–19. 9. 2016, Boston, Massachusetts, USA, pp. 43–50. ACM Press (2016)
7. Zhang, Y., Dai, H., Xu, C., et al.: Sequential click prediction for sponsored search with recurrent neural networks (2014)
8. Zhang, Z.Q., Zhou, Y., Xie, X.Q., et al.: Research on advertising click-through rate estimation based on feature learning. Chin. J. Comput. **39**(04), 780–794 (2016)
9. Dave, K.S., Varma, V.: Learning the click-through rate for rare/new ads from similar ads. In: International ACM SIGIR Conference on Research & Development in Information Retrieval (2010)
10. Yue, Q., Wang, C.L., Zhu, Y.L., et al.: Click-through rate prediction of online advertisements based on probabilistic graphical model. J. East China Normal Univ. (Nat. Sci.) **03**, 15–25 (2013)
11. Qin, C.Y.: Research on influencing factors of keyword conversion rate based on hierarchical Bayesian method. Stat. Decis. **19**, 64–67 (2016)
12. Xiao, Y., Bi, J.F., Han, Y., et al.: Click rate prediction research in online advertising. J. East China Normal Univ. (Nat. Sci.) **05**, 80–86 (2017)
13. Yang, C.C., Mei, J.J., Wu, Y., et al.: Advertising click-through rate prediction based on feature dimension reduction and deep belief network. Comput. Eng. Des. **39**(12), 3700–3704 (2018)
14. She, X.Y., Wang, S.P.: Research on advertising click through rate prediction model based on CNN-LSTM network. Comput. Eng. Appl. **55**(02), 193–197 (2019)

15. He, X., Pan, J., Jin, O., et al.: Practical lessons from predicting clicks on ads at facebook. In: Proceedings of 20th ACM SIGKDD Conference on Knowledge Discovery and Data Mining, pp. 1–9. ACM (2014)
16. Friedman, J.H.: Greedy function approximation: a gradient boosting machine. Ann. Stat. **29**(5), 1189–1232 (2001)
17. Wang, Z.N., Xiao, W.M., Wang, J.: A practical pipeline with stacking models for KKBOX's churn prediction challenge. In: Proceedings of the 11th ACM International Conference on Web Search and Data Mining (2018)
18. Tian, C.L., Zhang, X., Pan, B., Yang, C., Xu, Y.R.: Research and implementation of feature extraction methods on Internet CTR prediction model. Appl. Res. Comput. **34**(2), 334–338 (2017)

Health and Biological Data Science

An Integrated Robust Graph Regularized Non-negative Matrix Factorization for Multi-dimensional Genomic Data Analysis

Yong-Jing Hao, Mi-Xiao Hou, Rong Zhu, and Jin-Xing Liu$^{(\boxtimes)}$

School of Information Science and Engineering, Qufu Normal University, Rizhao 276826, China
yongjinghaozb@163.com, mixiaohou@163.com,
{zhurongsd,sdcavell}@126.com

Abstract. With the generation of "multi-dimensional genomic data", the multi-platform genomic analysis technology for simultaneous biological samples has been rapidly developed. However, the existing shortcomings of the models for comprehensive analysis of multi-dimensional genomic data are the lack of robustness. We use the integrated matrix factorization model by introducing $L_{2,1}$-norm to deal with above problem. In this paper, we propose an Integrated Robust Graph Regularization Non-negative Matrix Factorization for multi-dimensional genomic data analysis which named iRGNMF. In the formulation of the objective function, we introduced the local structure to maintain regularization and $L_{2,1}$-norm to consider the data geometry and model robustness. We also applied this model to three types of data of the same cancer from The Cancer Genome Atlas (TCGA). Experiments shown that iRGNMF has obtained considerable effects in sample clustering, and found suspicious disease genes, which are used to reveal hidden patterns and information of biology in multi-dimensional gene data.

Keywords: Integrated model · NMF · TCGA · $L_{2,1}$-norm · Network analysis

1 Introduction

Non-negative matrix factorization (NMF) is a common matrix factorization technique designed to decompose non-negative data matrices into two low rank non-negative matrices. NMF was first proposed in Paatero and Tapper's paper [1]. In the follow-up work, Lee and Seung's research made NMF [2] to be an active field. In past studies, many NMF-based variants were nominated by introducing various constraints, including discriminant constraints, network regularization or local retention constraints, sparse constraints, orthogonality constraints and other constraints. NMF and its improved models have been defined as one of the valuable exploratory analysis tools and have been successfully applied in many fields, including bioinformatics [3, 4], data mining [5], signal processing [6], pattern recognition [7], etc.

Most of the existing improved NMF-based models can only be applied to the analysis of a single data matrix. In most cases, the occurrence of cancer is the result of the action

© Springer Nature Singapore Pte Ltd. 2020
H. Han et al. (Eds.): IDMB 2019, CCIS 1099, pp. 97–111, 2020.
https://doi.org/10.1007/978-981-15-8760-3_7

of multiple genes, as well as the interaction between various genes. With the development of second-generation sequencing technology [8], multi-dimensional genomic data has been greatly developed. Different types of multi-dimensional genomic data for the same cancer can be gained in the published database TCGA (https://www.cancer.gov). Thence it is crucial for the design of multi-dimensional data models, which has aroused much attention of researchers. In the research work of multi-dimensional genomic data, the integration research of genomic data is more and more extensive. Joint analysis of data from the same set of samples from multiple omic sources may yield results that are not available in a single data analysis and provide more comprehensive biological implications. The key step in the general multi-dimensional data model is how to handle heterogeneous data. Data from different sources is difficult to compare due to inherent differences in data. Different genomic variables are measured and collected in different ways, and they are also different from different types of noise. To analyze different types of data for the same cancer, many researchers have proposed various integration models.

Recently, more and more matrix factorization models for multi-dimensional data have been put forward. For example, the improved matrix factorization-based algorithm integration matrix factorization (IMF) [9] proposed by Creene and Cunningham in 2009 which can learn the structure of embedded underlying clusters across multiple views, It is a post-finishing strategy, IMF combines the clustering solution for each view. A joint matrix factorization model called SNMFCA [10]: Supervised NMF-Based Image Classification and Annotation was proposed by Jing *et al.* in 2012. The purpose of this model is to make full use of known class relationships to determine potential class relationships. In order to further analyze the data, a Joint Non-negative Matrix Factorization (jNMF) [11] based on the original NMF improvement was proposed by Zhang *et al.* in 2012. This method allows multiple inputs and multiple outputs, which has the advantage of not only detecting partial-based fundamental patterns in each matrix, but also the potential relationship between different matrices can be fully considered, and jNMF has a good application for bioinformatics. Data Fusion by Matrix Factorization (DFMF) [12] with a penalty matrix was proposed by Zitnik and Zupan in 2015. This method aims to simultaneously decompose multiple relation matrices in the same framework, making excellent use of the abstract relation matrix. In 2016, a integration Orthogonal-regularized Non-negative Matrix Factorization (iONMF) [13] method based on jNMF improvement was proposed by Stražar *et al.* used to predict protein-RNA interactions. iONMF regularizes by integrating multiple types of data and orthogonality on the base matrix. The coefficient matrix of the model is learned from the training datasets, and the basic matrix is learned from the test datasets, there by completing the prediction of the interaction matrix. However, the jNMF and iONMF models do not take into account the inherent geometric relationship between the processed data and the lack of robustness of the model itself in solving the problem, which makes the part useful information ignored and the robustness of the model limited.

In this paper, we summarize and introduce the Integrated Robust Graph Regularization Non-negative Matrix Factorization (iRGNMF) for Multi-dimensional Genomic Data Analysis. We use the multiplication update algorithm and existing NMF pattern recognition and data integration techniques. In contrast, iRGNMF has the following characteristics:

(i) The regularity of the graph is introduced in the iRGNMF model. It can explore more valuable information through considering the geometry of the data itself.

(ii) The introduction of the $L_{2,1}$-norm in the iRGNMF model can improve the robustness of the model.

(iii) iRGNMF can be solved directly by the multiplication update algorithm, which proves that the process is simple and easy to understand.

The remainder of this paper is structured as follows: We describe the related work including NMF and RGNMF in Sect. 2; we present the methodology of iRGNMF in Sect. 3; the experiment of the bioinformatics datasets is listed in Sect. 4; last but not the least, we give a conclusion in Sect. 5.

2 Related Work

2.1 NMF

NMF model decompose one input matrix into two output matrices both of their elements are non-negative. As for the input data matrix $\mathbf{X} = [x_1, x_2, x_3, \ldots, x_M] \in R^{m \times n}$, where m is the number of features about the input data matrix. n is the number samples about of the input data matrix. The expression of NMF [14] can be described by:

$$\mathbf{X} = \mathbf{AY},$$
$$s.t. \ \mathbf{A} \geq 0, \ \mathbf{Y} \geq 0. \tag{1}$$

To reduce data dimensions, NMF minimizes the following objective function:

$$O^{NMF} = \|\mathbf{X} - \mathbf{AY}\|_F^2, \tag{2}$$

where $\|\cdot\|_F$ denotes the F-norm of the matrix. Lee and Seung proposed an iterative updates algorithm [15]:

$$a_{ik} \leftarrow a_{ik} \frac{(\mathbf{XY}^T)_{ik}}{(\mathbf{AYY}^T)_{ik}}. \tag{3}$$

$$y_{jk} \leftarrow y_{jk} \frac{(\mathbf{A}^T\mathbf{X})_{jk}}{(\mathbf{A}^T\mathbf{AY})_{jk}}. \tag{4}$$

It is proved that the above iterative update algorithm is convergent.

2.2 RGNMF

Compared with other improved NMF models, robust NMF with manifold learning not only considers the internal spatial structure of data but also has good stability, showing remarkable performance in experiments. Robust Graph Regularized Non-negative Matrix Factorization (RGNMF) [3] model is a classic NMF-based model that goes into a variety of learning. Minimize the following equation:

$$O^{RGNMF} = \|\mathbf{X} - \mathbf{AY}\|_{2,1} + \lambda \operatorname{tr}(\mathbf{YLY}^T),$$
$$s.t. \ \mathbf{A} \geq 0, \ \mathbf{Y} \geq 0, \tag{5}$$

where tr($\cdot$) is the trace of the matrix, and the regularization parameter λ controls the smoothness of the regularization of the graph. $\mathbf{L} = \mathbf{D} - \mathbf{W}$, which is named Graph Laplacian [16], $\mathbf{W}$ is the weight matrix of the graph, $\mathbf{D}$ is the diagonal matrix.

3 Methodology

In this section, we seek a new Integrated Robust Graph Regularized Non-negative Matrix Factorization (iRGNMF) model for multi-dimensional genomic data analysis by using the combination of Laplacian regularization and $L_{2,1}$-norm.

3.1 The $L_{2,1}$-Norm

The $L_{2,1}$-norm [17] of one matrix was originally employed as a rotational invariant L_1-norm [18], which was usually used fortensor factorization, multi-task learning and so on. The $L_{2,1}$-norm [19] mathematical definition is as follows:

$$\|\mathbf{P}\|_{2,1} = \sum_{i=1}^{m} \sqrt{\sum_{j=1}^{n} \mathbf{P}_{i,j}^2} = \sum_{i=1}^{m} \|p_i\|_2, \tag{6}$$

where m is the number of rows in the formula and n is the number of columns. When $m = 1$, it is equivalent to finding the L_2-norm for the entire column vector, so the understanding of the $L_{2,1}$-norm is the sum of the L_2-norm of the row vectors. The L_2-norm [20] refers to the sum of the squares of the elements of the vector and then the square root. The $L_{2,1}$-norm can produce the advantage of line sparsity relative to the L_2-norm and theL_1-norm [21].

3.2 Laplacian Regularization

Through the study of graph theory [22] and manifold learning theory [23], we can know that embedding the local geometry of data in high-dimensional data with nearest neighbor graph method can improve the effect of the model. For every data point x_i, we find its k nearest neighbors and place the edge between its neighbors. There are many options for defining a weight matrix $\mathbf{W}$ on a graph. w_{jl} is used to measure the proximity of two points x_j and x_l.

There are a lot of weighting schemes, like heat kernel weighting [24], Gaussian weighting [25], dot-product weighting [26] and zero-one weighting [27]. The method of defining the weight matrix is as follows: zero-one weighting: $w_{jl} = 1$, only in this status that nodes j and l are linked by one edge. This is the very wise weighting method and is easy enough to calculate. $w_{jl} = 1$ is only used to measure the proximity of two data points in the method, so we calculate the distance between the two data points by using of the Euclidean distance [28]. Next, we use the following formula to calculate the smoothness of the low-dimensional matrix:

$$\begin{aligned} R &= \tfrac{1}{2} \sum_{j,l}^{N} \|s_j - s_l\|^2 w_{jl} = \sum_{j=1}^{N} s_j^T s_j d_{jj} - \sum_{j,l=1}^{N} s_j^T s_l w_{jl} \\ &= \mathrm{Tr}\left(\mathbf{Y}\mathbf{D}\mathbf{Y}^T\right) - \mathrm{Tr}\left(\mathbf{Y}\mathbf{W}\mathbf{Y}^T\right) = \mathrm{Tr}\left(\mathbf{Y}\mathbf{L}\mathbf{Y}^T\right) \end{aligned} \tag{7}$$

3.3 Integrated Robust Graph Regularized Non-negative Matrix Factorization (iRGNMF)

The integrative model proposed in this paper which is designed for multi-dimensional genomic data. The formula of the model can be written as follows:

$$O^{iRGNMF} = \sum_{I=1}^{d} \|\mathbf{X}_I - \mathbf{A}_I \mathbf{Y}\|_{2,1} + \sum_{I=1}^{d} \lambda_I \text{tr}(\mathbf{Y}\mathbf{L}_I \mathbf{Y}^T),$$
$$s.t. \ \mathbf{A}_I \geq 0, \quad \mathbf{Y} \geq 0, \tag{8}$$

where d is the number of datasets. $\mathbf{X}$ and $\mathbf{L}$ are one-to-one correspondence, and λ is the regularization factor of the graph regular term.

3.4 An Efficient Algorithm of iRGNMF

The error function in Eq. (8) can be rewritten as:

$$\begin{aligned} f &= \text{tr}\big((\mathbf{X}_I - \mathbf{A}_I \mathbf{Y})\mathbf{G}_I(\mathbf{X}_I - \mathbf{A}_I \mathbf{Y})^T\big) + \lambda_I \text{tr}(\mathbf{Y}\mathbf{L}_I \mathbf{Y}^T) \\ &= \text{tr}(\mathbf{X}_I \mathbf{G}_I \mathbf{X}_I^T) - 2\text{tr}(\mathbf{X}_I \mathbf{G}_I \mathbf{Y}^T \mathbf{A}_I^T) + \text{tr}(\mathbf{A}_I \mathbf{Y}\mathbf{G}_I \mathbf{Y}^T \mathbf{A}_I^T) + \lambda_I \text{tr}(\mathbf{Y}\mathbf{L}_I \mathbf{Y}^T), \end{aligned} \tag{9}$$

where $\mathbf{G}$ is a diagonal matrix, which elements obtained by:

$$G_{jj} = 1/\sqrt{\sum_{m=1}^{i} (\mathbf{X} - \mathbf{A}\mathbf{Y})_{mj}^2 + \delta} = 1/\|x_j - \mathbf{F}g_j + \delta\|, \tag{10}$$

where δ is a positive number close to 0 but not equal to 0.

To resolve the problem in Eq. (7), we first give the Lagrangian multipliers Φ_{ik} and Ψ_{kj} to restricts $a_{ik} \geq 0$ and $y_{kj} \geq 0$. Let Φ_{ik} and Ψ_{kj}, the Lagrangian function L be written as [29]:

$$\begin{aligned} L &= \text{tr}(\mathbf{X}_I \mathbf{G}_I \mathbf{X}_I^T) - 2\text{tr}(\mathbf{X}_I \mathbf{G}_I \mathbf{Y}^T \mathbf{A}_I^T) + \text{tr}(\mathbf{A}_I \mathbf{Y}\mathbf{G}_I \mathbf{Y}^T \mathbf{A}_I^T) \\ &\quad + \lambda_I \text{tr}(\mathbf{Y}\mathbf{L}_I \mathbf{Y}^T) + \text{tr}(\Phi \mathbf{A}_I^T) + \text{tr}(\Psi \mathbf{Y}^T) \end{aligned} \tag{11}$$

Using the KKT [30], the iterative formula of the corresponding variable is shown as follows:

$$a_{ik} \leftarrow (a_I)_{ik} \frac{(\mathbf{X}_I \mathbf{G}_I \mathbf{Y}^T)_{ik}}{(\mathbf{A}_I \mathbf{Y}\mathbf{G}_I \mathbf{Y}^T)_{ik}}. \tag{12}$$

$$y_{jk} \leftarrow y_{jk} \frac{\sum_{I=1}^{d} (\mathbf{A}_I^T \mathbf{X}_I \mathbf{G}_I + \lambda_I \mathbf{Y}\mathbf{W}_I)_{jk}}{\sum_{I=1}^{d} (\mathbf{A}_I^T \mathbf{A}_I \mathbf{Y}\mathbf{G}_I + \lambda_I \mathbf{Y}\mathbf{D}_I)_{jk}}. \tag{13}$$

The summary of the steps of our algorithm is shown in Algorithm 1. Repeat the algorithm steps until convergence.

Algorithm 1: iRGNMF

Input: X and parameter λ.

Output: Y, **A** and **W**.

1: Initialize $\mathbf{A}_{I0}$, $\mathbf{Y}_0$.

2: Repeat

Update $\mathbf{A}_{r+1}$ as

$$\mathbf{A}_{r+1} \leftarrow \mathbf{A}_r \frac{(\mathbf{X}_I \mathbf{G}_I \mathbf{Y}^T)_{ik}}{(\mathbf{A}_I \mathbf{Y} \mathbf{G}_I \mathbf{Y}^T)_{ik}}$$

Update $\mathbf{Y}_{r+1}$ as

$$\mathbf{Y}_{r+1} \leftarrow \mathbf{Y}_r \frac{\sum_{I=1}^{d}(\mathbf{A}_I^T \mathbf{X}_I \mathbf{G}_I + \lambda_I \mathbf{Y} \mathbf{W}_I)_{jk}}{\sum_{I=1}^{d}(\mathbf{A}_I^T \mathbf{A}_I \mathbf{Y} \mathbf{G}_I + \lambda_I \mathbf{Y} \mathbf{D}_I)_{jk}}$$

Compute diagonal matrix $\mathbf{L}_I$ according to Eq. (7).

Compute diagonal matrix $\mathbf{G}_I$ according to Eq. (11).

$r = r + 1$

Until convergence

3.5 Computational Complexity Study

In this subsection, we discuss the computational complexity of our approach. Using the large O to represent the computational complexity of the model, the arithmetic operation of calculating the inner loop in iRGNMF is very similar to the multi-view update rule of the single view, assuming that the multiplication update stops after the tin iteration, and the time cost of the multiplication update become $O(dt_{in}MNK)$. In addition to the multiplication update, iRGNMF also requires $O(dNK)$ to calculate the Laplacian matrix from the equation. The total cost of the iRGNMF method is $O(dt_{in}t_{out}MNK)$ until the inner loop converges. It is worth noting that our method converges after 200 times and is proved in the experimental part.

4 Experiment

4.1 Datasets

There are four cancer datasets involved in this paper: Pancreatic Adenocarcinoma (PAAD), Esophageal Carcinoma (ESCA), Colon Adenocarcinoma (COAD) and Head and Neck Squamous Cell Carcinoma (HNSC), a multi-omics data from The Cancer Genome Atlas (TCGA: https://cancergenome.nih.gov/). Every dataset are contained three types of data: CNV, ME, and GE. The integrated model in our experiment is a three-dimensional integrated model. All three types of data were collected from the same sample. The status of the sample includes two types: tumor or normal. For different types of the same cancer, we retained the common genes of the three datasets to keep the size of the input matrix consistent. The significant statics of our datasets are listed in the following table (Table 1):

Table 1. Datasets details

Datasets	Data types	Size
PAAD	CNV, ME, GE	19877 * 180
ESCA	CNV, ME, GE	19877 * 192
COAD	CNV, ME, GE	22723 * 281
HNSC	CNV, ME, GE	19877 * 418

Note: The number of columns is represented (number of genes * samples)

4.2 Convergence Study

Equation (8) is the objective function of the iRGNMF model. It is iterative in nature, and it can be proved that the update rules of the method are convergent. Figure 1 shows the convergence curves for the four datasets, one representing a method. From the figure we can see that after about 200 iterations. The model will be convergence.

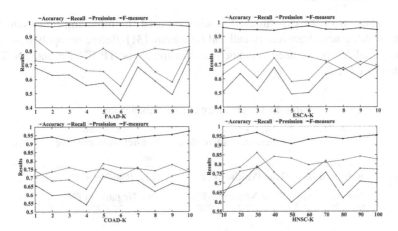

Fig. 1. Performance of iRGNMF with different K

4.3 Parameters Selection

In our experiments, the dimensionality parameter K and the regularization parameter λ are needs to choose. We use the control variable method to find the optimal value of each parameter. In the experiment, we can find that the value of the regularization parameter has little effect on the experimental result, and the dimension reduction parameter has a great influence on the experimental result. In order to obtain the best experimental results, we make the $\lambda_1 = \lambda_2 = \lambda_3 = 0.01$, then test the dimensionality reduction K from 1 to 120 ($K \in [1, 2, 3, \ldots, 10, 20, 30, \ldots 120]$), and finally choose the most reasonable K value. The specific experimental results are shown in Fig. 2: From the above figure, we can get the optimal dimension reduction for each cancer. For PAAD, ESCA, COAD and HNSC, the K values are 10, 5, 4 and 70 respectively.

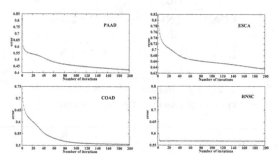

Fig. 2. Convergence performance on different datasets

4.4 Evaluation Metrics

In the evaluation of the experimental results, we used the following four evaluation indicators: Accuracy, Precision, Recall and F-measure [31]. Before we apply those four evaluation indicators, we need to define four classifications: TP, FN, FP, and TN. They are listed in Table 2.

Table 2. The definition of four classifications

	Relevant, positive class	Non-relevant, negative class
Retrieved	True Positives (TP)	False Positives (FP)
Not retrieved	False Negatives (FN)	True Negatives (TN)

The definition of Accuracy is that for a given test datasets, the ratio of the number of samples correctly classified by the classifier to the total number of samples. Recall is that how many positive examples in the sample are predicted correctly, which is also called sensitivity. Precision means that how many of the positively predicted samples are true positive, which is also called the positive prediction ratio. F-measure, also known as F-score, is a commonly used evaluation standard in the field of Information Retrieval.

$$\text{Accuracy} = \frac{TP + TN}{TP + FP + FN + TN} \tag{14}$$

$$\text{Recall} = \frac{TP}{TP + FN} \tag{15}$$

$$\text{Precision} = \frac{TP}{TP + FP} \tag{16}$$

$$\text{F-Measure} = \frac{2 \cdot \text{Recall} \cdot \text{Precision}}{\text{Recall} + \text{Precision}} \tag{17}$$

4.5 Clustering Results

In the clustering experiment results, we compare our method iRGNMF with the other three integration methods (jNMF, iNMF and iONMF). As for jNMF, which allows input from multiple data sources, simultaneously outputs a shared matrix and multiple special matrices based on heterogeneity of data to discover more useful information about multidimensional data; iONMF, which is On the basis of jNMF, orthogonal constraints are applied. The introduction of orthogonal constraints fully considers the orthogonal relationship between data. The iNMF utilizes the advantages of multiple data sources to obtain robustness to heterogeneous perturbations.

There are the results of comparing the different models of the four datasets. The experimental results are shown in Fig. 3. In most instances, we can see that our model can achieve higher Accuracy, Precision, Recall and F-Measure than other integrated models, it shows that our model is effective compared to other models.

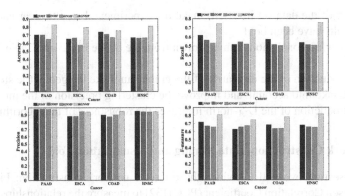

Fig. 3. Clustering performance on different datasets about integrated model

To demonstrate the superiority of the integrated model over a single model, we next compare our model to the single model RGNMF. The following are detailed experimental results:

As can be seen from Table 3, the integrated model can achieve good experimental results. For the datasets PAAD, ESCA, COAD, four of the three metrics exceed a single model, for the datasets HNSC only Precision exceeds a single model, probably due to

Table 3. Clustering performance on different datasets about RGNMF and iRGNMF

		Accuracy	Recall	Precision	F-measure
PAAD	GE	0.7144 ± 0.0000	0.5763 ± 0.0470	0.9779 ± 0.0001	0.6997 ± 0.0379
	ME	0.6336 ± 0.0051	0.5211 ± 0.0229	0.9729 ± 0.0005	0.6666 ± 0.0178
	CNV	0.8097 ± 0.0105	0.7128 ± 0.0035	**0.9806 ± 0.0000**	0.8038 ± 0.0968
	G-M-C	**0.8278 ± 0.1457**	**0.7460 ± 0.2950**	0.9716 ± 0.0072	**0.8104 ± 0.2270**
ESCA	GE	0.5449 ± 0.00001	0.5063 ± 0.0019	0.9537 ± 0.0001	0.6606 ± 0.0016
	ME	0.5410 ± 0.0002	0.5022 ± 0.0047	0.9576 ± 0.0016	0.6542 ± 0.0026
	CNV	0.6798 ± 0.0094	0.5521 ± 0.0530	0.962 ± 0.0009	0.6675 ± 0.0394
	G-M-C	**0.7958 ± 0.9653**	**0.6792 ± 0.3064**	0.9410 ± 0.0231	**0.7472 ± 0.2653**
COAD	GE	0.7286 ± 0.0038	0.6446 ± 0.0950	0.9538 ± 0.0041	0.7149 ± 0.0768
	ME	0.7330 ± 0.0066	0.5588 ± 0.0049	**0.9609 ± 0.0049**	0.6624 ± 0.0552
	CNV	0.6767 ± 0.0054	0.5334 ± 0.0463	0.9577 ± 0.0018	0.6560 ± 0.0343
	G-M-C	**0.7544 ± 0.1251**	**0.7080 ± 0.2477**	0.9502 ± 0.0316	**0.7843 ± 0.2041**
HNSC	GE	**0.9877 ± 0.0011**	**0.8528 ± 0.1143**	0.8807 ± 0.0915	**0.8596 ± 0.1122**
	ME	0.7347 ± 0.0091	0.5425 ± 0.0571	0.9062 ± 0.0159	0.6597 ± 0.0500
	CNV	0.6320 ± 0.0010	0.5063 ± 0.0123	0.9255 ± 0.0061	0.5016 ± 0.0126
	G-M-C	0.8093 ± 0.1100	0.7565 ± 0.2537	**0.9441 ± 0.0652**	0.8184 ± 0.2287

the structure of the data set itself. To make the experimental results clear enough, the optimal results have been bold.

The following conclusions can be drawn from the above Fig. 3 and Table 3:

1) The introduction of the regularity of the graph can make the geometric structure of the data take into account, and improve the clustering precision of the model.
2) The introduction of the $L_{2,1}$-norm improves the robustness of the algorithm.

4.6 Network Construction, Data Mining and Visualization Tool

In this subsection, we will introduce the rules for network construction: Firstly, we utilize Pearson Correlation Coefficient (PCC) [32] to measure the relationships of nodes in networks to gain the adjacency matrix. Secondly, we sort absolute values of PCC matrix and perform curve fitting. Finally, we choose the first inflection point as the filtering threshold of the matrix to obtain final networks.

About data mining, we consider several different important properties of nodes in the network: like degree, betweenness and closeness. In order to find abnormal nodes better, we define the score of every node x:

$$Score(x) = \frac{DB}{C} \tag{14}$$

where D is degree of x, B is betweenness of x and C is closeness of x. Degree represents local properties of the node in network, and the betweenness and closeness reflect global properties of the node in network.

Last but not the least, we use Cytoscape [33] software to visualize the data obtained. An open source software platform for visualizing molecular biology pathways and interaction networks is called Cytoscape, which integrates these gene expression profiles, networks and annotations with other state data. Cytoscape was originally designed for use in biological research, but is now more used for complex network visualization and analysis. More details on Cytoscape can be found at https://cytoscape.org/.

4.7 Network Analysis

In the network analysis section, we used two datasets (PAAD and ESCA) to demonstrate the effectiveness of our model. We reserved more than 20 nodes for each experimental result, 7 modules of the PAAD datasets, and 10 modules of the ESCA datasets.

For the PAAD datasets, the network modules are illustrated in Fig. 4. In the figure, the node with the larger area represents the degree of the gene, and the node with the darker color represents the higher the mediation centrality of the gene. We selected the highest-scoring genes in each module according to our scoring rules. The details are listed in Table 4. The black body is the gene that can be found in GeneCards (https://www.genecards.org/) and has been documented.

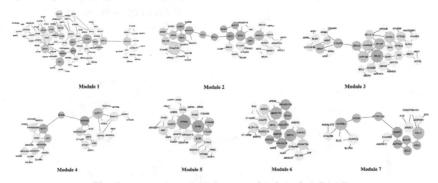

Fig. 4. Network modules construction based on PAAD

Pancreatic cancer is one of the most deadly human malignancies. Annexin A1 (ANXA1) is a Ca^{2+}-binding protein involved in pancreatic cancer (PC) progression. It mediates the maintenance of the cytoskeletal organization of the malignant phenotype [34]. The key role of the metabolic enzyme ACLY in the production of acetyl-CoA during pancreatic carcinogenesis was determined [35].

For the ESCA dataset, the network module is illustrated in Fig. 5. In the above figure, we can see that a total of ten modules of the datasets were mined according to the same structuring rule (Table 5).

Esophageal carcinoma is a cancer caused by the esophagus. The esophageal tube flows between the throat and the stomach. The reference indicates that the expression levels of AURKA, AURKB, and NEK6 are significantly elevated in esophageal

Table 4. Top one gene of every module (PAAD)

Gene	Description	Summary
ACSM6	Acyl-CoA synthetase medium chain family member 6	Protein coding gene
ANXA1	Annexin A1	This gene encodes a membrane-localized protein that binds phospholipids
ARL4A	ADP ribosylation factor like GTPase 4A	ADP-ribosylation factor-like 4A is a member of the ADP-ribosylation factor family of GTP-binding proteins
ARGFXP2	Arginine-fifty homeobox pseudogene 2	Homeobox genes encode DNA-binding proteins, many of which are thought to be involved in early embryonic development
ATP8B2	ATPase phospholipid transporting 8B2	The protein encoded by this gene belongs to the family of P-type cation transport ATPases, and to the subfamily of aminophospholipid-transporting ATPases
BSPH1	Binder of sperm protein homolog 1	Protein coding gene
ACLY	ATP citrate lyase	The primary enzyme responsible for the synthesis of cytosolic acetyl-CoA in many tissues

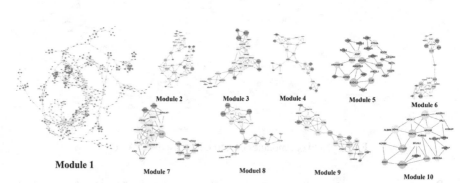

Fig. 5. Network modules construction based on ESCA

adenocarcinoma [36]. The study found that two novel lncRNAs (ADAMTS9-AS1 and AP000696.2) may be required for ectoderm and epithelial cell development, significantly stratifying ESCA patients into high-risk and low-risk groups, and far superior to traditional clinical Tumor markers [37].

There are some genes with higher scores have not been clinically confirmed such as ACSM6, ARL4A, ARGFXP2, ADSSL1, BUD31, ANGPT4 and so on. However, they need to be further clinical examination due to the potential significance is unpredictable.

Table 5. Top one gene of every module (ESCA)

Gene name	Description	Summary
ADSSL1	Adenylosuccinate synthase like 1	This gene encodes a member of the adenylosuccinate synthase family of proteins
BUD31	BUD31Homolog	Protein coding gene
AURKB	Aurora Kinase B	This gene encodes a member of the aurora kinase subfamily of serine/threonine kinases
AQP2	Aquaporin 2	This gene encodes a water channel protein located in the kidney collecting tubule
BOD1L1	Biorientation of chromosomes in cell division 1 like 1	Protein coding gene
C1orf100	Chromosome 1 open reading frame 100	Protein coding gene
ANGPT4	Angiopoietin 4	Angiopoietins are proteins with important roles in vascular development and angiogenesis
ADAMTS9	ADAM metallopeptidase with thrombospondin type 1 Motif 9	This gene encodes a member of the ADAMTS protein family
BUB1	BUB1 mitotic checkpoint serine/threonine kinase	This gene encodes a serine/threonine-protein kinase that play a central role in mitosis

5   Conclusion

In this paper, to model for multi-dimensional genomic data, we propose a novel matrix factorization model to make sure more robust, which is named integrated Robust Graph Regularization Non-negative Matrix Factorization. In spite of our put forward objective function is hard to deal with, we apply the MUR to solve it easily. Moreover, we provide the computational complexity and convergence of our model. And experiments demonstrate our model is more effective in multi-dimensional genomic clustering and network analysis. Experiments on our datasets demonstrate its effectiveness of our model both in multi-dimensional genomic clustering and network analysis.

Acknowledgments. This work was supported in part by the NSFC under grant Nos. 61872220 and 61572284.

References

1. Paatero, P., Tapper, U.: Positive matrix factorization: a non-negative factor model with optimal utilization of error estimates of data values. Environmetrics 5(2), 111–126 (2010)

2. Lee, D.D., Seung, H.S.: Learning the parts of objects by non-negative matrix factorization. Nature **401**(6755), 788–791 (1999)
3. Wang, D., Liu, J.X., Gao, Y.L., Zheng, C.H., Xu, Y.: Characteristic gene selection based on robust graph regularized non-negative matrix factorization. IEEE/ACM Trans. Comput. Biol. Bioinf. **13**(99), 1059–1067 (2016)
4. Zhang, S., Li, Q., Liu, J., Zhou, X.J.: A novel computational framework for simultaneous integration of multiple types of genomic data to identify microRNA-gene regulatory modules. Bioinformatics **27**(13), i401–i409 (2011)
5. Li, Y., Ngom, A.: The non-negative matrix factorization toolbox for biological data mining. Source Code Biol. Med. **8**(1), 1–15 (2013)
6. Grais, E.M., Erdogan, H.: Regularized nonnegative matrix factorization using Gaussian mixture priors for supervised single channel source separation. Comput. Speech Lang. **27**(3), 746–762 (2013)
7. Liu, W., Zheng, N., You, Q.: Nonnegative matrix factorization and its applications in pattern recognition. Chin. Sci. Bull. **51**(1), 7–18 (2006)
8. Li, X.L., Bai, B., Wu, J.: Transcriptome analysis of early interaction between rice and Magnaporthe oryzae using next-generation sequencing technology. Hereditas **34**(1), 102–112 (2012)
9. Greene, D., Cunningham, P.: A matrix factorization approach for integrating multiple data views. In: Buntine, W., Grobelnik, M., Mladenić, D., Shawe-Taylor, J. (eds.) ECML PKDD 2009. LNCS (LNAI), vol. 5781, pp. 423–438. Springer, Heidelberg (2009). https://doi.org/10.1007/978-3-642-04180-8_45
10. Liping, J., Chao, Z., Ng, M.K.: SNMFCA: supervised NMF-based image classification and annotation. IEEE Trans. Image Process. **21**(11), 4508–4521 (2012)
11. Zhang, S., Liu, C., Li, W., Shen, H., Peter, W.L., Zhou, X.J.: Discovery of multi-dimensional modules by integrative analysis of cancer genomic data. Nucleic Acids Res. **40**(19), 9379–9391 (2012)
12. Žitnik, M., Zupan, B.: Data fusion by matrix factorization. IEEE Trans. Pattern Anal. Mach. Intell. **36**(1), 41–53 (2015)
13. Stražar, M., Žitnik, M., Zupan, B., Ule, J., Curk, T.: Orthogonal matrix factorization enables integrative analysis of multiple RNA binding proteins. Bioinformatics **32**(10), 1527–1535 (2016)
14. Lee, D.D., Seung, H.S.: Algorithms for non-negative matrix factorization. NIPS **13**(6), 556–562 (2001)
15. Xie, W., Wang, G., Bindel, D.: Fast iterative graph computation with block updates. Proc. VLDB Endow. **6**(14), 2014–2025 (2013)
16. Hein, M., Audibert, J.-Y., von Luxburg, U.: From graphs to manifolds – weak and strong pointwise consistency of graph Laplacians. In: Auer, P., Meir, R. (eds.) COLT 2005. LNCS (LNAI), vol. 3559, pp. 470–485. Springer, Heidelberg (2005). https://doi.org/10.1007/11503415_32
17. Zhu, R., Liu, J.X., Zhang, Y.K., Guo, Y.: A robust manifold graph regularized nonnegative matrix factorization algorithm for cancer gene clustering. Molecules **22**(12), 2131–2142 (2017)
18. Shen, B., Liu, B.D., Wang, Q., Ji, R.: Robust nonnegative matrix factorization via L1 norm regularization by multiplicative updating rules. In: IEEE International Conference on Image Processing, pp. 5282–5286. IEEE, Paris (2014)
19. Geng, B., Tao, D., Xu, C., Yang, L., Hua, X.S.: Ensemble manifold regularization. IEEE Trans. Pattern Anal. Mach. Intell. **34**(6), 1227–1233 (2012)
20. Yang, S., Zhang, C., Yi, W.: Robust non-negative matrix factorization via joint sparse and graph regularization for transfer learning. Neural Comput. Appl. **23**(2), 541–559 (2013)

21. Nojun, K.: Principal component analysis based on l1-norm maximization. IEEE Trans. Pattern Anal. Mach. Intell. **30**(9), 1672–1680 (2008)
22. Diestel, R.: Graph theory. Math. Gaz. **173**(502), 67–128 (2000)
23. Tuia, D., Volpi, M., Trolliet, M.: Semisupervised manifold alignment of multimodal remote sensing images. IEEE Trans. Geosci. Remote Sens. **52**(12), 7708–7720 (2014)
24. Bueler, E.L.: The heat kernel weighted Hodge Laplacian on noncompact manifolds. Trans. Am. Math. Soc. **351**(2), 683–713 (1999)
25. Bonnet, L., Rayez, J.C.: Gaussian weighting in the quasiclassical trajectory method. Chem. Phys. Lett. **397**(1), 106–109 (2004)
26. Gao, L., Alibart, F., Strukov, D.B.: Analog-input analog-weight dot-product operation with Ag/a-Si/Pt memristive devices. In: IEEE/IFIP International Conference on VLSI and System-on-chip, pp. 1–6. IEEE, Santa Cruz (2015)
27. Cai, D., He, X., Han, J., Huang, T.S.: Graph regularized nonnegative matrix factorization for data representation. IEEE Trans. Pattern Anal. Mach. Intell. **33**(8), 1548–1560 (2011)
28. Danielsson, P.E.: Euclidean distance mapping. Comput. Graph. Image Process. **14**(3), 227–248 (1980)
29. Wu, G.C., Baleanu, D.: Variational iteration method for the Burgers' flow with fractional derivatives—new Lagrange multipliers. Appl. Math. Model. **37**(9), 6183–6190 (2013)
30. Facchinei, F., Fischer, A., Kanzow, C., Peng, J.M.: A simply constrained optimization reformulation of KKT systems arising from variational inequalities. Appl. Math. Optim. **40**(1), 19–37 (1999)
31. Nguyen, T., Duong, T., Phung, D., Venkatesh, S.: Autism blogs: expressed emotion, language styles and concerns in personal and community settings. IEEE Trans. Affect. Comput. **6**(3), 312–323 (2015)
32. Mudelsee, M.: Estimating Pearson's correlation coefficient with bootstrap confidence interval from serially dependent time series. Math. Geol. **35**(6), 651–665 (2003)
33. Saito, R., et al.: A travel guide to cytoscape plugins. Nat. Methods **9**(11), 1069–1076 (2012)
34. Belvedere, R., et al.: miR-196a is able to restore the aggressive phenotype of annexin A1 knock-out in pancreatic cancer cells by CRISPR/Cas9 genome editing. Int. J. Mol. Sci. **19**(7), 1967 (2018)
35. Alessandro, C., Sophie, T., Steven, Z.: Acetyl-CoA metabolism supports multistep pancreatic tumorigenesis. Cancer Discov. **9**(3), 416–435 (2019)
36. Kasap, E., et al.: Aurora kinase A (AURKA) and never in mitosis gene A-related kinase 6 (NEK6) genes are upregulated in erosive esophagitis and esophageal adenocarcinoma. Exp. Therap. Med. **4**(1), 33–42 (2012)
37. Zhen, L., Yao, Q., Zhao, S., Yin, W., Li, Y., Zhen, W.: Comprehensive analysis of differential co-expression patterns reveal transcriptional dysregulation mechanism and identify novel prognostic lncRNAs in esophageal squamous cell carcinoma. Oncotargets Ther. **10**, 3095–3105 (2017)

Hyper-graph Robust Non-negative Matrix Factorization Method for Cancer Sample Clustering and Feature Selection

Cui-Na Jiao, Tian-Ru Wu, Jin-Xing Liu, and Xiang-Zhen Kong[✉]

School of Information Science and Engineering, Qufu Normal University, Rizhao, China
jiaocuina123@163.com, {wutianru,sdcavell}@126.com,
kongxzhen@163.com

Abstract. Non-negative Matrix Factorization (NMF) algorithm is a useful method for data dimensionality reduction, which is performed with the Euclidean distance. However, the basic NMF only assumes that data will be destroyed by Gaussian noise. It ignores both the intrinsic geometrical structure and the influence of sparse noises existing in the gene expression data. To enhance the robustness of the NMF, a novel method called Hyper-graph Robust Non-negative Matrix Factorization (HRNMF) is proposed for cancer sample clustering and feature selection. The merits of the HRNMF mainly consist of two aspects. Firstly, the $L_{2,1}$-norm is combined with the objective function, which can effectively handle noise and outliers. Secondly, the manifold information and sparsity are also considered so we add the hyper-graph regularization term and sparse constraints to the error function. It can effectively preserve the geometric structure and enhance matrix sparsity. Experiments on Cancer Genome Atlas (TCGA) gene expression data have demonstrated that HRNMF performs better than other advanced methods in cancer sample clustering and feature selection.

Keywords: Non-negative Matrix Factorization · Hyper-graph regularization · $L_{2,1}$-norm · Sample clustering · Robustness

1 Introduction

Over the past few years, many dimensional reduction methods are proposed to find the useful genes. Locally Linear Embedding (LLE) [1], Nonlinear Principal Component Analysis (NPCA) [2], Sparse Principal Component Analysis (SPCA) [3] and ISOMAP [4] are traditional approaches for reducing data dimension. Later, a popular method called Non-negative matrix factorization (NMF) [5] is introduced to ensure that elements of matrices are non-negative when reducing dimensionality.

NMF is parts-based and linear representations of the non-negative data, which can factorize an original non-negative matrix into the product of two non-negative matrices [6]. Currently, lots of variants based on the NMF have been developed by modifying the objective function or the constraint conditions [7, 8]. To reduce the influence of noises

© Springer Nature Singapore Pte Ltd. 2020
H. Han et al. (Eds.): IDMB 2019, CCIS 1099, pp. 112–125, 2020.
https://doi.org/10.1007/978-981-15-8760-3_8

and outliers existing in real data, Kong *et al.* proposed Robust Non-negative Matrix Factorization using $L_{2,1}$-norm ($L_{2,1}$-NMF) [9]. By separately modeling the Gaussian noise and the sparse noise, Sparsity-Regularized Robust Non-negative Matrix Factorization (SRNMF) [10] is introduced by He *et al.*. Then, some manifold regularization theories have been extended. Maaten *et al.* introduced t-SNE to visualize high-dimensional data [11]. Cai *et al.* proposed Graph regularized Non-negative Matrix Factorization (GNMF) to encode the geometrical information of the data space [12]. With the maturity of hyper-graph technology, Zeng *et al.* proposed Hyper-graph regularized Non-negative Matrix Factorization (HNMF) for image clustering [13]. Furthermore, considering the manifold structure and the sparsity, Graph Regularized Robust Non-negative Matrix Factorization (GrRNMF) [14] is proposed by Yu *et al.*.

Although these methods mentioned above are effective in handling gene expression data, there are still many details should be considered. On the one hand, the most traditional NMF methods use squared loss function to minimize the objective function, which is sensitive to the noise and outliers. On the other hand, the low dimensional manifold structure exists in the high dimensional space should be preserved.

Inspired by the above points, an improved method called Hyper-graph Robust Non-negative Matrix Factorization (HRNMF) is proposed for cancer sample clustering and feature selection in this paper. This method considers the influence of noises or outliers and the intrinsic manifold information of real data simultaneously. So, using the $L_{2,1}$-norm based loss function replaces the L_2-norm-based function to enhance the robustness of our method. The hyper-graph regularization completely preserves the geometric structure of high dimensional data.

Our main contributions are listed as follows:

1) The squared loss of the $L_{1/2}$-RNMF is too sensitive to outliers and noises. So, we used $L_{2,1}$-norm to minimize the error function of $L_{1/2}$-RNMF, which is modeling the Gaussian noise and sparse noise to reduce influences of noises. The $L_{2,1}$-norm not only effectively handles noises and outliers, but also enhances the robustness of HRNMF method.
2) The hyper-graph regularization term is incorporated to the objective function. HRNMF not only exploits the intrinsic geometric structure of the data distribution, but also considers the multiple geometric relationships between samples. This makes sense for having better performance of our method, which has the advantages of $L_{1/2}$-RNMF method, $L_{2,1}$-norm and hyper-graph regularization at the same time.
3) Scientific and extensive experiments are designed on the gene expression data to assess the effectiveness of the HRNMF method.

The remainder of this paper are organized as follows: We give a brief introduction about the related work containing MMF, and $L_{1/2}$-RNMF in the second part. The third part introduces our proposed HRNMF method with its formulas and specific algorithms in detail. Experiments and analysis are in the fourth part of the paper. Finally, the conclusions are drawn in the fifth part.

2 Related Work

2.1 Non-negative Matrix Factorization

Non-negative Matrix Factorization is a useful dimensionality reduction method, which is used in many fields, such as pattern recognition and image processing [15, 16]. Given a non-negative decomposed matrix $\mathbf{X}$, which size is $m \times n$. And the m denotes the number of genes, n denotes the number of samples.

Through the NMF, we can get two low-dimensional matrices $\mathbf{U} \in R^{m \times k}$ and $\mathbf{V} \in R^{k \times n}$, i.e., $\mathbf{X} \approx \mathbf{UV}^T$. $\mathbf{U}$ is a basis matrix, and $\mathbf{V}$ is a coefficient matrix. To guarantee the error between the $\mathbf{X}$ and $\mathbf{UV}^T$ is minimum, using the Euclidean distance to minimize the optimization problem. So, we get the objective function of the NMF method

$$O_{NMF} = \left\| \mathbf{X} - \mathbf{UV}^T \right\|^2 = \sum_{ij} \left(x_{ij} - \sum_{k=1}^{r} u_{ik} v_{jk} \right)^2,$$
$$s.t.\ \mathbf{U} \geq 0, \mathbf{V} \geq 0, \tag{1}$$

where $\|\cdot\|_F$ is applying the Frobenius norm to the matrix. For the (1), whose noise obeys the Gaussian distribution. According to the iterative update algorithms, the multiplicative updating rules of $\mathbf{U}$ and $\mathbf{V}$ are as follows:

$$u_{ik} \leftarrow u_{ik} \frac{\left(\mathbf{XV}^T\right)_{ik}}{\left(\mathbf{UVV}^T\right)_{ik}}, \tag{2}$$

$$v_{jk} \leftarrow v_{jk} \frac{\left(\mathbf{U}^T\mathbf{X}\right)_{jk}}{\left(\mathbf{VUU}^T\right)_{jk}}. \tag{3}$$

2.2 $L_{1/2}$-Robust Non-negative Matrix Factorization

However, most of the existing NMF-based methods only assume that the real data are corrupted by Gaussian noise. In reality, the sparse noise also influences the high dimensional data. Therefore, the proposed $L_{1/2}$-RNMF approach can simultaneously model Gaussian noise and sparse noise [10]. Compared with other methods, this algorithm performs better. The $L_{1/2}$-RNMF model can be expressed as:

$$\mathbf{X} = \mathbf{UV} + \mathbf{S} + \mathbf{G}, \tag{4}$$

where $\mathbf{S}$ represents spare noises, and $\mathbf{G}$ represents Gaussian noises. Because the new emerging $\mathbf{U}$, $\mathbf{V}$, and $\mathbf{S}$ matrices are prepared to the sparse noise corruption, it is more robustness than other basic NMF methods.

Firstly, the RNMF model is:

$$\min_{\mathbf{U},\mathbf{V},\mathbf{S}} \frac{1}{2} \|\mathbf{X} - \mathbf{S} - \mathbf{UV}\|_F^2, \quad s.t.\ \mathbf{U} \geq 0, \mathbf{V} \geq 0. \tag{5}$$

It is same with the NMF, which adds sparsity regularizations to the objective function.

The error function ban be presented as follows:

$$\min_{\mathbf{U},\mathbf{V},\mathbf{S}} \frac{1}{2}\|\mathbf{X} - \mathbf{S} - \mathbf{UV}\|_F^2 + \lambda\|\mathbf{S}\|_{2,1} + \beta\|\mathbf{V}\|_{1/2},$$

$$s.t.\ \mathbf{U} \geq 0,\ \mathbf{V} \geq 0, \tag{6}$$

where $\|\mathbf{S}\|_{2,1} = \sum_{i=1}^{L}\|\mathbf{E}_{i,:}\|_2$ and $\|\mathbf{V}\|_{1/2} = \sum (\mathbf{V}_{i,j})^{1/2}$, $\lambda \geq 0$ and $\beta \geq 0$ are regularization parameters to control the sparsity of the matrix $\mathbf{S}$ and matrix $\mathbf{V}$, respectively. The update rules for (6) is similar to (1) because the noise matrix $\mathbf{G}$ must obey the Gaussian distribution.

3 Method

3.1 Definition of $L_{2,1}$-Norm

The error function of standard NMF model uses L_2-norm, thus a few noises or outliers with large errors will dominate the objective function easily because of the squared errors [9]. Therefore, the $L_{2,1}$-norm replaces the L_2-norm to minimize the objective function, which is not sensitive to outliers.

The $L_{2,1}$-norm was proposed in [9], which is computed the L_2-norm of rows m_i and then calculated the L_1-norm of the vector $b(M) = (\|m^1\|_2, \|m^2\|_2, \cdots, \|m^i\|_2)$. So, the definition of $L_{2,1}$-norm is:

$$\|\mathbf{M}\|_{2,1} = \sum_{i=1}^{n}\sqrt{\sum_{j=1}^{s} m_{ij}^2} = \sum_{i=1}^{n}\|m^i\|_2, \tag{7}$$

where m^i is the i-th row of matrix $\mathbf{M}$. The $L_{2,1}$-norm not only gets over the drawbacks of NMF-based methods, but also enhances the robustness of the method that uses it.

3.2 Hyper-graph Regularization

Hyper-graph regularization not only can consider the manifold geometric information of the real data, but also can connect more nodes among multiple data samples. In contrast to the simple graph, whose edge is only connected with two data samples, the hyper-graph performs more effective. And the simple graph will lose lots of important information about many crucial relationships because of that. Compared with the traditional NMF method, the method with the hyper-graph constraint, which preserves the similarity between the original data points after dimensionality reduction.

Hyper-graph $G = (V, E, \mathbf{W})$ makes up of multiple hyperedges subsets $E = \{e_i | i = 1, 2 \cdots \cdots n\}$, non-empty vertices set $V = \{v_j | j = 1, 2 \cdots \cdots p\}$ and a weights matrix $\mathbf{W}$ that is expressed by a diagonal matrix. In spatial structure, the hyper-graph can divide the dataset m into n hyper-edges e_i of the vertex interconnect [17, 18]. The hyper-graph structure and its incidence matrix as follows (Fig. 1):

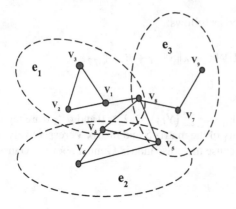

	e1	e2	e3
v1	1	0	0
v2	1	0	0
v3	1	0	0
v4	1	1	0
v5	0	1	1
v6	0	1	0
v7	0	0	1
v8	1	0	1
v9	0	0	1

Fig. 1. Illustration of the hyper-graph and its incidence matrix. The points in this figure denotes the spatial distribution of the data. Each hyper-edge consists of multiple data points.

Now, some definitions of the hyper-graph are given as below:

Every hyperedge will be represented as a positive value, which is the weight of each hyperedge. So, the initial weight for each hyperedge is calculated

$$\mathbf{W}_i = \sum_{v_j \in e_i} \mathbf{A}_{ij}, \tag{8}$$

where the affinity matrix $\mathbf{A}$ is defined as

$$\mathbf{A}_{ij} = \exp\left(-\frac{\|v_i - v_j\|^2}{\sigma^2}\right). \tag{9}$$

In which, the σ denotes the average distance among all vertices.

Generally speaking, the relationships between a vertex and a hyperedge use an incidence matrix $\mathbf{H}(v, e)$ to express. Then, we get the $\mathbf{H}$ as below:

$$\mathbf{H}(v, e) = \begin{cases} 1, & if\ v \in e \\ 0, & if\ v \notin e \end{cases}. \tag{10}$$

The above weights of hyperedges that v_j belongs to are calculated as the degree of v_j. The number of vertices of E represents the hyperedge degree.

$$d(v) = \sum_{\{e_i \in E | v \in e\}} w(e) = \sum_{e_i \in E} w(e)\mathbf{H}(v, e), \tag{11}$$

$$\delta(e) = |e| = \sum_{v_j \in V} \mathbf{H}(v, e). \tag{12}$$

The elements of D_v and D_e are vertex degrees and hyperedge degrees, which stand for a diagonal matrix. At last, the unnormalized hyper-graph Laplacian matrix $\mathbf{L}_{hyper} = \left[l_{ij}\right]$ can be calculated as $\mathbf{L}_{hyper} = \mathbf{D}_v - \mathbf{S}$, where $\mathbf{S} = \mathbf{H}\mathbf{W}\mathbf{D}_e^{-1}\mathbf{H}^T$ [13].

3.3   HRNMF Method

The $L_{1/2}$-RNMF is a popular method for ensuring dimensionality reduction. However, it is used squared residue error on the objective function, which is sensitive to noises or outliers. And the $L_{1/2}$-RNMF only considers the sparsity of error function, which ignores the internal geometry structure of high dimensional data space. For the sake of enhancing the robustness of the $L_{1/2}$-RNMF model, based on the $L_{2,1}$-norm and hyper-graph manifold learning theories, we proposed a novel method called Hypergraph Robust NMF (HRNMF) for cancer sample clustering and feature selection in this paper.

This algorithm considers two conditions, containing the noises or outliers existed in data samples and the geometric manifold structure of the high dimensional data space. Therefore, the hyper-graph Laplacian regularization term $\mathbf{L}_{hyper}$ and the $L_{2,1}$-norm are integrated into the objective function of popular $L_{1/2}$-RNMF model. The error function of HRNMF is defined as below:

$$O_{HRNMF} = \frac{1}{2}\|\mathbf{X} - \mathbf{S} - \mathbf{UV}\|_{2,1} + \frac{\alpha}{2}tr\left(\mathbf{VL}_{hyper}\mathbf{V}^T\right) + \beta\|\mathbf{S}\|_{2,1} + \gamma\|\mathbf{V}\|_{1/2},$$
$$s.t. \quad \mathbf{U} \geq 0, \mathbf{V} \geq 0 \tag{13}$$

where $\mathrm{Tr}(\cdot)$ represents the trace of the matrix and $\alpha \geq 0$ is a balancing parameter that controls the hyper-graph regularization term. Besides, $\beta \geq 0$ and $\gamma \geq 0$ as two regularization parameters are used to constrain the sparseness of $\mathbf{S}$ and $\mathbf{V}$.

HRNMF method has the merits of $L_{1/2}$-RNMF method, $L_{2,1}$-norm and hyper-graph regularization simultaneously.

3.4   Optimization and Updating Rules

Owing to the HRNMF method is non-convex, it is difficult to find its global optimal solution. In this paper, we used the multiplication update rule to iteratively update the error function and minimize O_{HRNMF} to find local optimal solution. The $\mathbf{X} - \mathbf{S}$ can be seen as a variable, so the objective function in (13) is rewritten as follows:

$$O_{HRNMF} = \frac{1}{2}tr\left[((\mathbf{X} - \mathbf{S}) - \mathbf{UV})\mathbf{G}((\mathbf{X} - \mathbf{S}) - \mathbf{UV})^T\right]$$
$$\frac{\alpha}{2}tr\left(\mathbf{VL}_{hyper}\mathbf{V}^T\right) + \beta\|\mathbf{S}\|_{2,1} + \gamma\|\mathbf{V}\|_{1/2}, \tag{14}$$

where $\mathbf{G}$ is a diagonal matrix with diagonal elements defined by

$$\mathbf{G}_{jj} = \frac{1}{\sqrt{\sum_{m=1}^i ((\mathbf{X} - \mathbf{S}) - \mathbf{UV})_{mj}^2 + \varepsilon}} = \frac{1}{\|x_j - Fg^j + \varepsilon\|}. \tag{15}$$

In which, the ε is a non-zero positive number, which infinitely closes to zero.

To solve this problem, we introduced $\Psi = [\Psi_{ik}]$ and $\varphi = [\varphi_{kj}]$ Lagrange multipliers, which are constrained to $\mathbf{U} \geq 0$ and $\mathbf{Z} \geq 0$, respectively [19]. Then, we got the Lagrange function that is defined:

$$f = \frac{1}{2}tr\left[(\mathbf{X} - \mathbf{S})\mathbf{G}(\mathbf{X} - \mathbf{S})^T\right] - tr\left[(\mathbf{X} - \mathbf{S})\mathbf{GV}^T\mathbf{U}^T\right]$$

$$+ \frac{1}{2}tr\left(\mathbf{UVGV}^T\mathbf{U}^T\right) + \frac{\alpha}{2}tr\left(\mathbf{VL}_{hyper}\mathbf{V}^T\right) + \beta\|\mathbf{S}\|_{2,1}$$
$$+ \gamma\|\mathbf{V}\|_{1/2} + tr\left(\Psi\mathbf{U}^T\right) + tr\left(\varphi\mathbf{V}^T\right). \tag{16}$$

Let $\Psi\mathbf{U} = 0$ and $\varphi\mathbf{V} = 0$ by using Karush-Kuhn-Tucher (KKT) [19] conditions. The iterative formulas of the objective function are expressed as follows:

$$u_{ik} \leftarrow u_{ik}\frac{\left((\mathbf{X} - \mathbf{S})\mathbf{GV}^T\right)_{ik}}{\left(\mathbf{UVGV}^T\right)_{ik}}, \tag{17}$$

$$v_{jk} \leftarrow v_{jk}\frac{\left(\mathbf{U}^T(\mathbf{X} - \mathbf{S})\mathbf{G}\right)_{jk}}{\left(\mathbf{U}^T\mathbf{UVG} + \alpha\mathbf{VL}_{hyper}^T + \frac{\gamma}{2}\mathbf{V}^{-1/2}\right)_{jk}}, \tag{18}$$

$$s_{ij} \leftarrow soft_\beta(\mathbf{X} - \mathbf{UV}), \tag{19}$$

where iterative formula of $\mathbf{S}$ can be solved through soft threshold in Lemma 1 [6]:

Lemma 1: Given a matrix $\mathbf{A}$, the solution to the following formula

$$\min_{\mathbf{X}} \frac{1}{2}\|\mathbf{X} - \mathbf{A}\|_F^2 + \beta\|\mathbf{X}\|_{2,1}, \tag{20}$$

is $\mathbf{X}^* = soft(\mathbf{A})$, then the i-th row of $\mathbf{X}^*$ is

$$\mathbf{X}_{i,:}^* = \begin{cases} \frac{\|\mathbf{A}_{i,:}\|_2 - \gamma}{\|\mathbf{A}_{i,:}\|_2}\mathbf{A}_{i,:}, & if \, \|\mathbf{A}_{i,:}\|_2 \geq \beta, \\ 0, & otherwise. \end{cases} \tag{21}$$

Through the detailed description of the above process, the algorithm of the HRNMF method can be organized as follows:

Algorithm 1: HRNMF

Data Input: $\mathbf{X} \in R^{m \times n}$ and $\mathbf{L}_{hyper} \in R^{n \times n}$

Parameters: α, β, γ

Output: $\mathbf{U} \in R^{m \times k}, \mathbf{V} \in R^{k \times n}, \mathbf{S} \in R^{m \times n}$

Initialization: $\mathbf{U} \geq 0$, $\mathbf{V} \geq 0$, $\mathbf{S} = 0$

Set $r = 1$.

Repeat

      Update $\mathbf{U}$ using Eq. (17)

      Update $\mathbf{V}$ using Eq. (18)

      Update $\mathbf{S}$ using Eq. (19)

      $r = r + 1$;

Until convergence

---

4   Results and Discussion

In this section, we assessed our proposed method with clustering and feature selection on four gene expression data. Besides, we compared our method with NMF [5], $L_{2,1}$-NMF [9], HNMF [13], $L_{1/2}$-RNMF [10], and GrRNMF [14] methods. Detailed information and related results are shown in the following subsections.

4.1 Datasets

The datasets are downloaded from TCGA (https://cancergenome.nih.gov/), which is established by the National Cancer Institute (NCI) and the National Human Genome Research Institute (NHGRI) [20]. It is designed to detect mutations in cancer cells and get approaches to diagnose and treat cancer, and ultimately to prevent cancer. In our experiments, we evaluated the HRNMF performance on four gene expression datasets, containing breast invasive carcinoma (BRCA_GE), lung adenocarcinoma (LUAD_GE), prostate adenocarcinoma (PRAD_GE), and ovarian serous cystadenocarcinoma (OV_GE). These gene expression data are introduced in Table 1.

Table 1. Summary of the four gene expression datasets.

Datasets	Normal samples	Tumor samples	Genes
BRCA_GE	120	1102	24991
LUAD_GE	61	533	24991
PRAD_GE	56	495	24991
OV_GE	5	374	24991

4.2 Convergence and Computational Time Analysis

Here, we iterated the update rules of HRNMF to get its local optimal values of the error function on the four gene expression datasets. The maximum number of iterations is set to 100 times during the experiment. The error value is the loss function value. And the error value will tend to zero with the increasing of convergence criterions. The convergence curves are shown in Fig. 2, which proves that our algorithm can converge better and quickly.

Furthermore, we also compared the execution time of these six methods. Experiments were running on a PC with 2.30 GHz Intel(R) Xeon(R) CPU and 4G RAM. To guarantee the fairness and save space, we get the execution time of six methods on the PRAD_GE datasets.

The statistics running times are shown in Fig. 3. NMF, HNMF, $L_{1/2}$-RNMF and GrRNMF approaches have satisfactory running times. But the $L_{2,1}$-NMF has the longest running time because it is imposed the $L_{2,1}$ norm constraint, which increases time complexity. HRNMF integrated the $L_{2,1}$ norm and hyper-graph regularization to the objective function, so the cost of time is also higher. But I think that it is worthwhile to use the time performance in exchange for the improvement of the experimental results. In the future, we will strive to make our method reduce the running time on the basis of achieving better performance results.

4.3 Parameters Setting

In the HRNMF model, we used four essential parameters, including p-nearest neighbours, hyper-graph regularization parameter α, sparsity regularization parameters β and

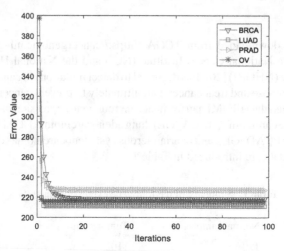

Fig. 2. Convergence curves of HRNMF method on BRCA, LUAD, PRAD, and OV datasets.

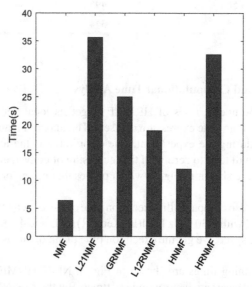

Fig. 3. Running time of these six methods on PRAD datasets.

γ. For convenience, we selected all these parameters using cross validations search. Finally, we set the same size of neighbourhood p to be 6 and $\alpha = 0.3$. The parameters β and γ are set to be 2 and 5 to achieve the better performance in the experiments.

4.4 Clustering Results and Analysis

Evaluation Metrics. We used K-means clustering approach to perform sample clustering on the coefficient matrix decomposed by HRNMF. The accuracy (AC) is the percentage of correct cluster samples, which is used as an evaluation criterion to assess the clustering performance. It is defined as follows:

$$AC = \frac{\sum_{i=1}^{n} \delta(s_i, \text{map}(r_i))}{n}, \tag{22}$$

where $\delta(\cdot)$ and $\text{map}(\cdot)$ denote the function permutation and delta mapping function, respectively. The actual sample label, the predicted sample label and the total number of samples are denoted by s, r and n, respectively. If x is equal to y, the $\delta(x, y) = 1$; otherwise $\delta(x, y) = 0$.

Cancer Sample Clustering Results. To assess the advantages of $L_{2,1}$-norm and the manifold structure in clustering, the five methods are compared with our method. In clustering process, each algorithm runs 20 times to get the averages values for comparison. Using the average results reduce the influence of random initialization on these initial methods results.

The detailed descriptions about the clustering accuracy of these six methods are shown in Table 2. From the Table 2, we can see our method has better performance than other advanced approaches. The specific analysis of the clustering results is given as follows:

Table 2. The clustering accuracy results of six methods.

Datasets	BRCA_GE	LUAD_GE	PRAD_GE	OV_GE	Average
NMF	66.64 ± 0.82	57.79 ± 0.16	67.32 ± 0.40	61.46 ± 0.32	63.30 ± 0.43
$L_{2,1}$-NMF	64.50 ± 0.62	57.13 ± 0.04	61.47 ± 0.09	57.22 ± 0.23	60.08 ± 0.25
$L_{1/2}$-RNMF	68.78 ± 0.01	72.53 ± 0.07	74.09 ± 0.14	76.65 ± 0.03	73.01 ± 0.06
HNMF	73.59 ± 0.71	74.06 ± 1.64	72.75 ± 0.81	83.88 ± 1.62	76.07 ± 1.20
GrRNMF	87.04 ± 0.15	83.88 ± 0.47	83.76 ± 0.89	95.18 ± 0.33	87.47 ± 0.46
HRNMF	$\mathbf{90.07 \pm 0.01}$	$\mathbf{89.36 \pm 0.03}$	$\mathbf{89.45 \pm 0.05}$	$\mathbf{98.36 \pm 0.01}$	$\mathbf{91.81 \pm 0.03}$

1) The clustering results of original NMF method is not the worst. This means the improvements in basic NMF may lose some important information and affect the accuracy results. Although clustering accuracy of $L_{2,1}$-NMF is lower than NMF, the multiple clustering variances of $L_{2,1}$-NMF method are smaller. Owing to the $L_{2,1}$-norm loss replaces the squared loss to improve robustness of the method.

2) In the above datasets, the clustering accuracy of HNMF is higher than the basic NMF method. The hyper-graph regularization term plays an important role in the HNMF, which preserves the geometrical structure of the high dimensional data points in dimensionality reduction process. And compared with the simple graph, which only gets the structure between two data points, hyper-graph considers multiple manifold relationships between the complex data points.

3) On the basis of $L_{1/2}$-RNMF, we can get GrRNMF and HRNMF methods. And the four gene expression datasets are larger, which have some noises and manifold structures. Compared with the GrRNMF, HRNMF integrated the hyper-graph regularization and $L_{2,1}$-norm to the objective function. Therefore, the performance of HRNMF is better than other methods. In contrast to the GrRNMF, HRNMF has improved by 3.03%, 5.48%, 5.39%, and 3.18% on the four gene expression datasets, respectively.

In summary, the clustering performance of our proposed method has improved on the four gene expression data. Compared with the $L_{1/2}$-RNMF, HRNMF has improved 18.8% on average. And compared with the GrRNMF, HRNMF has improved 4.34% on average. The above clustering results prove that our method is feasible and effective.

4.5 Feature Selection Results and Analysis

Feature selection is an important analysis tool to search differentially expressed genes for cancers. Nowadays, exploring the relations between genes and diseases are important. Thus, in this subsection, we also analyzed the performance of six methods to select differentially expressed genes.

Differentially Expressed Genes Selection Results. We used BRCA_GE, LUAD_GE, PRAD_GE datasets that have a better clustering effect to perform experiment. Then we selected 500 genes of every method for comparison. They are sent to the GeneCards (https://www.genecards.org/) to analyze. Through the analysis of the selected genes, we can obtain a corresponding relevance score, which is an indictor to assess the results of the feature selection. In which, the N represents the number of overlapping genes selected by each method in the associated gene pool. And the MRS is the maximum relevance score of the selected genes. With the higher maximum relevance score, the probability of the abnormal expressed genes causes cancers that is higher.

The detailed feature selection results of these six methods are shown in Table 3. Through this Table 3, we can see the differentially expressed genes selected by HRNMF have the higher maximum relevance scores and the number of genes. It means our proposed model can better find the genes associated with the cancer.

Analysis of Differential Genes. Table 4 lists the genes that have top ten highest correlation scores selected by the HRNMF model on the LUAD_GE dataset. In detail, we also specific introduced two genes with the highest relevance score selected by HRNMF on LUAD_GE dataset in Table 4.

The PTEN gene encodes a dual-specificity phosphatase mutated in a variety of human cancers [21]. Recently, a tumor suppressor from human chromosome 10q23, called PTEN or MMAC1, has been identified that shares homology with the protein tyrosine phosphatase family [22]. Germ-line mutations in PTEN give rise to several related neoplastic disorders, including Cowden disease (CD), Lhermitte-Duclos disease (LDD) and Bannayan-Zonana syndrome (BZS), characterized by tumor susceptibility and developmental defects [21, 22]. Loss of PTEN correlated with increased S6 kinase activity and phosphorylation of ribosomal S6 protein [23]. The important step to understand the function of PTEN as a tumor suppressor is to identify its physiological substrates.

Table 3. Feature selection performance of these six methods.

Datasets	BRCA_GE		LUAD_GE		PRAD_GE	
	N	MRS	N	MRS	N	MRS
NMF	166	29.89	151	**69.68**	108	24.10
$L_{2,1}$-NMF	171	29.89	168	**69.68**	114	24.10
$L_{1/2}$-RNMF	168	29.89	157	**69.68**	112	19.77
HNMF	169	29.89	163	**69.68**	112	24.10
GrRNMF	170	29.89	159	**69.68**	101	**34.85**
HRNMF	**185**	**33.21**	**175**	**69.68**	**122**	**34.85**

Table 4. The top ten genes selected by HRNMF on LUAD_GE dataset.

Gene name	Relevance score	Gene official name	Related diseases
PTEN	69.68	Phosphatase and TensinHomolog	Prostate cancer, lung and breast cancer
MLH1	55.50	Mut L homolog 1	Colorectal cancer, and mismatch repair cancer syndrome
STAT3	50.89	Signal transducer and activator of transcription 3	Autoimmune disease, and hyper-ige recurrent infection syndrome
BAX	47.06	BCL2 associated X, apoptosis regulator	Precursor T-cell acute lympho-blastic leukemia and acute t cell leukemia
APC	41.73	APC, WNT signaling pathway regulator	Familial adenomatous polyposis 1 and desmoid disease, hereditary
OGG1	38.36	8-oxoguanine DNA glycosylase	Clear cell renal cell carcinoma and renal cell carcinoma
CDKN1B	30.47	Cyclin dependent kinase inhibitor 1B	Multiple endocrine neoplasia, and primary hyperparathyroidism
MIR197	25.29	MicroRNA 197	Diabetes mellitus and thyroid cancer, nonmedullary, 2
BAP1	25.15	BRCA1 associated protein 1	Tumor predisposition syndrome and Bap1 tumor predisposition syndrome
PDPK1	15.35	3-phosphoinositide dependent protein kinase 1	Lung cancer

Persons who have hypermethylation of one allele of MLH1 in somatic cells throughout the body (a germ-line epimutation) have a predisposition for the development of cancer in a pattern typical of hereditary nonpolyposis colorectal cancer [24, 25]. The Mlh1 appears to localize to sites of crossing over on meiotic chromosomes. These findings suggest that Mlh1 is involved in DNA mismatch repair and meiotic crossing over [26].

5 Conclusion

In this paper, we have proposed the HRNMF based on the $L_{1/2}$-RNMF with the hyper-graph regularization and $L_{2,1}$-norm. This algorithm is effective for sample clustering and feature selection on several gene expression data. Firstly, the hyper-graph preserves the low-dimensional manifold structure existing in the high dimensional data space. Secondly, the $L_{2,1}$-norm is added to the objective function to reduce noises and outliers on the gene expression data. Furthermore, our method simulations the Gaussian noise and sparse noise, which can reduce the influence of the sparse noise. At last, we performed experiments on the gene expression datasets to analyze the feasible of our method. Compared with the other five advanced methods, our algorithm has achieved better results in clustering and feature selection.

Acknowledgments. This work was supported in part by the grants of the National Science Foundation of China, Nos. 61872220, and 61702299.

References

1. Roweis, S.T., Saul, L.K.: Nonlinear dimensionality reduction by locally linear embedding. Science **290**, 2323–2326 (2000)
2. Leeuw, J.D.: Nonlinear principal component analysis (2011)
3. Zou, H., Hastie, T., Tibshirani, R.: Sparse principal component analysis. J. Comput. Graph. Stat. **15**, 265–286 (2006)
4. Tenenbaum, J.B., De Silva, V., Langford, J.C.: A global geometric framework for nonlinear dimensionality reduction. Science **290**, 2319–2323 (2000)
5. Lee, D.D., Seung, H.S.: Learning the parts of objects by non-negative matrix factorization. Nature **401**, 788–791 (1999)
6. Zhu, R., Liu, J.X., Zhang, Y.K., Guo, Y.: A robust manifold graph regularized nonnegative matrix factorization algorithm for cancer gene clustering. Molecules **22**, 21–31 (2017)
7. Huang, Z., Zhou, A., Zhang, G.: Non-negative matrix factorization: a short survey on methods and applications. In: Li, Z., Li, X., Liu, Y., Cai, Z. (eds.) ISICA 2012. CCIS, pp. 331–340. Springer, Heidelberg (2012). https://doi.org/10.1007/978-3-642-34289-9_37
8. Wang, Y.-X., Zhang, Y.-J.: Nonnegative matrix factorization: a comprehensive review. IEEE Trans. Knowl. Data Eng. **25**, 1336–1353 (2013)
9. Kong, D., Ding, C., Huang, H.: Robust nonnegative matrix factorization using L21-norm. In: ACM International Conference on Information and Knowledge Management (2011)
10. He, W., Zhang, H., Zhang, L.: Sparsity-regularized robust non-negative matrix factorization for hyperspectral unmixing. IEEE J. Sel. Top. Appl. Earth Obs. Remote Sens. **9**, 4267–4279 (2017)
11. Maaten, L.V.D., Hinton, G.: Visualizing data using t-SNE. J. Mach. Learn. Res. **9**, 2579–2605 (2008)
12. Cai, D., He, X., Han, J., Huang, T.S.: Graph regularized nonnegative matrix factorization for data representation. IEEE Trans. Pattern Anal. Mach. Intell. **33**, 1548–1560 (2011)
13. Zeng, K., Yu, J., Li, C., You, J., Jin, T.: Image clustering by hyper-graph regularized non-negative matrix factorization. Neurocomputing **138**, 209–217 (2014)
14. Yu, N., Liu, J.X., Gao, Y.L., Zheng, C.H., Wang, J., Wu, M.J.: Graph regularized robust non-negative matrix factorization for clustering and selecting differentially expressed genes. In: IEEE International Conference on Bioinformatics and Biomedicine (2017)

15. Li, S.Z., Hou, X., Zhang, H., Cheng, Q.: Learning spatially localized, parts-based representation. In: CVPR, vol. 1, pp. 207–212 (2001)
16. Lee, D.D., Seung, H.S.: Algorithms for non-negative matrix factorization. In: Advances in Neural Information Processing Systems, pp. 556–562 (2001)
17. Konstantinova, E.V., Skorobogatov, V.A.: Application of hypergraph theory in chemistry. Discret. Math. **235**, 365–383 (2001)
18. Simovici, D.A., Djeraba, C.: Graphs Hypergraphs **34**, 1307–1315 (1973)
19. Qi, L., Jiang, H.: Semismooth Karush-Kuhn-Tucker equations and convergence analysis of Newton and quasi-Newton methods for solving these equations. Math. Oper. Res. **22**, 301–325 (1997)
20. Katarzyna, T., Patrycja, C., Maciej, W.: The cancer genome atlas (TCGA): an immeasurable source of knowledge. Contemp. Oncol. **19**, 68–77 (2015)
21. Cristofano, A.D., Pesce, B., Cordon-Cardo, C., Pandolfi, P.P.: PTEN is essential for embryonic development and tumour suppression. Nat. Genet. **19**, 348 (1998)
22. Myers, M.P.: The lipid phosphatase activity of PTEN is critical for its tumor supressor function. Proc. Natl. Acad. Sci. U. S. A. **23**, 13513–13518 (1998)
23. Neshat, M.S., et al.: Enhanced sensitivity of PTEN-deficient tumors to inhibition of FRAP/mTOR. Proc. Natl. Acad. Sci. **98**, 10314–10319 (2001)
24. Hitchins, M.P., et al.: Inheritance of a cancer-associated MLH1 germ-line epimutation. New Engl. J. Med. **356**, 697–705 (2007)
25. Edelmann, W., et al.: Meiotic pachytene arrest in MLH1-deficient mice. Cell **85**, 1125–1134 (1996)
26. Baker, S.M., et al.: Involvement of mouse Mlh1 in DNA mismatch repair and meiotic crossing over. Nat. Genet. **13**, 336 (1996)

Association of PvuII and XbaI Polymorphisms in Estrogen Receptor Alpha (ESR1) Gene with the Chronic Hepatitis B Virus Infection

Ke Men[1(✉)], Wen Ren[2], Xia Wang[3], Tianjian Men[3], Ping Li[2], Kejun Ma[2], and Mengyan Gao[1]

[1] Institute for Research on Health Information and Technology, School of Public Health Xi'an Medical University, Xi'an 710021, Shaanxi, China
menke@xiyi.edu.cn
[2] Lanzhou University Second Hospital, Lanzhou 730030, Gansu, China
[3] Department of Military Preventive Medical, The Air Force Military Medical University, Xi'an 710032, Shaanxi, China

Abstract. This study aims to investigate the correlation of PvuII (rs2234693) and XbaI (rs9340799) polymorphisms in estrogen receptor alpha (ESR1) gene with Chronic Hepatitis B Virus Infection, and provide new insight and scientific basis for controlling chronic infection of hepatitis B virus (HBV). We selected 107 healthy subjects as the control group and 107 patients with chronic hepatitis B (CHB) as the treated group. The molecular diagnosis methods of PCR-HRM based on the high-resolution melting curve (HRM) technique were employed to determine the gene polymorphisms of two SNPs: rs2234693 (T > C) and rs9340799 (A > G). The correlations between these two single nucleotide polymorphisms (SNP) sites and HBV chronic were further verified by gel electrophoresis and following sequencing. It was found that the genotype frequencies of rs2234693 (T > C) site were significantly different between CHB cases and the controls (P < 0.05), the allele frequency of rs2234693 showed no difference between two groups. The genotype frequencies of rs9340799 site was significantly different between CHB cases and the controls (P < 0.05). The GG genotype was significantly increased in the chronic hepatitis B group. The allele frequency of rs9340799 site was significantly different between the two groups (P < 0.05). According to Chi square analysis, the G gene of rs9340799 could increase the risk of chronic infection of HBV and the A gene could reduce the risk of chronic infection of HBV. The GG genotype and G allele of rs9340799 (A > G) of estrogen-receptor gene ESR1 may be the genetic susceptibility gene of HBV infection. The GG genotype is associated with chronic infection of HBV.

Keywords: Hepatitis B · Estrogen receptor · Gene polymorphism · High resolution melting

1 Introduction

Chronic hepatitis B is a worldwide widespread infectious disease caused by Hepatitis B virus [1, 2]. It is ranked as the tenth lethal factor in the world according to WHO [3]. The

H. Han et al. (Eds.): IDMB 2019, CCIS 1099, pp. 126–136, 2020.
https://doi.org/10.1007/978-981-15-8760-3_9

number of deaths associated with HBV infection is about a half to 1.2 million per year. Although the neonatal immunization program has successfully reduced the infection rate of hepatitis B virus surface antigen (HBsAg) carriers from 9.8% in 1992 to 7.18% in 2006, HBV still remains as one of the major infectious diseases for developing countries [4]. For example, China is a high-incidence area of HBV infection, HBV has become one of the major infectious diseases threatening Chinese people health due to the lack of effective treatment. The national HBV serological epidemiological survey showed that HBV carriers in China were up to 93 million and chronic hepatitis B patients reached 25 million in 2006 [5]. The incidence of chronic hepatitis B in male is generally higher than female. This is mainly because HBV is a sex hormone response virus and liver is a sexual dimorphic organ [6]. Previous studies have reported that the gender difference is due to the decrease of estrogen secretion or decrease of estrogen response, and the methylation of ESR1 gene promoter exists in hepatitis B disease [7]. The gender factor also plays an important role in HBV chronic infection, because liver is a metabolic organ of sex hormone and human body can make sex hormone disorder after infection with HBV, which has influence on the progress of chronic hepatitis B. As a result, the prevalence rate of male viral hepatitis is higher than that of female one [8], so do acute hepatitis [9].

A certain level of estrogen can prevent liver from hepatic fibrosis and avoid inflammatory injury of hepatocyte besides inhibiting apoptosis of hepatocytes [10]. Recent studies have shown that the interaction between estrogen and estrogen receptor plays an important role. The effect of estrogen is mainly regulated by two known estrogen receptors: estrogen receptor α (ESR1) and estrogen receptorβ (ESR2). In fact, most of the biological effects are mainly mediated by ESR1, and estrogen plays a role that mainly binds specifically to estrogen receptorα (ESR1) [11, 12]. Many studies have reported that ESR1 gene polymorphism may be associated with tumor risk-related in various organs, such as prostate [13], endometrium [14], mammary gland [15], ovary [16], uterine leiomyoma [17], and colorectum [18]. Estrogen plays an anticancer role by binding to estrogen receptor α (ESR1) [19]. The gene polymorphism of estrogen receptor ESR1 may also be one of the important causes of HBV chronic infection. The aim of this study is to explore the association between estrogen receptor gene ESR1 polymorphism and HBV chronic infection in order to provide a new scientific basis and clue for the control of HBV chronic infection.

Estrogen receptor (ER) is a protein molecule that exists in target cells and can bind specifically with estrogen to form a hormone-receptor complex that makes estrogen play a biological role. Estrogen receptor is a member of nuclear receptor superfamily and mediates many effects of estrogen, including classical nuclear receptor and membrane receptor. Among them, nuclear receptors include ERα (ESR1) and ERβ (ESR2) subtypes. Estrogen is regulated by ERα and ERβ, which are nuclear transcription factor [20]. They play a genotypic regulatory effect by regulating the transcription of specific target genes. The effects of estrogen are mainly regulated by estrogen receptors ERα (ESR1) and ERβ (ESR2), and most of the biological effects are mediated by ESR1. The human ESR1 gene is located in the 6q24–27 region of chromosome and is composed of 8 exons and 7 introns with the length of 140 kb. The two SNP sites: rs2234693 and rs9340799 of estrogen receptor gene ESR1, are the two most widely studied polymorphism sites in ESR1 gene. Some scholars screened single nucleotide polymorphisms of ERa and determined that two SNP sites of 29T/C and 252966A/G were associated with HBV infection [21]. Therefore, it is valuable to further study the polymorphism of two SNP loci in ESR1 gene.

2 Method

2.1 Data Acquisition

According to the serological markers of HBV, the quantitative results of HBV DNA and indicators of liver function, 107 cases of chronic hepatitis B patients hospitalized in Lanzhou University Second Hospital from October 2016 to October 2017 were selected as case group. All patients in the case group with hepatitis B history or hepatitis B surface antigen HBsAg positive for more than six months. The patients with HBsAg positive and/or HBV DNA positive were diagnosed as chronic HBV infection. The 107 patients consist of 70 males and 37 females with their ages 9–68 (mean: 36.9, std: 12.4) years. On the other hand, another 107 healthy people consisting of 70 males and 37 females with ages 14–79 (mean: 42.3, std: 14.5) years, were selected as control group from October 2016 to October 2017 in the physical examination center of Lanzhou University Second Hospital. It is noted that the diagnosis of chronic hepatitis B patients accords with the diagnostic criteria of chronic hepatitis B as revised by the Chinese Society of Hepatology, CMA and Society of Infectious Diseases, CMA in 2015. The subjects of the two groups were not related to each other and all of them were Han nationalities.

2.2 Extraction and Quantification of Genomic DNA

We employed the blood genomic DNA extraction kit from Tigen Biochemical Technology Co. Ltd. to extract genomic DNA from 214 blood samples collected in the case and control group. We further employed NanoDrop-2000 trace nucleic acid protein detector to extract 1 µl from the genomic DNA extracted above, and put it into the instrument to detect the concentration and purity of DNA. The unit of concentration detected was ng/µl in our experiment. It is noted that OD 260/280 ≈ 1.8 and OD 260/280 > 1.9 indicates RNA pollution. Similarly, OD 260/280 < 1.6 indicates there are protein and phenol pollution. The purity of OD 260/280:1.6–2.0, OD 260/230:1.5–2.0 was used as purity requirement, and the reserve was stored at −40 °C.

2.3 Primer Design and PCR Reaction Conditions

According to the two SNP sites of estrogen receptor gene ESR1, rs2234693 (T > C) & rs9340799 (A > G), the upstream and downstream primers from two SNP sites were designed as follows.

The rs2234693 (T > C)'s upstream is 5'-ACATGTTCTGTTGTCCATC-3' and downstream is 5'-CCAACTCTAGACCACACTCAGG-3'. The length of the PCR product is102bp. On the other hand, the upstream primer of rs9340799 (A > G) is 5'-CATCTGAGTTCCAAATGTCCC-3', and the downstream primer is 5'-TTTCAGAACCATTAGAGACCAATG-3'.

The length of the PCR product is 107 bp. The primers were further synthesized by Beijing Zixi Biotechnology Co., Ltd. The reaction was amplified by PCR three-step. The optimized reaction conditions were as follows: initial denaturation at 95 °C for 2 min, one cycle; denaturation at 95 °C for 10 s, annealing at 58 °C for 15 s, extending at 72 °C for 15 s and amplifying for 40 cycles. The PCR products were detected by 4% agarose gel electrophoresis.

2.4 HRM Assay

The assay was performed on Rotor-gene6000 fluorescence quantitative PCR instrument. The HRM Analysis PreMix enzyme system was used to amplify and melt, which contained EvaGreen saturated dyes, and could not inhibit the PCR reaction at a high concentration. It makes combining weight of the double-stranded PCR products could reach saturation state, and distinguishes the difference of single base between amplification products.

The melting conditions were denatured at 96 °C for 1 min and annealed at 40 °C for 1 min. The fluorescence signals were collected from 71 °C to 83 °C, at the speed of 0.2 °C/s. HRM analysis was performed according to the characteristic peaks of the samples melting curve of each SNP site.

2.5 Methodology Verification

Genotyping is combined with the position and shape of the melting curve. The 2–3 samples with different peak types of melting curves were randomly selected from those that have been genotyped by HRM as well as the samples with excursion of melting curves that cannot be genotyped directly. The PCR products of these samples were identified as a single bright band by 4% agarose gel electrophoresis for 15 min at 100 v voltage. Then, the products were sent to the biological company for gene sequencing to verity the correctness of genotyping in this experiment.

2.6 Statistical Analysis

The Hardy-Weinberg balance test is used to determine whether the study population is representative or not. When the $p > 0.05$, it means that the studied population conforms to the Hardy-Weinberg genetic balance law, i.e. the population is representative. The Hardy-Weinberg balance test was carried out by using SHEsis online analysis software, and the allele frequency and genotype frequency of the studied gene loci were statistically analyzed. The odds ratio (OR) and its 95% confidence interval (95% CI) indicated the risk assessment of each locus genotype. When $P < 0.05$, the difference is statistically significant.

3 Results

3.1 Electrophoretic Analysis of PCR Products at rs2234693 and rs9340799 Sites of ESR1 Gene

The integrity of genomic DNA was identified by 4% agarose gel electrophoresis after PCR reaction at two sites of estrogen receptor ESR1 gene. The following is the electrophoretogram of partial genomic DNA of two sites. The result of electrophoresis showed that the length of the amplified target gene product was about 110 bp, which was a single bright band with no heterologous band. It indicated that there was no non-specific amplification of the PCR product, the integrity of genomic DNA extracted was good. As shown in Fig. 1.

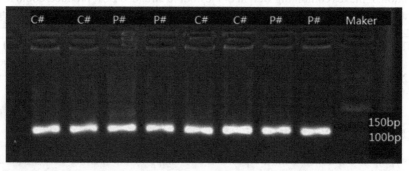

Fig. 1. Electrophoresis of PCR product. Note. The right end band is DNA Maker, and the remaining bands are PCR amplification products of randomly selected SNP sites. The bands are all located between 100 bp–150 bp which is consistent with the length of PCR products designed by our primer.

3.2 Genotyping and Validation of SNP Locus

3.2.1 HRM Genotyping and Sequencing of rs2234693 (T > C)

Three genotypes of rs2234693 (T > C) locus were found in this study, which were wild-type TT, mutation heterozygous CT, mutation homozygous CC. The HRM genotyping map and corresponding gene sequencing of melting curve were shown in Figs. 2-1 and 2-2 respectively. The result showed that concordance rate of HRM genotyping and sequencing was 99%.

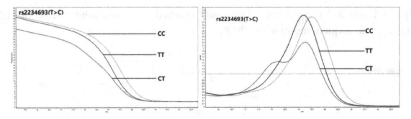

Fig. 2-1. Genotyping map of rs2234693 (T > C)

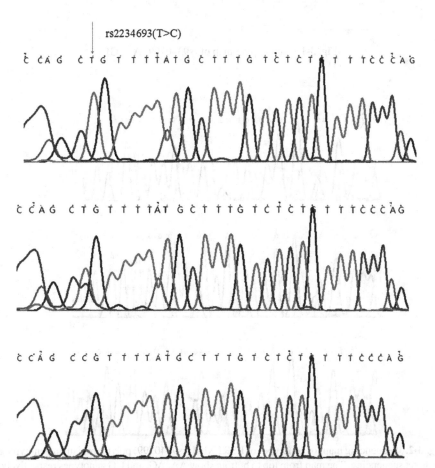

Fig. 2-2. Sequencing map for genotypes of ESR1 PvuII(rs2234693) polymorphism in genotyping by direct sequencing. The map from top to bottom show TT, TC, and CC genotypes, respectively.

3.2.2 HRM Genotyping and Sequencing of rs9340799 (a > G)

Through the detection, three genotypes of rs9340799 (A > G) locus were found. They were wild-type AA, mutation heterozygous AG, mutation homozygous GG. The HRM genotyping map and corresponding gene sequencing of melting curve were shown in Figs. 3-1 and 3-2 respectively. The result showed that concordance rate of HRM genotyping and sequencing reached 99%.

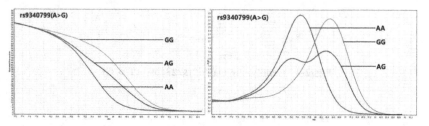

Fig. 3-1. Genotyping map of rs9340799 (A > G)

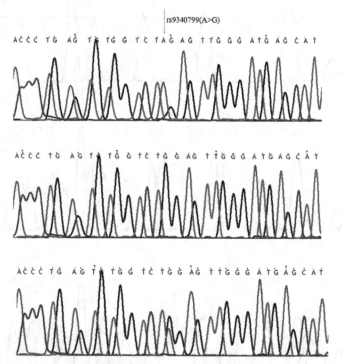

Fig. 3-2. Sequencing map for genotypes of ESR1 XbaI (rs9340799) polymorphism in genotyping by direct sequencing. The map from top to bottom show AA, AG, and GG genotypes respectively.

3.3 Hardy-Weinberg Balancing Test

The two SNP sites rs2234693 (T > C) and rs9340799 (A > G) of the case group and the healthy control were tested for Hardy-Weinberg balance test. The test result of rs2234693 (T > C) locus showed that p-$value = 0.216$ in the case group. Such a result was consistent with the genetic balance study and had population genetic representativeness. Alternatively, we have *p-value* = 0.000897 in the healthy control group. It was not consistent with genetic balance study and had no population genetic representativeness. The reason for the result is probably due to the sample size, which leads to the imbalance of Hardy-Weinberg test.

The results of rs9340799 (A > G) at SNP locus tested by Hardy-Weinberg balance test were as follows. The p-values were 0.342 and 0.177 for the case and control group respectively. They indicated that both case group and healthy control group were consistent with the genetic balance study and had population genetic representation.

3.4 Polymorphism Distribution of Two SNP Loci in ESR1 Gene

We further detected the genotypic frequency and allele frequency of rs2234693 (T > C) locus in ESR1. The results showed that there was some statistical difference in the distribution of genotypic frequency between HBV case group and healthy controls group (p < 0.01), and there was no statistical difference in allele frequency between HBV case group and healthy controls (P > 0.05). The chi square analysis was performed by SHEsis online software, OR is 1.0 (95% CI 0.67–1.47), as shown in Table 1.

Table 1. ESR1 PvuII(rs2234693) polymorphism between CHB and healthy group

groups	n	Genotype frequency (n, %)			Gene frequency (n, %)	
		CC	CT	TT	C	T
CHB	104	20(19.2)	44(42.3)	40(38.5)	84 (40.4)	124 (59.6)
Control	105	9(8.6)	67(63.8)	29(27.6)	85 (40.5)	125 (59.5)
χ^2	10.687				0.000364	
P	0.004778				>0.05	
OR					0.99	
95%CI					(0.67-1.47)	

Genotypic frequency and allele frequency of rs9340799 (A > G) locus in ESR1 were detected. It was found that they demonstrated statistically significant difference (P < 0.01) between the HBV case group and healthy control group. Furthermore, A/G OR value was 0.27 (95% CI 0.16–0.46). The distribution GG genotype G allele in HBV case group was significantly higher than in healthy control group (Table 2). The result indicated that allele in the rs9340799 (A > G) locus was a protective gene of chronic hepatitis B, and G allele was a risk gene for chronic hepatitis B, and the GG genotype was a susceptible genotype of chronic HBV infection that can increase the risk of HBV.

4 Discussion

In this study, the relationship between polymorphism of two loci gene PvuII (rs2234693) and XbaI (9340799) of estrogen receptor ESR1 and chronic HBV infection was studied.

Table 2. ESR1 XbaI(rs9340799) polymorphism between CHB and healthy group

groups	n	Genotype frequency (n, %)			Gene frequency (n, %)	
		AA	AG	GG	A	G
CHB	105	45(42.9)	44(41.9)	16(15.2)	134（63.8）	76（36.2）
Control	80	59(73.8)	21(26.3)	0(0.00)	139（86.9）	21（13.1）
χ^2	23.066				24.976	
P	<0.01				<0.01	
OR					0.27	
95%CI					(0.16-0.46)	

We employed case-control methods, and analyzed clinical phenotypes from subjects infected with HBV, such as chronic virus carrying, hepatitis B, hepatic injury, hepato-cirrhosis, primary hepatic carcinoma. The patients with chronic hepatitis B virus (CHB) were selected as the case group and the healthy people as the control group. Due of time and funding constrains, there was no further study of genetic polymorphism in hepatic insufficiency and hepatocirrhosis.

The result of rs2234693 (T > C) polymorphism in estrogen receptor ESR1 gene in this work showed that the proportion of TT genotypes in chronic HBV case group was slightly higher than that in healthy control group. There were significantly statistical differences in genotype frequencies between HBV case group and healthy control group, but there were no significant statistical differences in allele frequencies in HBV distribution. Deng found that the ESR1 29T/T genotype of individuals with at least one 29C allele gene compared to the susceptibility of persistent HBV infection significantly, through comparing of 1277 persistent HBV infected subjects with 748 spontaneously recovered subjects [21]. Furthermore, other investigators studied the polymorphism of rs2234693 loci in Guangxi Zhuang people, and showed that susceptibility of CC genotype to CHB was significantly higher than that of TT genotype. It was suggested that the CC genotype was associated with increased risk of CHB [22]. Liu reported Zhuang population which is minor population and may be different from Han population. It suggests that there may be differences in HBV infection among different nationalities. The new study reported that the rs2234693 locus of ESR1 gene is located at the potential binding site of b-Myb transcription factor, and the correlation between ESR1 overexpression and C allele is significantly higher than of T allele [23]. It is worthwhile to point out that and the sample size of the study is also relatively small that may limit the generalization of our conclusion. However, We plan to expand the sample size to further study the locus in the future to further enhance robustness of our findings.

For the rs9340799 (A > G) locus, the GG genotype appeared only in the HBV case group, and no GG genotype was found in the healthy control group. It may be related to the low frequency of the genotype itself because of the relatively small sample size in our study. In the HBV case group, pure mutation was significantly higher than that in the healthy control group. Logistic regression analysis showed that A allele gene in rs9340799 (A > G) site was a protective gene for chronic hepatitis B, and allele gene G was the susceptible genotype of chronic HBV infection. The GG genotypes and G allele gene at rs9340799 (A > G) site may be the genetic susceptibility genes for chronic HBV infection.

Chronic HBV infection can be the result of multiple factors. In addition to the variation of HBV, host age, sex and immune status, etc., genetic polymorphisms may be an important cause of HBV chronic infection [24]. There are few studies on estrogen receptor gene polymorphism and HBV chronic infection. Association of PvuII and XbaI polymorphisms with diseases such as prostate cancer susceptibility and systemic lupus erythematosus were studied [25, 26]. The reason may be the complexity of the interaction between the virus and the host with multiple allele genes, and the genetic susceptibility mechanism may involve the interaction between multiple gene loci and haploid variation. Therefore, we should further study the relationship between HBV chronic infection and gene polymorphism from many aspects, large samples, and multiple regions to explore the pathologic mechanisms of genetic susceptibility to HBV infection and to provide new insight for anti-HBV therapy and hepatitis B prevention.

Acknowledgment. This work was partially supported by National Natural Science Foundation of China (Grant No. 81773496, 2018PT01) and Natural Science Basic Research Plan in Shaanxi Province of China (Grant No. 2016JM8103, 2017PT22).

References

1. Mahoney, F.J.: Update on diagnosis, management, and prevention of hepatitis B virus infection. Clin. Microbiol. Rev. **12**(2), 351–366 (1999)
2. Ocama, P., Opio, C.K., Lee, W.M.: Hepatitis B virus infection: current status. Am. J. Med. **118**(12), 1413 (2005)
3. Zampino, R., Boemio, A., Alessio, L., et al.: Hepatitis B virus burden in developing countries. World J. Gastroenterol. **21**, 11941–11953 (2015)
4. Lesmana, L.A., Leung, N.W., Mahachai, V., et al.: Hepatitis B: overview of the burden of disease in the Asia-Pacific region. Liver Int. **26**(Suppl 2), 3–10 (2006)
5. Ministry of Health of the PRC: National hepatitis B prevention and control plan from 2006 to 2010. Chin. Pract. Rural Doctor **13**(08), 1–4 (2006)
6. Wang, S.H., Chen, P.J., Yeh, S.H.: Gender disparity in chronic hepatitis B: mechanisms of sex hormones. J. Gastroenterol. Hepatol. **30**, 1237–1245 (2015)
7. Dou, C.Y., Fan, Y.C., Cao, C.J., et al.: Sera DNA methylation of CHD1, DNMT3b and ESR1 promoters as biomarker for the early diagnosis of hepatitis B virus-related hepatocellular carcinoma. Dig. Dis. Sci. **61**, 1130–1138 (2016)
8. Chan, S.L., Wong, V.W., Qin, S., et al.: Infection and cancer: the case of hepatitis B. J. Clin. Oncol.: Off. J. Am. Soc. Clin. Oncol. **34**, 83–90 (2016)
9. Miao, N., Zhang, G., Zheng, H., et al.: Analysis of the hepatitis B report data pilot surveillance in 200 counties in China, 2013. Chin. J. Prev. Med. **49**, 766–770 (2015)
10. Dan, L., X-H, L., C-G, G., et al.: The role of sex hormones in the progression of chronic hepatitis B. J. Hepatol. **17**(11), 821–823 (2012). (in Chinese)
11. Ahlbory-Dieker, D.L., Stride, B.D., Leder, G., et al.: DNA binding by estrogen receptor-alpha is essential for the transcriptional response to estrogen in the liver and the uterus. Mol. Endocrinol. **23**(10), 1544–1555 (2009)

12. Yan, Z., Tan, W., Xu, B., et al.: A cis-acting regulatory variation of the estrogen receptor alpha (ESR1) gene is associated with hepatitis B virus-related liver cirrhosis. Hum. Mutat. 32(10), 1128–1136 (2011)

13. Li, L., Zhang, X., Xia, Q., Ma, H., Chen, L., Hou, W.: Association between estrogen receptor alpha PvuII polymorphism and prostate cancer risk. Tumor Biol. 35(5), 4629–4635 (2014). https://doi.org/10.1007/s13277-014-1606-9

14. Zhou, X., Gu, Y., Wang, D.N., et al.: Eight functional polymorphisms in the estrogen receptor 1 gene and endometrial cancer risk: a meta-analysis. PLoS ONE 8(4), e60851 (2013)

15. Chattopadhyay, S., Siddiqui, S., Akhtar, M.S., et al.: Genetic polymorphisms of ESR1, ESR2, CYP17A1 and CYP19A1 and the risk of breast cancer: a case control study from North India. Tumour Biol. 35(5), 4517–4527 (2014). https://doi.org/10.1007/s13277-013-1594-1

16. Doherty, J.A., Rossing, M.A., Cushing-Haugen, K.L., et al.: Polymorphism and invasive epithelial ovarian cancer risk an ovarian cancer association consortium study. Cancer Epidemiol. Biomarkers Prev. 19, 245–250 (2010)

17. Feng, Y., Lin, X., Zhou, S., et al.: The associations between the polymorphisms of the ER-alpha gene and the risk of uterine leiomyoma (ULM). Tumour Biol. 34(5), 3077–3082 (2013)

18. Slattery, M.L., Sweeney, C., Murtaugh, M., et al.: Associations between ERalpha, ERbeta, and AR genotypes and colon and rectal cancer. Cancer Epidemiol. Biomarkers Prev. 14(12), 2936–2942 (2005)

19. Hishida, M., Nomoto, S., Inokawa, Y., et al.: Estrogen receptor 1 gene as a tumor suppressor gene in hepatocellular carcinoma detected by triple-combination array analysis. Int. J. Ooncol. 43, 88–94 (2013)

20. Perissi, V., Rosenfeld, M.G.: Controlling nuclear receptors: the circular logic of cofactor cycles. Nat. Rev. Mol. Cell Biol. 6(7), 542–554 (2005)

21. Deng, G., Zhou, G., Zhai, Y., et al.: Association of estrogen receptor alpha polymorphisms with susceptibility to chronic hepatitis B virus infection. Hepatology 40(2), 318–326 (2004)

22. Liu, Y., Liu, Y., Huang, X., et al.: Association of PvuII and XbaI polymorphisms in estrogen receptor alpha gene with the risk of hepatitis B virus infection in the Guangxi Zhuang population. Infect. Genet. Evol. 27, 69–76 (2014)

23. Herrington, D.M., Howard, T.D., Brosnihan, K.B., et al.: Common estrogen receptor polymorphism augments effects of hormone replacement therapy on E-selectin but not C-reactive protein. Circulation 105(16), 1879–1882 (2002)

24. Thursz, M.: Genetic susceptibility in chronic viral hepatitis. Antiviral Res. 52(2), 113–116 (2001)

25. Zhao, Y., Zheng, X., Zhang, L., et al.: Association of estrogen receptor alpha PvuII and XbaI polymorphisms with prostate cancer susceptibility and risk stratification: a meta-analysis from case-control studies. Onco Targets Ther. 10, 3203–3210 (2017)

26. Xie, Q., Hu, H., Li, S.S., et al.: Association of oestrogen receptor alpha gene polymorphisms with systemic lupus erythematosus risk: an updated meta-analysis. Microb. Pathog. 127, 352–358 (2019)

Inferring Communities and Key Genes of Triple Negative Breast Cancer Based on Robust Principal Component Analysis and Network Analysis

Qian Ding[1], Yan Sun[1(✉)], Junliang Shang[1(✉)], Yuanyuan Zhang[2], Feng Li[1], and Jin-Xing Liu[1]

[1] School of Information Science and Engineering, Qufu Normal University, Rizhao 276826, China
{dingqian19,sunyan225,sdcavell}@126.com,
shangjunliang110@163.com, lifeng_10_28@163.com
[2] School of Information and Control Engineering,
Qingdao University of Technology, Qingdao 266000, China
yyzhang1217@163.com

Abstract. Triple negative breast cancer (TNBC), one subtype of breast cancer, has become the leading cause of cancer-associated death in women worldwide. In this paper, we provided a framework to identify communities and key genes of TNBC based on robust principal component analysis (RPCA) and network analysis. Firstly, RPCA was employed to detect differentially expressed genes of TNBC in an integrated dataset that collected from three profile datasets, i.e., GSE38959, GSE45827 and GSE65194. Secondly, these differentially expressed genes were used to construct a weighted correlation gene network, from which five communities were identified. Then, three traditional topological properties, as well as a novel developed one named Normalized Score which takes into account both local and global features of nodes, were applied to identify key genes in these communities. Finally, both GO enrichment analysis and KEGG pathway analysis were performed on these communities to detect significant biological pathways. Results not only proved that this framework is effective in inferring communities and key genes of complex diseases like TNBC, but also captured several useful clues for better uncovering genetic architecture of TNBC.

Keywords: Triple negative breast cancer · Topological property · Community · Gene network · Key genes

1 Background

Triple negative breast cancer (TNBC), which with aggressive tumor behavior and distinct disease features, represents 10% to 17% [1] of all diagnosed breast tumors. Owing to the lack of effective targeted medicine and the high recurrence rates, identifying communities and key genes of breast cancer has become a hot topic. There are various

© Springer Nature Singapore Pte Ltd. 2020
H. Han et al. (Eds.): IDMB 2019, CCIS 1099, pp. 137–151, 2020.
https://doi.org/10.1007/978-981-15-8760-3_10

methods have been proposed to identify the key genes of TNBC. For example, Wang *et al.* [2] used t-test method to identify key genes from breast cancer dataset. R. Bell *et al.* [3] obtained the significant differential expression in breast neoplasms by comparing the results of the data meta-analysis with literature mining. Li *et al.* [4] used GEO2R tool to identify key genes between TNBC and normal tissue. However, most of these methods focused on analyzing a single dataset or multiple datasets independently, which ignored the interaction between genes from different datasets. Therefore, we analyzed the integrated dataset collected from GSE38959, GSE45827 and GSE65194 to quantify the interaction between genes. Identifying interactions between key genes from high dimensional integrated datasets makes the advantages of feature selection and network mining methods more promising.

In recent years, the network methods were used to identify communities which include novel biomarkers associated with TNBC. Yang *et al.* [5] identified communities of an integrated gene regulatory network for studying the aggressive behavior of breast cancer. Tang *et al.* [6] used weighted gene co-expression network analysis (WGCNA) to identify significant communities of breast cancer dataset. Tang *et al.* [7] established protein-protein interaction (PPI) network and identified communities of the molecular complex detection (MCODE) plugin. However, these network methods, which were based on WGCNA and MCODE, only focused on analyzing communities but ignored the role of individual genes played in the network.

To reduce the complexity of network analysis, we used feature selection methods to select genes for network construction. Principal component analysis (PCA) is one of the most common methods for feature selection. There are so many peer methods in PCA such as, sparse principal component analysis (SPCA) [8], nonnegative principal component analysis (NPCA) [9], supervised discriminative sparse PCA (SDSPCA) [10]. Since robust principal component analysis (RPCA) can recover the essentially low rank data from the observation data of the large and sparse noise pollution, we identified significant communities and key genes from network which was constructed by integrated dataset based on RPCA feature selection. First, RPCA was performed on the integrated dataset which collected from GSE38959, GSE45827 and GSE65194 to select differentially expressed genes. These selected genes were used to construct network and to identify significant communities. Then, a new topological property Normalized Score (*NS*) was proposed to identify key genes from each community. Finally, we applied the Gene Ontology (GO) and Kyoto Encyclopedia of Genes and Genomes pathway (KEGG) enrichment to verify the experimental results. The experimental results of our framework provided a better understanding of the underlying mechanisms and biomarkers for TNBC.

2 Method

2.1 Datasets

Three profile datasets, GSE38959, GSE45827 and GSE65194, were downloaded from the National Center for Biotechnology Information Gene Expression Omnibus (GEO) database (http://www.ncbi.nlm.nih.gov/geo/) in this study. The array data of GSE38959, which were based on GPL4133 platform (Agilent-014850 Whole Human Genome

Microarray 4x44K G4112F), had 30 TNBC samples and 13 normal samples [11]. GSE45827 was based on GPL570 platform (Affymetrix Human Genome U133 Plus 2.0 Array) including 41 TNBC samples and 11 normal samples [12]. The array data of GSE65194 were based on GPL570 platform (Affymetrix Human Genome U133 Plus 2.0 Array). This dataset had 55 TNBC samples and 11 normal samples [13].

To make these three datasets conform to a normal distribution, after converting the probe names of the three data sets into the corresponding gene symbols and log 2 transformation, z-score normalization was performed by calculating the mean and standard deviation [14]. After aligning the genes of these three datasets, we removed the under-expressed genes in multiple samples, that is, rows with most values of zero were removed, and we integrated them into a matrix (11796 × 161) which covers 161 samples with 11796 genes for our subsequent processing and analysis.

2.2 Robust Principal Component Analysis (RPCA)

Robust principal component analysis (RPCA) [15] decomposes large data matrix M (m × n) as perturbation signals S and low-rank matrix A. The differentially expressed genes were viewed as the sparse perturbation signals S and the differentially expressed genes as the low-rank matrix A. [16]:

$$M = A + S \qquad (1)$$

where m represents the number of genes, and n represents the number of samples. By using RPCA, perturbation signals S was recovered from the gene expression data to select the differentially expressed genes. Assuming that the observation matrix M is given by Eq. (1), RPCA is used to solve the following optimization problem:

$$\min \|A\|_* + \lambda \|S\|_1$$
$$\text{subject to } M = A + S \qquad (2)$$

where $\|A\|_*$ represents the nuclear norm of the matrix A: $\|A\|_* = \sum_i \sigma_i(A)$; $\|S\|_1$ represents L1-norm of S: $\|S\|_1 = \sum_{ij} |S_{ij}|$; $\lambda = (c * \max(m, n))^{-\frac{1}{2}}$, c is set to 0.3 [16].

For solving the RPCA problem in Eq. (2), we introduced the inexact Augmented Lagrange Multipliers ALM (IALM) method [17] to eliminate equality constraints by introducing a Lagrange multiplier. The details of this algorithm are referred to [17].

After decomposing the data matrix M by using RPCA, sparse perturbation matrix S was obtained to select differentially expressed genes. In matrix S, each row represents one gene, and each column represents one sample:

$$S = \begin{pmatrix} s_{11} & \cdots & s_{1n} \\ \vdots & \ddots & \vdots \\ s_{m1} & \cdots & s_{mn} \end{pmatrix} \qquad (3)$$

The differentially expressed genes can be determined by analyzing matrix S: after calculating the absolute values of factors in the sparse matrix S, we calculated the sum of each row in S to get the evaluation vector S_0, which is defined as:

$$S_0 = \left[\sum_{i=1}^{n} |s_{1i}| \cdots \sum_{i=1}^{n} |s_{mi}| \right]^T \tag{4}$$

We sorted values of S_0 and performed a curve fit to select the value of inflection point, which is the number of differentially expressed genes.

2.3 Weighted Gene Co-expression Network Analysis (WGCNA)

2.3.1 Network Construction

In co-expression network, which is also called weighted gene network of differentially expressed genes, a node represents a gene; an edge represents an association between any two genes. The weighted coefficient matrix a_{xy} is used to quantify the correlation between these two genes, which could strengthen strong correlations and weaken weak correlations between genes:

$$a_{xy} = |S_{xy}|^{\beta} \tag{5}$$

where S_{xy} is the Pearson Correlation Coefficient (PCC) between any two genes, and the soft-thresholding parameter β was used to approximate the distribution of scale-free networks.

2.3.2 Communities Identification

The weighted coefficient matrix a_{xy} was transformed into topological overlap matrix w_{xy} to measure the network connectivity of a gene:

$$w_{xy} = \frac{l_{xy} + a_{xy}}{\min\{k_x, k_y\} + 1 - a_{xy}} \tag{6}$$

where $l_{xy} = \sum_{u} a_{xu} a_{yu}$, u is a node that is connected to both gene x and y. k_x represents the sum of the weighted coefficient matrix of nodes connected to gene x: $k_x = \sum_{u} a_{xu}$. Similarly, k_y represents the sum of the weighted coefficient matrix of nodes connected to gene y: $k_y = \sum_{u} a_{yu}$. Based on topological overlap coefficient, genes with different expression patterns are classified into different communities.

2.4 Topological Properties Analysis

2.4.1 Traditional Properties

After using WGCNA to construct a gene interaction network, we analyzed the topology properties of the network. There are a lot of topology properties to identify key nodes. The following three traditional properties were considered in the network: degree centrality (DC), betweenness centrality (BC) and closeness centrality (CC). The definitions of them are as follows.

Degree centrality of node v is defined as the number of edges which are directly connected to v in the network:

$$DC_v = \deg(v) \tag{7}$$

Degree centrality is the most basic topological property, which reflects local features of nodes in the network. Betweenness centrality of node v is used to quantify the role of node v plays in connecting to other nodes:

$$BC_v = \sum_{i \neq j \neq v \in V} \frac{N_{ivj}}{N_{ij}} \tag{8}$$

where N_{ij} represents the number of the shortest path from node i to node j, in which N_{ivj} represents the number of paths through the node v. The betweenness value represents the connection tightness of each node in the global network. Closeness centrality represents the average shortest distance from node v to all other nodes in the network, which is defined as:

$$CC_v = \frac{1}{n} \sum_{j \neq v \in V} d_{vj} \tag{9}$$

where d_{vj} represents the shortest distance from node v to node j. The smaller the value of closeness centrality, the more important the node in the network. Closeness centrality reflects the global features of nodes.

2.4.2 Normalized Score

There are a host of traditional properties that quantify the topological properties of the nodes in network. However, most of these properties only focus on one side of local or global features. To find key genes better, we proposed a novel topological property, namely Normalized Score (NS) by combining the above three traditional properties: degree centrality, betweenness centrality and closeness centrality:

$$NS_v = \frac{DC_v/med(DC_v) + BC_v/med(BC_v)}{CC_v/med(CC_v)} \tag{10}$$

where DC_v, BC_v and CC_v represent the degree centrality, betweenness centrality and closeness centrality of node v, respectively. $med(DC_v)$, $med(BC_v)$ and $med(CC_v)$ represent the median of degree centrality, betweenness centrality and closeness centrality of node v. NS property considers both local features and global features, the calculation results of NS may solve the problem of one-sided.

2.5 GO Enrichment Analysis and KEGG Pathway Analysis

Gene ontology (GO) enrichment analysis was used to annotate genes and products of key genes. GO enrichment was performed on the communities by using DAVID online tool (https://david.ncifcrf.gov/) [18], which was an essential foundation for identifying biological characteristic of genes. What's more, Kyoto Encyclopedia of Genes and Genomes (KEGG) pathway enrichment analysis was performed to detect functional attributes of five communities, by using the KOBAS online analysis database (version 3.0) (http://kobas.cbi.pku.edu.cn/) [19]. P-value < 0.05 was set as the cut-off criterion.

3 Results and Discussions

3.1 RPCA for Identifying Differential Expressed Genes

After using RPCA method for feature selection, finally, a matrix of differentially expressed genes including 1267 rows and 161 columns was obtained to construct the network. To further demonstrate the advantages of selecting network construction genes by using RPCA, we compared the results of RPCA with PCA. Firstly, the same number of differentially expressed genes identified by the RPCA and PCA methods were used for GO enrichment analysis. Compared with the PCA method, RPCA matched significantly more GO items (RPCA: 1077, PCA: 766). In addition, to further verify the biological significance of the differentially expressed genes selected by the RPCA method, we compared the number of pathways matched by two networks of the same scale (RPCA: 124, PCA: 55). The network construction genes in these two networks were selected by RPCA method and PCA method, respectively. It is obvious that the network constructed by differentially expressed genes selected by the RPCA can find more pathways.

3.2 WGCNA for Constructing Network and Identifying Communities

3.2.1 Selecting Threshold for Constructing Network

We selected a suitable weighted parameter β of adjacency matrix from 1–30 to determine the correlation coefficient between two genes, which makes the connections of genes approximately in the distribution of scale-free. That is, only a small number of nodes in a network is related to most nodes, and most of them have no association. The scale free property meets the following rule: the node $\log(i)$ and frequency of occurrences about the node $\log(p(i))$ were negatively correlated. As is shown in Fig. 1, we chose the first correlation coefficient close to 0.9 as the soft threshold ($\beta = 3$) to construct the network by using the WGCNA package [20].

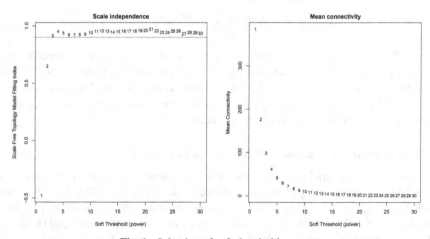

Fig. 1. Selection of soft-threshold power.

3.2.2 Identifying Communities by WGCNA

A system clustering tree of differentially expressed genes was constructed by a correlation coefficient matrix S_{xy}, which was shown as Fig. 2. Different branches of the clustering tree corresponding to different colors, represent different gene communities. Genes were clustered to different expression patterns based on topological overlap matrix w_{xy}. Then, genes with similar patterns were grouped into one community, the minimum number of genes in each community was 30. From Fig. 2, we can see that there were six communities identified. After merging these communities, a total of five communities were recognized by the dynamic tree cut algorithm. The number of genes included in community 1 to community 5 is 237, 83, 59, 50, and 838.

Cluster Dendrogram

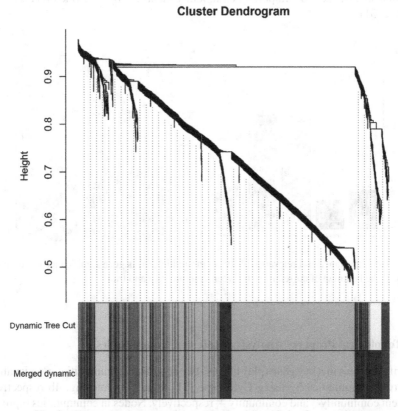

Fig. 2. Cluster dendrogram of gene communities. (Color figure online)

We calculated the gene significance by WGCNA to represent the expression value of these five communities which was shown in Fig. 3a. From Fig. 3a we can see that, the significance of community 3 and community 4 were more prominent than other communities. Figure 3b shows the correlation values between these five communities and trait, which are between −1 and 1. In addition, the numbers in parentheses represent the P-values of the different communities. In these communities, the greater the abso-lute values of correlation values and the smaller the P-values, the more important the community is. From this heatmap we can see, the correlation values of community 3 and community 4 are higher as well as the P-values of them are lower. We will analyze these two most relevant communities below. Meanwhile, to verify the efficient of iden-tifying communities by using WGCNA, we compared the number of matched pathways from the most important communities which were identified by WGCNA and NCMine. NCMine software package is wildly used for community analysis. We used NCMined to identify communities from gene network which was constructed by Pearson correla-tion coefficient. We found that the number of pathways matched by the most important community identified by WGCNA was significantly greater than the number of path-ways matched by the most important community identified by NCMine (WGCNA: 79, NCMine: 31).

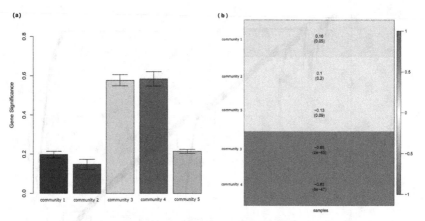

Fig. 3. The importance and relevance of five communities. (a) is gene significance of communi-ties; (b) is the heatmap of correlations between communities and trait.

3.3 Topological Property Analysis of Inferred Communities

To verify the roles that key genes play in these two significant communities, we visualized these two communities by using Cytoscape [21]. Figure 4a and Fig. 4b respectively represent community 3 and community 4, respectively. Nodes in communities represent genes, labels of nodes represent the names of genes, edges represent associations between two genes, and the weight of the edge is determined by the value of weighted coefficient matrix. The larger the node and the darker the color, the more neighbors connected to this node. What's more, the topological properties DC, BC, CC and NS of each node in these two communities were calculated respectively. In Fig. 5, we listed the proportions of the top10, 20, 30 and 40 genes that found in the GeneCard (http://www.genecards.org/). From Fig. 5, we can see that the accuracy of using NS property to detect genes is the highest overall.

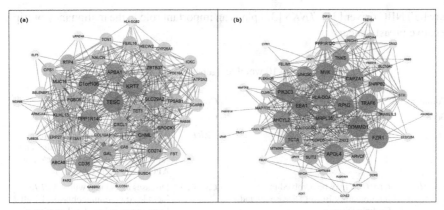

Fig. 4. Two significant communities. (a) is the community 3; (b) is the community 4. (Color figure online)

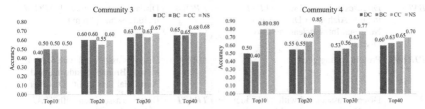

Fig. 5. Accuracy of detecting genes by different topological properties.

Refer to annotations on GeneCard (https://www.genecards.org/) for top 10 genes of these two communities, their annotations are listed in Table 1. Among them, genes with bold font have been recorded in GeneCard, which are related to TNBC: *SPOCK1* can enhance invasion and correlate with poor prognosis, which is a novel transforming growth factor-beta-induced myoepithelial marker in TNBC [22]. Gene *PPP1R14C* has demonstrated that it overexpressed in TNBC tumors, and gene *KRT7* acts as a down-regulated gene in TNBC [23]. *CD36* affects the process of proliferation, migration and tamoxifen-inhibited growth in breast cancer cells [24]. Though *PGBD* appears to be novel, with no obvious relationship to other transposases or other known protein families, it has a high *NS* score, and we suspect that it may have some connection with TNBC. *FZR1* resulted in stabilization of *DNMT1* [25], and a highly significant correlation with *DNMT1* immunohistochemical staining was observed in TNBC. [26] revealed the over-expression of chemokine *CXCL13* in TNBC specimens. [27] constructed a breast cancer risk predictive model including the *CAPZA1* gene. [28] demonstrated that the regulation of TWIST1 by *MAOA* is engaged in promoting EMT, and persistent TWIST1 expression induced growth arrest in EMT-like breast cancer cells. *PIK3CA* amplification related to *TRAT1* in the carcinomas is obvious, somatic missense mutations of the *PIK3CA* gene has been reported [29] in carcinomas of the breast. Silencing of *GDPD5* [30] alters the phospholipid metabolite profile in a TNBC model, which decreases triple negative breast cancer cell proliferation, migration, and invasion. *SNRPB2* acts as an amplified gene that may be TNBC driver [31]. *TNKS* [32] plays an important role in the malignancy of triple negative breast cancer.

Table 1. Top 10 weighted genes of communities.

Community 3		Community 4	
Gene	Comments	Gene	Comments
TESC	Protein coding gene	*FZR1*	Among its related pathways are CDK-mediated phosphorylation and removal of Cdc6 and Cell cycle role of APC in cell cycle regulation
C1orf106	*C1orf106* may be involved in Inflammatory Bowel Disease 29 and Crohn's Disease	*CXCL13*	Diseases associated with *CXCL13* include Angioimmunoblastic T-Cell Lymphoma and Burkitt Lymphoma
SPOCK1	The function of *SPOCK1* may be related to protease inhibition	*CAPZA1*	Diseases associated with *CAPZA1* include Cleft Palate with or without Ankyloglossia, X-Linked.
APBA1	Diseases associated with *APBA1* include Alzheimer Disease and Neuronal Intranuclear Inclusion Disease	*APOL4*	Diseases associated with *APOL4* include Schizophrenia
PPP1R14C	Protein coding gene	*MAOA*	Diseases associated with *MAOA* include Brunner Syndrome and Antisocial Personality Disorder
KRT7	Diseases associated with *KRT7* include Bile Duct Adenoma and Cystadenoma	*TRAT1*	Diseases associated with *TRAT1* include Bone Cancer
CHML	Diseases associated with *CHML* include Choroideremia and Usher Syndrome	*GDPD5*	Among its related pathways are PI Metabolism and Glycerophospholipid biosynthesis
CD36	Diseases associated with *CD36* include Platelet Glycoprotein Iv Deficiency and Coronary Heart Disease 7	*MAP7D1*	Protein Coding gene
KLHL13	Diseases associated with *KLHL13* include Cholestasis, Benign Recurrent Intrahepatic, 1	*SNRPB2*	Diseases associated with *SNRPB2* include Systemic Lupus Erythematosus
PGBD5	Protein Coding gene	*TNKS*	Diseases associated with *TNKS* include Lung Acinar Adenocarcinoma and Cherubism

3.4 GO Enrichment Analysis and KEGG Pathway Analysis of Communities

To investigate possible biological functions of these two communities expressed in TNBC, we applied the Gene Ontology (GO) enrichment analysis to verify the importance of communities, in which the threshold was set at P-value < 0.05. The genes of these two communities were classified into three functional groups: molecular function, biological process and cellular component. The significant enriched GO terms with the largest number of genes in these two communities are listed in Table 2. From the table, we can see that genes are mainly enriched in extracellular region, extracellular space, immune response, integral component of plasma membrane, and response to lipopolysaccharide.

Table 2. GO enrichment analysis of communities.

	Term	Category	Count	Genes
Community 3	Extracellular region	Cellular component	10	CXCL1, F13A1, OBP2B, FST, SUSD4, IGKC, TPSAB1, TCN1, COL10A1
	Extracellular space	Cellular component	9	CXCL1, CD36, SELENBP1, SPOCK1, IGKC, GAL, TPSAB1, TCN1, MUC16
	Immune response	Biological process	5	CXCL1, HLA-DQB2, CD36, CD274, IGKC
	Transporter activity	Molecular function	4	ABCA8, XK, SCARB1, SLCO5A1
Community 4	Extracellular region	Cellular component	12	PLAT, AZGP1, LAMB3, CD163L1, LIPG, TFF3, FGF13, LOX, TREM1, LBP, IGFBP2, CXCL11
	Extracellular space	Cellular component	11	PLAT, AZGP1, CST6, SCGB1D2, LIPG, TFF3, LOX, LBP, IGFBP2, CXCL11, SELE
	Integral component of plasma membrane	Cellular component	10	P2RX4, TMPRSS2, KISS1R, SLC26A7, SLC1A6, CDON, LIFR, FADS2, SLC6A16, HS3ST3B1
	Response to lipopolysaccharide	Biological process	5	MAOB, LBP, CXCL11, SELE, CITED1

To further elucidate the biological pathway of communities, we analyzed genes in these two communities by using the KOBAS online analysis database. P-value < 0.05 is selected as the cut-off criterion to identify enrichment pathways, and significant canonical pathways are listed in Table 3. From Table 3, we can see that the pathways of these two communities are major related to Nitrogen metabolism, Fat digestion and absorption, Phagosome, Morphine addiction, Arginine biosynthesis, Vitamin digestion and absorption, Glycine, serine and threonine metabolism, Hedgehog signaling pathway, Fatty acid metabolism, Transcriptional mis-regulation in cancer, PPAR signaling pathway, Toll-like receptor signaling pathway, Platelet activation, etc. Among these pathways, nitrogen metabolism and glycine, serine and threonine metabolism were identified to be enriched in breast cancer cell lines [33]. Mutation or misexpression of Hedgehog signaling can lead to the development of cancers in multiple organs. The potential role of Hedgehog signaling pathway has been confirmed to promote the progression of breast carcinogenesis [34]. Fatty acid metabolism pathway, and dysfunctional of PPAR signaling pathway, which were down regulated differential expressed genes, can greatly affect the progression of breast cancer [35]. The results of preclinical, clinical, and population-based studies has been used to explore the Toll-like receptor signaling pathway on breast cancer [36], which will prompt in the exploration of more effective breast cancer therapies.

Table 3. KEGG pathway analysis of DEGs associated with TNBC.

	Term	p-value	Genes
Community 3	Nitrogen metabolism	0.0006	*CPS1, CA8*
	Fat digestion and absorption	0.0030	*CD36, SCARB1*
	Phagosome	0.0032	*CD36, TUBB2B, SCARB1*
	Morphine addiction	0.0137	*PDE10A, GABBR2*
	Arginine biosynthesis	0.0430	*CPS1*
	Vitamin digestion and absorption	0.0467	*SCARB1*
Community 4	Glycine, serine and threonine metabolism	0.0042	*MAOB, GLDC*
	Hedgehog signaling pathway	0.0055	*PTCH1, CDON*
	Fatty acid metabolism	0.0062	*FADS2, ACADL*
	Transcriptional mis-regulation in cancer	0.0086	*PLAT, TMPRSS2, SLC45A3*
	PPAR signaling pathway	0.0130	*FADS2, ACADL*
	Toll-like receptor signaling pathway	0.0267	*LBP, CXCL11*
	Platelet activation	0.0348	*PRKG1, APBB1IP*

4 Conclusions

In this paper, we proposed a framework to identify significant communities and key genes of integrated datasets which were collected from triple negative breast cancer. We used RPCA feature selection method to select differentially expressed genes based on the integrated dataset. These differentially expressed genes were put into WGCNA software package for constructing network and identifying significant communities. Considering the role of individual genes play in communities, we proposed a new topological property NS to identify the key genes in each community considering both local and global features. In addition, GO terms and pathways which may be involved in the progression of TNBC were detected in this paper. We demonstrated the biological significance of these important communities and key genes through searching for relevant database and literatures. This paper provided a framework to better understand the molecular mechanisms underlying TNBC by detecting a set of useful clinic relevant genes. However, further experiments are required to confirm the function of these key genes.

Our method has several advantages. Firstly, differently from traditional methods of using WGCNA to process single dataset, this paper analyzed the integrated dataset which may produce more reliable results. Secondly, using differentially expressed genes identified by RPCA to construct network, which reduced the complexity of network construction. Thirdly, analyzing significant communities as well as the important individual gene simultaneously may provide a comprehensive bioinformatics analysis of TNBC.

Although this method is a beneficial exploration in identifying the communities and key genes, several limitations remain. First, the method of dataset integration needs to be improved. Second, there are many studies to select differentially expressed genes at present, and no reasonable description to explain that RPCA is the most appropriate method.

Acknowledgments. This work was supported by the National Science Foundation of China (61972226, 61902216, 61902430, 61872220) and the China Postdoctoral Science Foundation (2018M642635).

References

1. Reis-Filho, J.S., Tutt, A.N.J.: Triple negative tumours: a critical review. Histopathology **52**, 108–118 (2008)
2. Wang, Y.W., Zhang, W., Ma, R.: Bioinformatic identification of chemoresistance-associated microRNAs in breast cancer based on microarray data. Oncol. Rep. **39**, 1003–1010 (2018)
3. Bell, R., Barraclough, R., Vasieva, O.: Gene expression meta-analysis of potential metastatic breast cancer markers. Curr. Mol. Med. **17**, 200–210 (2017)
4. Li, M.X., et al.: Identification of potential core genes in triple negative breast cancer using bioinformatics analysis. Oncotargets Ther. **11**, 4105–4112 (2018)
5. Yang, X., et al.: Bioinformatics analysis of aggressive behavior of breast cancer via an integrated gene regulatory network. J. Cancer Res. Ther. **10**, 1013–1018 (2014)
6. Tang, J., et al.: Prognostic Genes of breast cancer identified by gene co-expression network analysis. Front. Oncol. **8**, 374 (2018)
7. Tang, D., Zhao, X., Zhang, L., Wang, Z., Wang, C.: Identification of hub genes to regulate breast cancer metastasis to brain by bioinformatics analyses. J. Cell. Biochem. **120**, 1–10 (2018)
8. Zou, H., Hastie, T., Tibshirani, R.: Sparse principal component analysis. J. Comput. Graph. Stat. **15**, 265–286 (2006)
9. Han, H.: Nonnegative principal component analysis for mass spectral serum profiles and biomarker discovery. BMC Bioinform. **11**, S1 (2010)

10. Feng, C., Xu, Y., Liu, J., Gao, Y., Zheng, C.: Supervised discriminative sparse PCA for com-characteristic gene selection and tumor classification on multiview biological data. In: IEEE Transactions on Neural Networks and Learning Systems, pp. 1–12 (2019)

11. Komatsu, M., et al.: Molecular features of triple negative breast cancer cells by genome-wide gene expression profiling analysis. Int. J. Oncol. **42**, 478–506 (2013)

12. Gruosso, T., et al.: Chronic oxidative stress promotes H2AX protein degradation and enhances chemosensitivity in breast cancer patients. EMBO Mol. Med. **8**, 527–549 (2016)

13. Maire, V., et al.: Polo-like kinase 1: a potential therapeutic option in combination with conventional chemotherapy for the management of patients with triple-negative breast cancer. Can. Res. **73**, 813–823 (2013)

14. Smyth, G.K.: LIMMA: linear models for microarray data. In: Gentleman, R., Carey, V.J., Huber, W., Irizarry, R.A., Dudoit, S. (eds.) Bioinformatics and Computational Biology Solutions Using R and Bioconductor. Statistics for Biology and Health. Springer, New York (2005). https://doi.org/10.1007/0-387-29362-0_23

15. Cand, E.J., Li, X., Ma, Y., Wright, J.: Robust principal component analysis? J. ACM **58**, 1–37 (2011)

16. Liu, J.-X., Wang, Y.-T., Zheng, C.-H., Sha, W., Mi, J.-X., Xu, Y.: Robust PCA based method for discovering differentially expressed genes. BMC Bioinform. **14**, S3 (2013). https://doi.org/10.1186/1471-2105-14-S8-S3

17. Lin, Z., Chen, M., Yi, M.: The Augmented Lagrange Multiplier Method for Exact Recovery of Corrupted Low-Rank Matrices, vol. 9. Eprint Arxiv (2010)

18. Dennis, G., et al.: DAVID: database for annotation, visualization, and integrated discovery. Genome Biol. **4**, R60 (2003)

19. Morishima, K., Tanabe, M., Furumichi, M., Kanehisa, M., Sato, Y.: KEGG: new perspectives on genomes, pathways, diseases and drugs. Nucleic Acids Res. **45**, D353–D361 (2016)

20. Langfelder, P., Horvath, S.: WGCNA: an R package for weighted correlation network analysis. BMC Bioinform. **9**, 559 (2008). https://doi.org/10.1186/1471-2105-9-559

21. Shannon, P., et al.: Cytoscape: a software environment for integrated models of biomolecular interaction networks. Genome Res. **13**, 2498–2504 (2003)

22. Fan, L.-C., Jeng, Y.-M., Lu, Y.-T., Lien, H.-C.: SPOCK1 is a novel transforming growth factor-β–induced myoepithelial marker that enhances invasion and correlates with poor prognosis in breast cancer. PLoS One **11**, e0162933 (2016)

23. Chen, D., Li, Y., Wang, L., Jiao, K.: SEMA6D expression and patient survival in breast invasive carcinoma. Int. J. Breast Cancer **2015**, 539721 (2015)

24. Liang, Y., et al.: CD36 plays a critical role in proliferation, migration and tamoxifen-inhibited growth of ER-positive breast cancer cells. Oncogenesis **7**, 98 (2018)

25. Agoston, A.T., Argani, P., De Marzo, A.M., Hicks, J.L., Nelson, W.G.: Retinoblastoma pathway dysregulation causes DNA methyltransferase 1 overexpression in cancer via MAD2-mediated inhibition of the anaphase-promoting complex. Am. J. Pathol. **170**, 1585–1593 (2007)

26. Panse, J., et al.: Chemokine CXCL13 is overexpressed in the tumour tissue and in the peripheral blood of breast cancer patients. Br. J. Cancer **99**, 930 (2008)

27. Huang, C.-C., et al.: Concurrent gene signatures for Han Chinese breast cancers. PLoS One **8**, e76421 (2013)

28. Wu, J.B., et al.: Monoamine oxidase A mediates prostate tumorigenesis and cancer metastasis. J. Clin. Investig. **124**, 2891–2908 (2014)

29. Fenic, I., Steger, K., Gruber, C., Arens, C., Woenckhaus, J.: Analysis of PIK3CA and Akt/protein kinase B in head and neck squamous cell carcinoma. Oncol. Rep. **18**, 253–259 (2007)

30. Cao, M.D., et al.: Targeting choline phospholipid metabolism: GDPD5 and GDPD6 silencing decrease breast cancer cell proliferation, migration, and invasion. NMR Biomed. **29**, 1098–1107 (2016)
31. Zhang, Q., Burdette, J.E., Wang, J.-P.: Integrative network analysis of TCGA data for ovarian cancer. BMC Syst. Biol. **8**, 1338 (2014). https://doi.org/10.1186/s12918-014-0136-9
32. Yang, H.-Y., Shen, J.-X., Wang, Y., Liu, Y., Shen, D.-Y., Quan, S.: Tankyrase promotes aerobic glycolysis and proliferation of ovarian cancer through activation of Wnt/β-catenin signaling. Biomed. Res. Int. **2019**, 14 (2019)
33. Bhute, V.J., Ma, Y., Bao, X., Palecek, S.P.: The poly (ADP-ribose) polymerase inhibitor veliparib and radiation cause significant cell line dependent metabolic changes in breast cancer cells. Sci. Rep. **6**, 36061 (2016)
34. Hatsell, S., Frost, A.R.: Hedgehog signaling in mammary gland development and breast cancer. J. Mammary Gland Biol. Neoplasia **12**, 163–173 (2007). https://doi.org/10.1007/s10911-007-9048-2
35. Zhu, X., et al.: Identification of collaboration patterns of dysfunctional pathways in breast cancer. Int. J. Clin. Exp. Pathol. **7**, 3853–3864 (2014)
36. Kidd, L.C.R., Rogers, E.N., Yeyeodu, S.T., Jones, D.Z., Kimbro, K.S.: Contribution of toll-like receptor signaling pathways to breast tumorigenesis and treatment. Breast Cancer (Dove Med. Press) **5**, 43–51 (2013)

A Novel Multiple Sequence Alignment Algorithm Based on Artificial Bee Colony and Particle Swarm Optimization

Fangjun Kuang[(✉)] and Siyang Zhang

School of Information Engineering, Wenzhou Business College, Wenzhou 325035, China
kfjztb@126.com

Abstract. Multiple sequence alignment (MSA), known as an NP-complete combinatorial optimization problem, is one of the most important and challenging tasks in bioinformatics. A novel hybrid algorithm of artificial bee colony and particle swarm optimization (HABC-PSO) for MSA is proposed in this paper to tackle this problem. The proposed method innovatively integrates Tent chaos search, opposition-based learning and recombination operator techniques. The best solutions obtained by the recombination operator are further set as the neighbor food source for onlooker bees and the global best of particle swarm, respectively. In addition, a new method of HABC-PSO algorithm for determining a food source or particle swarm in the neighborhood is proposed by exploiting the discreteness of the MSA problem. The experiment results strongly demonstrate its robustness and effectiveness, as well as achieve better performance and biological quality in comparison with its peers.

Keywords: Multiple sequence alignment · Artificial bee colony · Particle swarm optimization · Tent chaotic Opposition-based learning · Recombination operator

1 Introduction

Multiple sequence alignment (MSA) is one of the hot issues in bioinformatics. It is widely applied in sequence assembly, sequence annotation, the prediction of gene and protein's structure. MSA is an NP-complete combinatorial optimization problem in the sense of the sum-of-pairs scoring (SPS) [1], which mainly focused on discovering biological relations among different sequences, including nucleotide or amino acid, to investigate their underlying main characteristics or functions. The existing algorithms solving the MSA problem are roughly divided into four different types, which include exact alignment, evolutionary alignment algorithms, graph based algorithms and the iterative algorithms. The exact alignment algorithm is completely based on the dynamic programming. The evolutionary alignment algorithm uses evolutionary algorithm for solving MSA, such as the software package CLUSTALW. The alignment algorithm based on graphical model is partial order alignment. The iterative alignment algorithms is based on the algorithm which can produce alignment, and can improve the multiple

© Springer Nature Singapore Pte Ltd. 2020
H. Han et al. (Eds.): IDMB 2019, CCIS 1099, pp. 152–169, 2020.
https://doi.org/10.1007/978-981-15-8760-3_11

sequence alignment through a series of iterations until the results don't become better any longer.

In recent years, heuristic search, stochastic optimization and evolutionary algorithms have been increasingly used to tackle the MSA problem. The approaches can improve the multiple sequence alignment through a series of heuristic search until a global optimal solution. For example, Gupta et al. [2] proposed a multi-sequence alignment method based on genetic algorithms (MSA-GA); Gao et al. presented an inertial weight particle swarm optimization algorithm in [3]; Tsvetanov et al. initialized another MSA method based on an improved ant colony algorithm [4]; Öztürk et al. [5] proposed a new method by using artificial bee colony algorithm; In addition, Zhu et al. [6] proposed a novel multi-objective evolutionary algorithm to solve the MSA problem, where the initial population is generated by space insertion operation, and evolutionary operation is realized by crossover and mutation operators of genetic operation. Liu et al. employed Hidden Mark models and calculated a posterior probability assignment function, called MSAProbs to solve the MSA problem [7].

On the other hand, different hybrid algorithms have been proposed by integrating genetic algorithms and evolutionary algorithms in some recent work. Rani et al. [8] proposed two hybrid algorithms: Hybrid genetic algorithm with artificial bee colony (GA-ABC), and multi-objective bacterial foraging optimization algorithm (MO-BFO) for the MSA solving. Sun et al. [9] developed an MSA algorithm using quantum particle swarm optimization and hidden Markov model (HMM). Rubio-Largo et al. [10] proposed a hybrid frog leaping algorithm for MSA by integrating multi-objective and evolutionary heuristic algorithm. Moreover, Kuang et al. proposed a multi-strategy artificial bee colony (MS-ABC) for MSA [11], which consists of multiple strategies, such as Tent chaotic initialization population, different neighborhood search and tournament selection strategies. In addition, Chowdhury and Garai [12] reviewed recent trends in the multi-objective genetic algorithm for MSA solving. Zambrano-Vega et al. [13] made a comparative study of MSA methods for different multi-objective heuristic algorithms besides reviewing related heuristic methods.

A remaining challenge in the existing hybrid methods is how to enhance MSA accuracy by integrating different evolutionary and heuristic search algorithms. In this work, we propose a novel hybrid algorithm of artificial bee colony and particle swarm optimization (HABC-PSO) for MSA to handle this challenge. The hybrid algorithm integrates artificial bee colony (ABC) and particle swarm optimization (PSO) to optimize the MSA solving problem, by welding Tent chaos search, opposition-based learning and recombination operator techniques. HABC-PSO innovatively integrates artificial bee colony (ABC) and particle swarm optimization (PSO) by exploiting their complementary advantages in the optimization involved in MSA solving.

Artificial bee colony (ABC) is a swarm intelligent optimization algorithm proposed by Karaboga in 2005 [14]. It has strong and robust global exploration ability and attains a fast convergence with only few parameters [15]. Different improved ABC algorithms can be found in [16–20]. Particle swarm optimization (PSO) is also a swarm intelligent optimization algorithm, different improved PSO algorithms can be found in [21–27]. The ABC algorithm has effective global search ability but poor local search ability. Alternatively, the PSO has powerful local search ability but poor global search ability.

HABC-PSO considers two objective functions with the aim of preserving the quality and consistency of the alignment: the weighted sum-of-pairs function with affine gap penalties (WSP) and the number of totally conserved (TC) columns score. In order to assess the accuracy of HABC-PSO, we employ a BAliBASE benchmark (version 3.0) system, which presents more challenging test cases that represent the real problems encountered, when aligning large sets of complex sequences. The proposed algorithm is compared with 5 well-known peer MSA methods in bioinformatics field, as well as 8 evolutionary MSA algorithm peers.

The rest of this paper is organized as follows: Sect. 2 gives a description of the MSA problem. Hybridization algorithm of artificial bee colony and particle swarm optimization (HABC-PSO) is proposed in Sect. 3. Section 4 describes HABC-PSO algorithm for the MSA problem in details. Computational results of 14 methods for MSA are presented and some analysis is provided in comparison to them in Sect. 5. Section 6 presents the conclusions and the future work.

2 Multiple Sequence Alignment (MSA)

Multiple sequence alignment (MSA) reflects the evolutionary relationship among given sequences. It can be formulated mathematically as follows. Given a sequence group consisting of $N(N \geq 2)$ sequences $S = (s_1, s_2, \cdots, s_N)$, where $s_i = s_{i1}, s_{i1}, \cdots, s_{il_i}$, $s_{ij} \in \sum, 1 \leq i \leq N, 1 \leq j \leq l_i, l_i$ is defined as the length of the i sequence. There are some different objective functions formulated for multiple sequence alignment, such as hidden Markov model, sum-of-pairs (SP) function and COFFEE function [1, 28–31]. In this paper, the SP function is used as the objective function for its efficiency. Suppose the length of each sequence is L. The j character of the i sequence is denoted as $c_{ij}(1 \leq j \leq L)$. The sum-of-pairs scores of all the j character for all other sequences is defined as,

$$Score(j) = \sum_{i=1}^{N-1} \sum_{k=i+1}^{N} w_{ij} p(c_{ij}, c_{kj}) \tag{1}$$

where $p(c_{ij}, c_{kj})$ denotes the sum-of-pairs scores of the characters c_{ij} and c_{kj}. w_{ij} is the weight of sequences s_i and s_j, which depends on the distance between the two sequences. It is noted that w_{ij} calculated by Levenshtein distance (LD), i.e. the minimum number of single-character edits [10, 32, 33], $w_{ij} = 1 - \frac{LD(s_i, s_j)}{\max(|s_i|, |s_j|)}$, where $\max(|s_i|, |s_j|)$ is the maximum length of two sequences s_i and s_j. Thus, the sum-of-pairs score $p(c_{ij}, c_{kj})$ between c_{ij} and c_{kj} is described as follows,

$$p(c_{ij}, c_{kj}) = \begin{cases} 0, & \text{if } c_{ij} =' \_' \text{and } c_{kj} =' \_'; \\ -g_{open}, & \text{if } (c_{ij} =' \_' \text{ and } c_{kj} \neq' \_') \text{ or } (c_{ij} \neq' \_' \text{ and } c_{kj} =' \_') \text{ and it is a gap opening}; \\ -g_{extend}, & \text{if } (c_{ij} =' \_' \text{ and } c_{kj} \neq' \_') \text{ or } (c_{ij} \neq' \_' \text{ and } c_{kj} =' \_') \text{ and it is a gap extension}; \\ Blosum(c_{ij}, c_{kj}), & \text{if } (c_{ij} \neq' \_' \text{ and } c_{kj} \neq' \_'). \end{cases} \tag{2}$$

where g_{open} and g_{extend} are the penalty cost of a gap opening and extension, respectively. $Blosum(c_{ij}, c_{kj})$ is the score of matched residues characters c_{ij} and c_{kj} [6]. Then the total score of all characters in the sequence group is:

$$Score(S') = \sum_{j=1}^{L} Score(j) \qquad (3)$$

In this work, we employ multi-objective optimization for solving MSA problem. That is, we aim to maximize the number of matching base pairs among the sequences and minimize the score of gap penalties by inserting or deleting gaps via multi-objective optimization. The MSA multi-objective optimization function is formulated as:

$$\begin{aligned} \max &imize \; F(S') = (f_1(S'), f_2(S'))^T \\ & f_1(S') = Score(S') \\ & f_2(S') = -g(S') \\ & S' \in \Omega \end{aligned} \qquad (4)$$

where $g(S') = n_1 \times g_{open} + n_2 \times g_{extend}$ is the affine gap penalty score of sequence S', where n_1 and n_2 are the number of gap opening and extension respectively. Ω denotes multi-sequence alignment spaces. In this work, the BLOSUM62 substitution is used for protein sequences and we have parameters: $g_{open} = 6$ and $g_{extend} = 0.85$.

3 Hybridization Algorithm of ABC and PSO

ABC algorithm has effective global search ability but a poor local search ability, where PSO has powerful local search ability but poor global search ability [17, 34, 35]. Thus, it is desirable to integrate them to tackle the optimization in MSA. We invent a hybrid algorithm of Tent chaotic artificial bee colony and particle swarm optimization (HABC-PSO) to meet this demand. The proposed HABC-PSO algorithm, which is integrated by Tent chaos search, opposition-based learning, and recombination operator techniques, owns the advantages of ABC in global search and PSO's local serial search.

The proposed algorithm has an initialization strategy driven by the chaotic opposition-based learning is applied to diversify the initial individuals in the search space. All individuals are divided into two sub-swarms, one sub-swarm searches via Tent chaotic ABC (TCABC), and the other via Tent chaotic particle swarm optimization (TCPSO). The global optimal solutions obtained by the TCABC and TCPSO are then used for further recombination. The solution obtained from this recombination is given to the populations of the TCABC and TCPSO as the neighbor food source for onlooker bees and the global optima, respectively. We introduce relevant techniques involved in HABC-PSO as follows briefly.

3.1 Tent Chaotic Artificial Bee Colony Algorithm

Artificial Bee Colony Algorithm

The swarm of ABC algorithm is divided into employed bees, scouts and onlookers. In the initialization phase, the algorithm generates a group of food sources corresponding to the solutions in the search space. The food sources are produced randomly within the range of the boundaries of the variables. The fitness of food sources will be evaluated. In the employed bees' phase, a number of employed bees, set as the number of the food sources and half the colony size, are used to find new food sources. Onlooker bees next choose a random food source according to the selection probability. The employed bee that was exploiting this food source becomes a scout that looks for a new food source by randomly searching.

Tent Chaotic Opposition-Based Learning Initialization Strategy

Population initialization is a crucial task in evolutionary algorithms because it can affect the convergence speed and the quality of the final solution. If no information about the solution is available, then random initialization is then most commonly used method to generate initial population. Owing to the randomness and sensitivity dependence on the initial conditions, the chaotic maps have been used to initialize the population so that the search space information can be extracted to increase the population diversity. At the same time, the Tent-map shows outstanding advantages and higher iterative speed than the Logistic map. Therefore, Tent chaotic opposition-based learning strategy [17] is used to initialize the population, so that the initial population can increase diversity and preserve individual randomness.

Tournament Selection Strategy

Onlooker bees in the basic ABC algorithm select food source by using the proportion selection strategy, which leads to the consistency of individual adaptation in the later period, but it is not easy to jump out of local optimal states. Therefore, the tournament selection strategy [17] is employed to select the food source. This strategy is a selection process based on a local competition, which only refers to the relative value of individuals. Tournament selection probability is as follow:

$$P_i(t) = \frac{c_i(t)}{\sum_{i=1}^{N} c_i(t)} \tag{5}$$

where c_i is the score of an individual.

Tent Chaotic Artificial Bee Colony Algorithm

The procedure of TCABC algorithm is described as following steps.

Step 1: Initial the food sources and computation conditions include population of bee colony N, number of employed bees $SN = (N/2)$, upper and lower boundaries of every decision variable, maximum iteration G_{max}, $Limit$ and chaotic local search iteration number C_{max}.

Step 2: Set iteration $iter = 0$, generate SN vectors X_i with D dimensions as food sources according to Tent chaotic opposition-based learning initialization strategy.

Step 3: Sent SN employed bees to food sources. Initialize the flag vector $trial(i) = 0$, which is recorded the cycle number of a food source.

Step 4: Produce new solutions V_i using employed bees, and calculate the fitness value $fit(V_i)$.

Step 5: If $fit(V_i) > fit(X_i)$, then $X_i = V_i$, $trial(i) = 0$; Else X_i is maintained, $trial(i) = trial(i) + 1$.

Step 6: Calculate the probability values P_i of food sources by applying tournament selection using (5).

Step 7: Onlooker bees choose the food sources by probabilities P_i until all of them have a corresponding food source, and produce new solutions V_i. Calculate the fitness value $fit(V_i)$.

Step 8: If $fit(V_i) > fit(X_i)$, then $X_i = V_i$, $trial(i) = 0$; Else X_i is maintained, $trial(i) = trial(i) + 1$.

Step 9: If $trial(i) > Limit$, then there is an abandoned solution for the scout then replace it with a new food source V_i, which will be reinitialized by carrying out self-adaptive Tent chaotic search algorithm [17].

Step 10: Memorize the best solution found so far.

Step 11: Update $iter = iter + 1$. If the maximum iteration cycle is not reached yet, then go to step 4. Otherwise, return best solution.

3.2 Tent Chaotic Particle Swarm Optimization Algorithm

Particle Swarm Optimization Algorithm
Particle swarm optimization (PSO) is a computation intelligence technique, which was motivated by the organisms' behavior such as schooling of fish and flocking of birds. PSO can solve a variety of difficult optimization problems. The major advantage is that PSO uses the physical movements of the individuals in the swarm and has a flexible and well-balanced mechanism to enhance and adapt to the global and local exploration abilities. Another advantage of PSO is its simplicity in coding and consistency in performance.

Premature Judgment and Processing Mechanism
The position of particles determines the fitness of the particle, so we can track the status of particle swarms by the overall changing of all particles fitness. Group fitness variance δ^2 is defined as follows:

$$\delta^2 = \sum_{i=1}^{N} \frac{F_i - F_{avg}}{F} \qquad (6)$$

where N is population size; F_i denotes the fitness of the ith particle; F_{avg} is the average fitness of the particle swarms; F denotes normalization factor for limiting the size of δ^2, which is expressed as follows:

$$F = \begin{cases} \max_{1 \leq i \leq m} |F_i - F_{avg}|, & \max_{1 \leq i \leq m} |F_i - F_{avg}| > 1 \\ 1, & else \end{cases} \tag{7}$$

If $\delta^2 < H$ (H is a given constant), the premature processing is applied. Particles trapped in premature are employed Tent chaotic optimization according to the following equations.

$$V_i^d(t) < (2V_i^d(t-1)) \bmod 1 \tag{8}$$

$$V_i^d(t) = V_{min} + (V_{max} - V_{min})V_i^d(t) \tag{9}$$

where $[V_{min}, V_{max}]$ is the velocity range of the particles. The premature particles are updated according to the following equations.

$$\begin{cases} V_i^d(t+1) = W(t) \cdot V_i^d(t) + C_1(t) \cdot R_1(t) \cdot (P_{best}^d(t) - P_i^d(t)) + C_2(t) \cdot R_2(t) \cdot (G_{best}^d(t) - P_i^d(t)) \\ P_i^d(t+1) = P_i^d(t) + V_i^d(t+1) \end{cases} \tag{10}$$

where t is the current generation, T denotes the maximum number of generations.

Tent Chaotic Particle Swarm Optimization Algorithm
The procedure of Tent chaotic particle swarm optimization algorithm (TCPSO) algorithm is described as following steps.

Step 1: Initialize the swarm size S, maximum of generations T, setting $t = 1$, $[W_{min}, W_{max}], [C_{min}, C_{max}], [V_{min}, V_{max}], [P_{min}^d, P_{max}^d]$ is the value range of parameters, where $d = 1, 2, \cdots, D$. C_1, C_2, R_1, R_2 and W are generated by Tent chaotic optimization.

Step 2: Produce the positions and velocity of particles by chaos initialization.

Step 3: If the convergence criteria or one of the stopping criteria (Generally, a sufficiently good fitness or maximum iteration is met) is satisfied, go to step 10.

Step 4: Update the velocity V_i and position P_i of each particle according to Eq. (10), respectively. And C_1, C_2, R_1, R_2 and W are Obtained by Step 1.

Step 5: Compare the fitness value of each particle to its individual best P_{best}^d, if current value is better than P_{best}^d, then update P_{best}^d as current position.

Step 6: Compare the fitness value of each particle to the global best G_{best}^d. If current value is better than G_{best}^d, then update G_{best}^d as current position.

Step 7: If the convergence criteria or one of the stopping criteria (Generally, a sufficiently good fitness or maximum iteration is met) is satisfied, go to step 10.

Step 8: Calculate the group fitness variance δ^2 by (6)–(7). If $\delta^2 < H$ is not satisfied, let $t = t + 1$ and go back to step 4.

Step 9: Update the velocity and position of the premature particles according to (8)–(10), let $t = t + 1$, and go back to step 3.

Step 10: Obtain the optimal result.

3.3 Recombination Operator

In this paper, a new optimal solution is generated by recombination operator, which is used as the optimal solution of TCABC and the global optimal solution of TCPSO. The fitness values of the optimal solution of TCABC algorithm and the global optimal solution of TCPSO algorithm are F_{Abest} and F_{Gbest}, respectively. To obtain the optimal solution, the optimal solution selection probability expression is as follows:

$$P_{best} = \frac{F_{Abest}}{F_{Gbest} + F_{Abest}} \tag{11}$$

where P_{best} is the probability of the optimal solution A_{best} of TCABC algorithm, F_{Abest} is the fitness value of the optimal solution A_{best} of TCABC algorithm and F_{Gbest} is the fitness value of the global optimal solution G_{best} of TCPSO algorithm.

A random number r of [0, 1] is generated randomly for each dimension of the problem. If the random number r generated by the j dimension is less than or equal to P_{best}, the optimal value of the dimension is replaced by the optimal solution A_{best}^{j} obtained by ABC algorithm; otherwise, the optimal value of the dimension is replaced by the global optimal solution G_{best}^{j} obtained by PSO algorithm. The optimal solution is defined as follows:

$$Best_j = \begin{cases} A_{best}^{j} & if\,(r < P_{best}) \\ G_{best}^{j} & otherwise \end{cases} \tag{12}$$

where $Best_j$ is the j dimensional optimal solution of the optimal solution $Best$, $j = 1, 2, \cdots, D$. A_{best}^{j} and G_{best}^{j} represent the j dimensional optimal solution obtained by TCABC and TCPSO algorithm, respectively.

By using recombination operator, the relationship between particles and bees can be deepened and the information exchange between them can be enhanced. Thus, the ABC algorithm improves its local search ability by directly using the global optimal information, and the global search ability of PSO algorithm is improved, which can effectively jump out of the local optimum, thus balancing the exploitation and exploration of the algorithm.

4 Novel STOC-ABC Algorithms for MSA

4.1 Encoding Design

The STOC-ABC algorithm for the MSA problem uses Eq. (4) as the object function (fitness function). The main problems of the STOC-ABC algorithm for MSA problem are described below.

Initialization

It is supposed that there are N sequences to be aligned, and these sequences are generated with various lengths $l_1, l_2, \cdots, l_N$. Parent alignments are presented as matrix where each sequence is encoded as a row with the considered alphabet set. The length L of each row in the initialized matrix is from l_{max} to $1.2l_{max}$, where $l_{max} = \max(l_1, l_2, \cdots, l_N)$, because the number of the gaps is always less than 20% of the current sequence's length [1, 7].

Individual Encoding

In view of the characteristics of multiple sequence alignment, two-dimensional encoding method is used. It is simple and intuitive, easy to operate. However, this method always takes up a lot of memory space. Supposed N sequences to be aligned, and these sequences are generated with various lengths $l_1, l_2, \cdots, l_N$, the length of sequence after alignment is L. According to the mathematical model of sequence alignment, the number $L - l_i (1 \leq i \leq N)$ of spaces is inserted in each sequence. Therefore, randomly generated one dimensional matrix $a_i (1 \leq i \leq N)$. The code of bee source is $\gamma = [a_1, a_2, \cdots, a_N]$.

Alignment Results Expression

In the process of encoding implementation, the location of the matrix is stored in the inserted spaces. Therefore, the input of alignment results needs to be coded into the corresponding alignment. Alignment result is taken as matrix, which insert spaces after each sequence as a row of the matrix. E.g.: There are three sequences $s_1 = ydgeilyqskrf$, $s_2 = adesvynpgr$, and $s_3 = y\det pikqser$. Generated matrix $a_1 = [2, 8, 13]$, $a_2 = [1, 4, 8, 11, 14]$, $a_3 = [2, 4, 9, 13]$. The encoded alignment result can be as follows.

$$\begin{bmatrix} y\_d & g\,e\,i & l\_y & q\,s\,k & \_r\,f \\ \_a\,d & \_e\,s & v\_y & n\_p & g\_r \\ y\_d & \_e\,t & p\,i\_ & k\,q\,s & \_e\,r \end{bmatrix}$$

Produce Chaotic Sequence With the Chaotic Mapping

In this paper, using the Tent chaotic sequence produced in a chaotic mapping said inserted into the space position, must be an integer. Therefore, after Tent chaotic map sequences do integer operations, and determine whether to insert a space position already exists, if any, in the rest of the position from a position of insert, if does not exist, insert the location.

Neighborhood Search of the Population

MSA is a discrete optimization problem. Therefore, in the MSA problem, an interior dimension of one solution will be removed randomly and in its place a new encoded dimension of the solution will be inserted. This will help in producing a new and good neighboring solution. There are three sequences $s_1 = ydgeilyqskrf$, $s_2 = adesvynpgr$ and $s_3 = y\det pikqser$. Let $position\_old = [a_1, a_2, a_3]$, where $a_1 = [2, 8, 13]$, $a_2 = [1, 4, 8, 11, 14]$, $a_3 = [2, 4, 9, 13]$. If the third sequence is randomly selected as the neighborhood, four numbers between 1 and L are randomly generated, which are inserted into the sequence s_3 as new space. If $a_3' = [1, 7, 8, 12]$, then $position\_new = [a_1, a_2, a_3']$. The alignment results corresponding to $position\_old$ and $position\_new$ are shown in Fig. 1 and Fig. 2, respectively.

$$\begin{bmatrix} y & \_ & d & g & e & i & l & \_ & y & q & s & k & \_ & r & f \\ \_ & a & d & \_ & e & s & v & \_ & y & n & \_ & p & g & \_ & r \\ y & \_ & d & \_ & e & t & p & i & \_ & k & q & s & \_ & e & r \end{bmatrix} \quad \begin{bmatrix} y & \_ & d & g & e & i & l & \_ & y & q & s & k & \_ & r & f \\ \_ & a & d & \_ & e & s & v & \_ & y & n & \_ & p & g & \_ & r \\ \_ & y & d & e & t & p & \_ & \_ & i & k & q & \_ & s & e & r \end{bmatrix}$$

Fig. 1. Alignment results of *position_old*   **Fig. 2.** Alignment results of *position_new*

4.2 Main Procedure of the HABC-PSO for MSA

In this section, we mainly focus on the procedure of the HABC-PSO algorithm for MSA. The proposed HABC-PSO algorithm is novelly integrated by Tent chaos search, opposition-based learning, and recombination operator techniques. The flowchart of the proposed algorithm is illustrated in Fig. 3. The procedure of HABC-PSO algorithm for MSA is described as Algorithm 1.

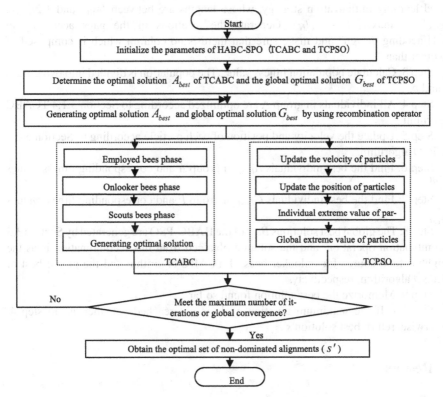

Fig. 3. The flowchart of algorithm HABC-PSO

Algorithm 1. HABC-PSO Algorithm for MSA

Input: Set of k unaligned sequences (s)

Penalty cost of a gap opening (g_{open})

Penalty cost of a gap extension (g_{extend})

Blosum scoring matrix ($Blosum(c_{ij}, c_{kj})$)

Population size N, maximum iteration G_{max}, $Limit$

Chaotic local search iteration number C_{max}, and iteration $iter$

Stopping Criterion

Output: Set of non-dominated alignments (s')

Step 1: Initialize the parameters of the HABC-PSO algorithm include.

Step 2: Generate the initial population solution by using Tent chaotic opposition-based learning initialization strategy, whose lengths are between l_{max} and $1.2l_{max}$, $l_{max} = \max(l_1, l_2, \cdots, l_N)$. Generate the locations of the gaps according to "4.1Encoding design" and make sure that there is no column which is composed of no other than "_".

Step 3: Calculate population fitness value. The population is divided into two groups A and p.

Step 4: All individuals in group A are optimized according to Section 3.1.4 TCABC algorithm.

Step 5: Update the velocity and position of each particle according to Section 3.2.3 TCPSO algorithm.

Step 6: Find the best individuals A_{best} in group A and corresponding fitness values F_{best}^{ABC}.

Step 7: Find the best individuals G_{best} in group P and corresponding fitness values F_{best}^{PSO}.

Step 8: The global best solutions $Best$ of the HABC-PSO are generated by Section 3.3 recombination operation, and the solution obtained from this recombination is as the neighbor food source for onlooker bees of TCABC algorithm and the global best of TCPSO algorithm, respectively.

Step 9: Memorize the best solution found so far.

Step 10: If the maximum iteration cycle is not reached yet, then go to step 4. Otherwise, return best solution s'.

5   Results

5.1   Datasets

In order to verify the performance and effectiveness of HABC-PSO algorithm for solving multi-sequence alignment problem, this paper uses BAliBASE 3.0 database [36, 37]. This database is based on three-dimensional overlapping standard alignment database, which contains 6255 sequences and 218 test cases. It is divided into six test sets: BB11, BB12, BB20, BB30, BB40 and BB50. Each set represents different biological characteristics and sequence similarity is also different. BB11 sequence set contains 38 test cases with sequence similarity less than 20%; BB12 sequence set contains 44 test cases with sequence similarity between 20% and 40%; BB20 sequence set contains 41 test cases with sequence similarity greater than 40%; BB30 sequence set contains 30 test cases with sequence similarity greater than 40%; BB40 sequence set contains 49 test cases, and the similarity between sequences is more than 20%; BB50 sequence set contains 16 test cases, and the similarity between sequences is more than 20%.

5.2 Performance Analysis

In order to verify the performance of HABC-PSO algorithm, almost all peer MSA algorithms are included for comparisons in this work. They include MSA-GA [2], MS-PSO [3], MS-ACO [4], ABC-Aligner [5], MOMSA [6], MSAProbs [7], MO-BFO [8], MS-ABC [11], MUSCLE [38, 39], Clustal Ω [38, 40], Kalign [38, 41], MAFFT [38, 42] and T-Coffee [38, 43]. Furthermore, we use MSA evaluation indicators: Sum of Pairs score (SP) and TC measure (TC). SP refers to the proportion of residue pairs aligned with standard reference alignment, while TC refers to the proportion of correctly aligned columns in all sequences.

In the experiment, the population size of all the algorithms is selected as follows. For examples, ABC algorithm population contains 50 employed bees and 50 onlooker bees; PSO algorithm population consists of 100 particles; ACO population has 100 ants, and GA population has 100 chromosomes. The maximum iterations number of all algorithms is $G_{max} = 5000$, and the food sources limited number of all ABC algorithms is $Limit = 25$. All algorithms are implemented on a computer with main frequency of 4 GHz and memory of 16G on the platform of MATLAB R2014b. Under fixed iteration times, the average SP and TC values obtained by running the algorithm 30 times are used to evaluate the performance of the algorithm by solving the matching of the same set of sequences.

Table 1 shows the average SP scores and the corresponding standard deviation of these algorithms on six subsets of BAliBASE 3.0 [36, 37]. Table 2 illustrates the average TC scores and the corresponding standard deviation of these algorithms on six subsets of BAliBASE 3.0. The best results are emphasized in bold in two tables. Figure 4 and 5 compare the average TC and SP scores of all algorithms across all datasets. It is easy to see that proposed HABC-PSO outperforms all its peers for almost all datasets.

Table 1 and Table 2 show the average SP scores and corresponding standard deviation, TC average scores and corresponding standard deviation of all 14 MSA algorithms respectively. From the analysis of Table 1, Table 2, Fig. 4 and 5, it can be seen that the performance of the HABC-PSO algorithm in BB50 sequence alignment group is slightly lower than the SP average performance of the *MSAProbs* algorithm, while in BB40 sequence alignment group is slightly lower than the TC average performance of the *MS-ABC* algorithm, and other situations have obvious advantages.

At the same time, regardless of the difference of sequence length and similarity, the proposed HABC-PSO algorithm can get better sequence alignment with smaller standard deviations, besides achieving the best performance for all alignments. These results show that the HABC-PSO algorithm has better robustness and stability, and the alignment results for BB12 and BB50 sequence alignment groups. It demonstrates

Table 1. Average SP scores and corresponding standard deviation of algorithms

Algorithm	SP						
	BB11(38)	BB12(44)	BB20(41)	BB30(30)	BB40(49)	BB50(16)	Overall(218)
HABC-PSO	**0.7408 ± 0.0125**	**0.9589 ± 0.0558**	**0.9457 ± 0.0925**	**0.8905 ± 0.0452**	**0.9308 ± 0.0073**	0.9015 ± 0.0132	**0.9004 ± 0.0101**
MS-ABC	0.7284 ± 0.1081	0.9516 ± 0.0112	0.9345 ± 0.0038	0.8873 ± 0.0133	0.9308 ± 0.0073	0.8996 ± 0.0115	0.8907 ± 0.0108
MSA-GA	0.4628 ± 0.2012	0.8389 ± 0.1205	0.8147 ± 0.0196	0.7886 ± 0.0824	0.8045 ± 0.1052	0.7746 ± 0.0624	0.7615 ± 0.0938
MSA-PSO	0.6175 ± 0.1534	0.9268 ± 0.1018	0.9069 ± 0.0986	0.8398 ± 0.1216	0.9063 ± 0.1008	0.8345 ± 0.0782	0.8467 ± 0.1585
MSA-ACO	0.6051 ± 0.1678	0.9134 ± 0.0936	0.9009 ± 0.1284	0.8246 ± 0.0989	0.8813 ± 0.1052	0.8234 ± 0.1205	0.8247 ± 0.0948
ABC-Aligner	0.6950 ± 0.0945	0.9425 ± 0.0656	0.9247 ± 0.0302	0.8532 ± 0.1079	0.9226 ± 0.0864	0.8886 ± 0.0158	0.8759 ± 0.1008
MOMSA	0.5678 ± 0.1247	0.8665 ± 0.0984	0.8925 ± 0.0083	0.7595 ± 0.1830	0.8241 ± 0.0098	0.7658 ± 0.1043	0.7546 ± 0.1584
MSAProbs	0.6821 ± 0.1005	0.9564 ± 0.0687	0.9287 ± 0.0098	0.8657 ± 0.0314	0.9236 ± 0.0805	**0.9016 ± 0.0198**	0.8754 ± 0.0245
MO-BFO	0.6354 ± 0.1058	0.9348 ± 0.0896	0.9024 ± 0.1278	0.8435 ± 0.0867	0.8945 ± 0.1002	0.8467 ± 0.0846	0.8349 ± 0.0887
MUSCLE	0.6826	0.9447	0.9284	0.8755	0.9254	0.8944	0.8752
Clustal Ω	0.5901	0.9060	0.9116	0.8624	0.9010	0.8620	0.8389
Kalign	0.6053	0.9121	0.9008	0.8126	0.8833	0.8201	0.8224
MAFFT	0.6712	0.9363	0.9262	0.8555	0.9191	0.8999	0.8680
T-Coffee	0.6572	0.9447	0.9157	0.8369	0.8964	0.8949	0.8576

Table 2. Average TC scores and corresponding standard deviation of algorithms

Algorithm	TC						
	BB11(38)	BB12(44)	BB20(41)	BB30(30)	BB40(49)	BB50(16)	Overall(218)
HABC-PSO	**0.6057 ± 0.1029**	**0.8872 ± 0.0246**	**0.5623 ± 0.0534**	**0.6621 ± 0.0326**	0.6549 ± 0.0213	**0.6465 ± 0.0468**	**0.6684 ± 0.0587**
MS-ABC	0.5912 ± 0.1254	0.8769 ± 0.0302	0.5157 ± 0.0705	0.6405 ± 0.0986	**0.6584 ± 0.0189**	0.6225 ± 0.0347	0.6506 ± 0.0865
MSA-GA	0.3167 ± 0.3023	0.6358 ± 0.1951	0.2458 ± 0.1158	0.6081 ± 0.1904	0.3587 ± 0.2087	0.2989 ± 0.0974	0.3564 ± 0.1293
MSA-PSO	0.3846 ± 0.1935	0.8368 ± 0.0906	0.3564 ± 0.1876	0.4892 ± 0.1075	0.5834 ± 0.0973	0.5095 ± 0.1003	0.5489 ± 0.1821
MSA-ACO	0.3685 ± 0.1734	0.8031 ± 0.0783	0.3987 ± 0.1632	0.4864 ± 0.0773	0.5081 ± 0.0856	0.4398 ± 0.0723	0.5164 ± 0.0984
ABC-Aligner	0.4864 ± 0.1123	0.8702 ± 0.0927	0.4469 ± 0.1276	0.5687 ± 0.1285	0.6258 ± 0.1124	0.5584 ± 0.0998	0.5846 ± 0.1240
MOMSA	0.3895 ± 0.1405	0.8143 ± 0.1129	0.3638 ± 0.1192	0.3841 ± 0.0982	0.5346 ± 0.0098	0.5146 ± 0.1870	0.5618 ± 0.1014
MSAProbs	0.4842 ± 0.1271	0.8715 ± 0.0985	0.4698 ± 0.0948	0.6118 ± 0.1009	0.6109 ± 0.0109	0.6112 ± 0.0695	0.6058 ± 0.0953
MO-BFO	0.4321 ± 0.2082	0.8148 ± 0.1009	0.4013 ± 0.1311	0.5013 ± 0.2106	0.5459 ± 0.0945	0.4867 ± 0.1854	0.5784 ± 0.1528
MUSCLE	0.4409	0.8619	0.4794	0.6227	0.6039	0.5931	0.6003
Clustal Ω	0.3622	0.7938	0.4529	0.5791	0.5826	0.5374	0.5513
Kalign	0.3687	0.7934	0.3625	0.4799	0.5078	0.4396	0.4920
MAFFT	0.4500	0.8423	0.4574	0.5733	0.6009	0.5661	0.5816
T-Coffee	0.4141	0.8593	0.4060	0.4776	0.5538	0.5911	0.5503

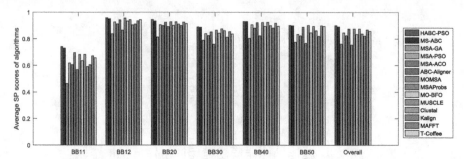

Fig. 4. Average SP scores of HABC-PSO and its comparison peer algorithms on each sequence alignment set

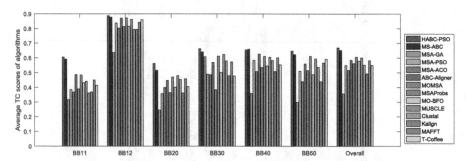

Fig. 5. Average TC scores of HABC-PSO and its comparison peer algorithms on each sequence alignment set

alignment advantages in MSA than *Clustal* Ω and MUSCLE professional software, especially when 218 test cases of 6255 sequences are aligned.

The reason why HABC-PSO outperforms other peers is because it optimizes the global and local exploration abilities in the process of neighborhood search by integrating ABC and PSO. For test dataset BB11 whose similarity between sequences is less than 20%, the HABC-PSO algorithm obtains SP average and TC average better than other non-professional alignment software, but only 0.7408 and 0.6057, respectively. The main reason is sequence similarity has a greater impact on sequence alignment results. Thus, dataset BB11 is still a challenge for the proposed algorithm, though it has an overall leading performance for almost all datasets.

Table 3 shows running time of all 14 algorithms for 218 test cases on BAliBASE 3.0 database. From Table 3, we can see that the runtime of *Kalign, Clustal* Ω and *MUSCLE* sequence alignment algorithms is shorter than those of other heuristic multi-sequence alignment algorithms, which are 1 min 1 s, 9 min 16 s and 13 min 29 s, respectively. The main reason is that *Kalign, Clustal* Ω and *MUSCLE* are highly specialized software for MSA. Before alignment sequence, they compare the evolutionary distance of two sequences in all sequences to construct phylogenetic tree. MOMSA algorithm is the slowest one, mainly because it fails to consider the optimization of parameters but only using default parameters. Alternatively, other heuristic sequence alignment algorithms actually optimize their parameters. The proposed HABC-PSO algorithm runs relatively fast compared with its peers, but it can be further accelerated by using multi-CPU or GPU optimization algorithms

Table 3. Algorithms runtime of 218 test cases on BAliBASE 3.0 (hh: mm: ss)

Algorithm	HABC-PSO	MS-ABC	MSA-GA	MSA-PSO	MSA-ACO	ABC-Aligner	MOMSA
Runtime	02:04:52	01:59:23	01:47:34	02:58:52	02:46:25	03:06:42	23:37:56
Algorithm	**MSAProbs**	**MO-BFO**	**MUSCLE**	**Clustal Ω**	**Kalign**	**MAFFT**	**T-Coffee**
Runtime	03:28:21	02:48:17	00:13:29	00:09:16	00:01:01	03:50:23	14:32:48

6 Conclusions

A novel hybridization algorithm of artificial bee colony and particle swarm optimization (HABC-PSO) is proposed to solve the MSA problem. The proposed method innovatively integrates Tent chaos search, opposition-based learning, and recombination operator techniques. Tent chaotic opposition-based learning initialization strategy is used to improve the diversity of algorithm population and preserve individual randomness. Employed bees use the tournament selection strategy to recruit onlooker bees for avoiding the influence of super individuals on the algorithm. Introduce food source coding method to adapt to the discretization of multiple sequence alignment. By using recombination operator, the relationship between particles and bees can be deepened and the information exchange between them can be enhanced, the algorithm can effectively jump out of the local optimal and quickly search the optimal solution, thus balancing the exploitation and exploration of the algorithm. To summarize, HABC-PSO is helpful to provide more meaningful MSA for user selection.

The experimental results show that the proposed algorithm can not only solve MSA problems effectively, but also obtain better alignment results than its peers. Therefore, the proposed method provides a novel way in MSA solving, and will inspire future new methods in related research fields.

However, it is worthwhile to point out that the proposed algorithm would encounter bottleneck and result in a slow running speed because of huge memory demand when the input data size increases, because the NP-hard nature of MSA. Our ongoing work is to consider how to tackle this challenge via novel parallel computing ways using multiple CPU or GPU and Spark platforms. Research on the initialization strategy optimization, balance mechanism for global exploration and local development ability, fitness design and convergence analysis of artificial bee colony will also be considered in our to-do list.

References

1. Wang, Y.X., Wang, Z.H.: Introduction to bioinformatics: algorithms and applications for high performance computing. Tsinghua University press, pp. 90–94 (2011)

2. Gupta, R., Pankaj, A., Soni, A.K.: MSA-GA: multiple sequence alignment tool based on genetic approach. Int. J. Soft Comput. Softw. Eng. **3**(8), 1–11 (2013)
3. Gao, Y.X.: A multiple sequence alignment algorithm based on inertia weights particle swarm optimization. J. Bionanosci. **8**(5), 400–404 (2014)
4. Tsvetanov, S., Ivanova, D., Zografov, B.: Ant colony optimization applied for multiple sequence alignment. Biomath Commun. **2**(1), 800–806 (2015)
5. Öztürk, C., Aslan, S.: A new artificial bee colony algorithm to solve the multiple sequence alignment problem. Int. J. Data Min. Bioinform. **14**(4), 332–353 (2016)
6. Zhu, H.Z., He, Z.S., Jia, Y.Y.: A novel approach to multiple sequence alignment using multi-objective evolutionary algorithm based on decomposition. IEEE Journal of Biomedical and Health Informatics **20**(2), 717–727 (2016)
7. Liu, Y.C., Schmidt, B., Maskell, D.L.: MSAProbs: Multiple sequence alignment based on pair hidden Markov models and partition function posterior probabilities. Bioinformaics **26**(16), 1958–1964 (2010)
8. Rani, R.R., Ramyachitra, D.: Multiple sequence alignment using multi-objective based bacterial foraging optimization algorithm. Biosystems **150**(10), 177–189 (2016)
9. Sun, J., Wu, X.J., Fang, W., et al.: Multiple sequence alignment using the Hidden Markov Model trained by an improved quantum-behaved particle swarm optimization. Inf. Sci. **182**, 93–114 (2012)
10. Rubio-Largo, A., Vega-Rodriguez, M.A., Gonzalez-Alvarez, D.L.: A hybrid multiobjective memetic metaheuristic for multiple sequence alignment. IEEE Trans. Evol. Comput. **20**(4), 499–514 (2016)
11. Kuang, F.J., Zhang, S.Y., Liu, C.C.: Multiple sequence alignment algorithm based on multi-strategy artificial bee colony. Control Decision **33**(11), 1990–1996 (2018)
12. Chowdhury, B., Garai, G.: A review on multiple sequence alignment from the perspective of genetic algorithm. Genomics **109**(5–6), 419–431 (2017)
13. Zambrano-Vega, C., Nebro, A.J., Durillo, J.J., et al.: Multiple sequence alignment with multiobjective metaheuristics: a comparative study. Int. J. Intell. Syst. **32**(2), 843–861 (2017)
14. Karaboga, D.: An idea based on honey bee swarm for numerical optimization. Technical Report-TR06, Erciyes University, Kayseri/Turkey (2005)
15. Öztürk, C., Hancer, E., Karaboga, D.: Dynamic clustering with improved binary artificial bee colony algorithm. Appl. Soft Comput. **28**(3), 69–80 (2015)
16. Mao, M.X., Duan, Q.C.: Modified artificial bee colony algorithm with self-adaptive extended memory. Cybern. Syst. Int. J. **47**(7), 585–601 (2016)
17. Kuang, F.J., Jin, Z., Xu, W.H., et al.: Hybridization algorithm of Tent chaos artificial bee colony and particle swarm optimization. Control Decision **30**(5), 839–847 (2015)
18. Gao, W.F., Huang, L.L., Liu, S.Y., et al.: Artificial bee colony algorithm based on information learning. IEEE Trans. Cybern. **45**(12), 2827–2839 (2015)
19. Saad, A., Khan, S.A., Mahmood, A.: A multi-objective evolutionary artificial bee colony algorithm for optimizing network topology design. Swarm Evol. Comput. **38**, 187–201 (2018)
20. Wang, S.H., Zhang, Y.D., Dong, Z.C., et al.: Feed-forward neural network optimized by hybridization of PSO and ABC for abnormal brain detection. Int. J. Imag. Syst. Technol. **25**(2), 153–164 (2015)
21. Wang, D., Tan, D., Liu, L.: Particle swarm optimization algorithm: an overview. Soft. Comput. **22**(2), 387–408 (2017). https://doi.org/10.1007/s00500-016-2474-6
22. Banka, H., Dara, S.: A hamming distance based binary particle swarm optimization (HDBPSO) algorithm for high dimensional feature selection, classification and validation. Pattern Recogn. Lett. **52**, 94–100 (2015)
23. Beheshti, Z., Shamsuddin, S.M.: Non-parametric particle swarm optimization for global optimization. Appl. Soft Comput. **28**, 345–359 (2015)

24. Kuang, F., Zhang, S., Jin, Z., Xu, W.: A novel SVM by combining kernel principal component analysis and improved chaotic particle swarm optimization for intrusion detection. Soft. Comput. **19**(5), 1187–1199 (2014). https://doi.org/10.1007/s00500-014-1332-7

25. Chang, W.D.: A modified particle swarm optimization with multiple subpopulations for multimodal function optimization problems. Appl. Soft Comput. **33**, 170–182 (2015)

26. Fang, W., Sun, J., Chen, H., et al.: A decentralized quantum-inspired particle swarm optimization algorithm with cellular structured population. Inf. Sci. **330**, 19–48 (2016)

27. Liu, T., Jiao, L., Ma, W., et al.: A new quantum-behaved particle swarm optimization based on cultural evolution mechanism for multiobjective problems. Knowl. Based Syst. **101**, 90–99 (2016)

28. Wang, X.-D., Liu, J.-X., Xu, Y., Zhang, J.: A survey of multiple sequence alignment techniques. In: Huang, D., Bevilacqua, V., Prashan, P. (eds.) ICIC 2015. LNCS, vol. 9225, pp. 529–538. Springer, Cham (2015). https://doi.org/10.1007/978-3-319-22180-9_52

29. Sara, S., Sameh, S., Arabi, K.: PoMSA: an efficient and precise position-based multiple sequence alignment technique. In: 2nd International Conference on Advanced Technology and Applied Science September 12–14, Alexandria, Egypt Arab Academy for Science, and Technology and Maritime Transport (2017)

30. Sara, S., Sameh, S., Arabi, K.: Parallel PoMSA for aligning multiple biological sequences on multicore computers. In: 13th International Conference on Computer Engineering and Systems (ICCES), Cairo, Egypt, 18–19 December 2018, 8639444 (2018)

31. Zambrano-Vega, C., Nebro, A.J., García-Nieto, J., Aldana-Montes, J.F.: Comparing multiobjective metaheuristics for solving a three-objective formulation of multiple sequence alignment. Progress Artif. Intell. **6**(3), 195–210 (2017). https://doi.org/10.1007/s13748-017-0116-6

32. Navarro, G.: A guided tour to approximate string matching. ACM Comput. Surv. **33**(1), 31–88 (2001)

33. Black, P.E., Levenshtein, distance.: Dictionary of Algorithms and Data Structures, U.S. National Institute of Standards and Technology, retrieved 2 November (2016)

34. Shi, X.H., Li, Y.W., Li, H.J., et al.: An integrated algorithm based on artificial bee colony and particle swarm optimization. Int. Conf. Nat. Comput. Yantai: IEEE Press. pp. 2586–2590 (2010)

35. El-Abd, M.: A hybrid ABC-SPSO algorithm for continuous function optimization. In: IEEE Symposium on Swarm Intelligence (SIS). pp. 1–6, Paris: IEEE Press (2011)

36. Thompson, J.D., Koehl, P., Poch, O.: BAliBASE 3.0: latest developments of the multiple sequence alignment benchmark. Proteins **61**, 127–136 (2005)

37. Edgar, R.C.: BENCH: a collection of protein sequence alignment benchmarks including BALIBASE v3, PREFAB v4, OXBENCH, and SABRE. http://www.drive5.com/bench (2016)

38. Multiple Sequence Alignment tools. http://www.ebi.ac.uk/Tools/msa/. Accessed 20 Dec 2018

39. Edgar, R.C.: MUSCLE: a multiple sequence alignment method with reduced time and space complexity. BMC Bioinform. **5**, 113 (2004)

40. Sievers, F., Wilm, A., Dineen, D., et al.: Fast, scalable generation of high-quality protein multiple sequence alignments using clustal omega. Molecular Syst. Biol. **7**(1), 539 (2011)

41. Lassmann, T., Frings, O., Sonnhammer, E.L.L.: Kalign2: high-performance multiple alignment of protein and nucleotide sequences allowing external features. Nucleic Acids Res. **37**(3), 858–865 (2009)

42. Katoh, K., Kuma, K., Toh, H., Miyata, T.: MAFFT version 5: improvement in accuracy of multiple sequence alignment. Nucleic Acids Res. **33**(2), 511–518 (2005)

43. Notredame, C., Higgins, D.G., Heringa, J.: T-coffee: a novel method for fast and accurate multiple sequence alignment. J. Molecular Biol. **302**(1), 205–217 (2000)

A Flexible and Comprehensive Platform for Analyzing Gene Expression Data

Bolin Chen[1,2] 📷, Chenfei Wang[1], Li Gao[3], and Xuequn Shang[1,2(✉)] 📷

[1] School of Computer Science, Northwestern Polytechnical University, Xi'an, China
npu_bioinf@hotmail.com
[2] Key Laboratory of Big Data Storage and Management, Ministry of Industry and Information Technology, Northwestern Polytechnical University, Xi'an, China
[3] School of Software, Northwestern Polytechnical University, Xi'an, China

Abstract. Studying the original gene expression dataset is one of the essential methods for analyzing biological processes. Many platforms were developed to conduct this kind of study, such as GSEA, and the online gene list analysis portal Metascape. However, these well-known platforms sometimes are not friendly enough for inexperienced users due to the following reasons. Firstly, many biological experiments only have three duplicates, which make classical statistical methods lack of efficient and accuracy. Secondly, different experiments could result in different gene expression profiles, where standard differential expressed gene identification methods still have room to be further improved. Thirdly, many platforms work only for specific experimental conditions based on their default parameters, where users are not easily setup parameters for their own studies. In this study, we designed a comprehensive and flexible gene expression data analysis tool, where six novel differential expressed gene identification methods and three functional enrichment analysis methods were proposed. Majority parameters can be friendly setting by users and a variety of algorithms can be 9 according to the user's own study designing. Experiments show that our platform provides an effective way for gene set series analysis, and has great performance in both practicality and convenience.

Keywords: Gene expression · Differentially expressed genes · Microarray data · Functional enrichment analysis

1 Introduction

Detecting differentially expressed genes aims to find the classes of genes that are significantly expressed or depressed. In many biological experiments, taking drug discovery as an example, researchers need to find genes that have significant changes in expression levels due to drug action. Furthermore, biological changes typically involve a gene set in which multiple genes are associated with individual biological pathways or GO terms.

B. Chen and C. Wang—Equal contributors.

© Springer Nature Singapore Pte Ltd. 2020
H. Han et al. (Eds.): IDMB 2019, CCIS 1099, pp. 170–183, 2020.
https://doi.org/10.1007/978-981-15-8760-3_12

Therefore, researchers normally follow the following two steps to analyze gene expression datasets: i) identify differentially expressed genes, ii) perform functional enrichment analysis based on differentially expressed genes, such as pathway enrichment analysis or GO analysis.

Although many methods and tools have been widely developed for analyzing gene expression data sets, there are still some shortcomings. Considering the flexibility and practicality of the tool, we think that the current analysis tools can be further improved in the following three aspects.

Firstly, due to the limited budgets or low sample availability, many biological experiments have only three duplications for each condition. These limited number of samples makes many standard statistical methods inefficient, and the results based on these methods contain many false positives, which is not reliable.

Secondly, Fig. 1(a) shows gene expression dataset that the majority of genes were normally expressed, and only a small number of genes were expressed with very low or very high levels. The standard fold-change method [1] works very well for this kind of dataset. However, when it comes to the dataset like Fig. 1(b), this method may not work well. The standard fold-change method could only detect genes with small expression levels, because a slight change of the gene expression could result in at least two-fold changes between the test and control groups, but a larger expression change is needed for a gene if the its expression level is high [2]. A scatter plot of this kind of gene expression with 3 test and 3 control samples was shown in Fig. 1(b). The simple fold change method is not sensitive to expression noises [3], which not only results in lots of false positives, but reduces the possibility of detecting highly expressed genes.

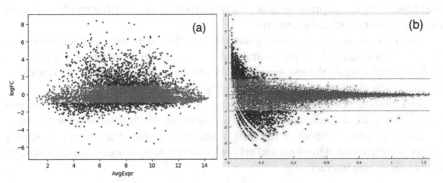

Fig. 1. The scatter plot of gene expression dataset. (a) A typical gene expression scatter plot where majority of genes were expressed normally, with only small number of genes were expressed with very low or very high levels. (b) Another gene expression scatter plot where the expression level of majority of genes were very low.

Thirdly, many gene expression analysis tools are designed only for general experiments and are not effective enough for other kinds of datasets. Many other tools only use default values to encapsulate the parameters of the differential expressed gene detection algorithms, and do not allow users to set parameters based on their own experimental conditions. Moreover, some platforms only focus on either differentially expressed genes or functional enrichment analysis, which makes users have to combine at least two different platforms to complete the whole analysis processes. Since one platform generally is designed only for specific kind of biological condition [4], the mixed combination of different platform may introduce new biases. Besides, some online analysis tools only allow limited gene expression data, for example, 3000 genes [5], which results in large volumes of data that cannot be processed.

In this study, we designed a comprehensive and flexible platform for analyzing gene expression data that allows for custom parameter settings for almost all-important parameters. In addition, we proposed six new methods for identifying differential expressed genes and two functional enrichment analysis methods. The platform is applicable to different types of data, making the design platform more versatile and effective.

The structure of the article is arranged as follows. Section 2 introduces methods for data cleaning and differentially expressed identification. Section 3 compares the results of the proposed methods. Section 4 draws conclusions and discussions.

2   Methods

2.1   The Overall Framework of the Gene Expression Analysis Platform

The overall framework of the proposed platform is shown in Fig. 2.

A: Input file types. The designed platform allows for the entire process of gene expression data analysis, from data cleansing to data normalization, from identification of expressed genes to functional enrichment analysis. Thus, it can support users either to upload the raw gene expression data sets or provide a list of differentially expressed genes for further enrichment analysis.

Two kinds of raw gene expression datasets are allowed as inputs, either in the form of one file that contains both the control and the test dataset together, or in the form of two files which contain the test group of data and the control group of data separately. This makes the platform has high compatibility and the scalability.

The proposed platform also allows users to define differentially expressed gene list as input, since in many cases, one only interest in the functional related analysis of a group of genes, without re-selecting genes from a given file. Several promising functional enrichment algorithms and network layout processes are employed to display the analysis results interactively.

B: Data cleaning. Data cleaning is one of the most overlooked but also the most critical steps in gene expression data analysis. In general, external experimental conditions affect gene expression data. For example, the drug injection dose of a sample may be proportional to the gene expression levels.

More importantly, gene expression data is affected by biological and technological variations, and a certain gene may correspond to multiple expression level. Therefore, the necessary data cleaning process must be performed prior to genetic analysis, which is an essential step in any data analysis.

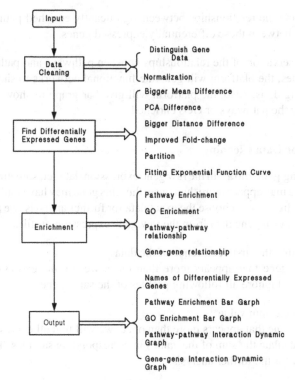

Fig. 2. The framework of the gene expression dataset analysis.

C: Identifying differentially expressed genes. Although many traditional methods to find differentially expressed genes in a given dataset, such as fold-change method [2] and t-test method [6, 7], these tradition methods are not proper for certain situations. For example, the fold-change based methods are not appropriate to this situation where there are lots of low expression values (or zero values) in the raw data, and the t-test will not perform well when the number of samples is small. Our platform improves existing traditional methods with new five algorithms to identify the differentially expressed genes in a given expression data set. The platform also allows users to set parameters according to their own experimental conditions.

D: Functional enrichment analysis. Once obtaining a list of significantly differentially expressed genes, the functional enrichment analysis will be conducted by integrating pathways, Gene Ontology (GO) and some biological networks. The platform implements different algorithms to functional enrichment analysis, as well as find the relationships between pathways and genes, between GOs and genes, between pathways and pathways, or between genes and genes.

E: Output information and results. The platform will finally prepare the outcomes drew from the above results. Users can get the following information: (1) significantly differentially expressed genes list, (2) pathways these genes enriched, (3) GOs that these genes enriched, (4) the relationships between significantly enriched pathways and (5) the relationships between these differentially expressed genes.

For better observation of the relationships between pathways and pathways as well as genes and genes, the platform will generate functional networks to show these relationships by using d3.js. Also, the platform will give bar graph to show the degree of how significantly the pathways or GOs enriched.

2.2 Methods for Data Cleaning

A: Distinguishing gene data. In the raw gene expression data set, sometimes we notice that a certain gene may appear more than once, thus this gene may have multiple different expression level. In order to choose the exact data for further analysis, we propose three methods to distinguish gene data from the raw gene expression profiles.

Method 1. selecting the first occurrence of the data
For a certain gene that appears more than once, we use the expression data that appears earliest and ignore all following profiles of the same gene.

Method 2. using the sum value
For a certain gene that appears more than once, we will find all expression levels of this gene, and calculate the sum of the values in corresponding samples. Then, the new data will be used for the further analysis.

Method 3. using the mean value
The average of all expression values is calculated for a particular gene as its new profile.

B: Data normalization. To eliminate the effect of different concentration degrees of the raw samples brought to data, the platform also provides normalization based on the sum of columns.

2.3 Algorithms for Identifying Differentially Expressed Genes

Giving two sets of m control samples and n test samples, respectively. Let $c_1, c_2, ..., c_m$ be the expression value of control samples and $t_1, t_2, ..., t_n$ be expression value in test samples. Let

$$\bar{c} = \frac{\sum_{i=1}^{m} c_i}{m}$$

represents the mean value of the expression data in control samples, and

$$\bar{t} = \frac{\sum_{i=1}^{n} t_i}{n}$$

represents the mean value of the expression data in test samples, respectively.

The traditional fold-change based method can be describe briefly as the following formula:

$$s = log_2 \frac{\bar{t}}{\bar{c}} \tag{1}$$

If s is larger than 1 or smaller than -1 [8, 9], the gene will be regarded as significantly highly expressed or depressed. However, there are some potential problems when using this method. Firstly, a lower expressed gene is easier to obtain a two-fold change compared to a higher expressed gene in term of their change requirement. Hence, this method is biased to low expressed genes, and may misclassify differentially expressed genes with small ratios but large differences, leading to poor identification of genes at high expression levels [2]. Secondly, when the expression values in either the denominator or numerator are close or equal to zero, the ratio is not stable and even cannot be calculated, and the fold change value can be disproportionately affected by measurement noise. To solve these specific problems, we provide the following six methods to find differentially expressed genes.

A: Improved fold-change method. This method is designed for the situation where there are zero values in the raw expression data, because zero cannot be denominator. Traditional methods normally filtered those genes, but we think those genes still have valuable information to interpret gene expressions.

Case 1. For normal gene expression values
This is the normal case where the gene expression data can be used to do fold-change. For all the genes to be analyzed, we take the gene as differentially expressed gene when the calculated s in (1) is larger than a fold threshold. Users can set the fold parameter or choose the genes with top $n\%$ s as differentially expressed genes, where n can be set by themselves according to requirements.

Case 2. data For gene expression values where $\bar{c} = 0$ or $\bar{t} = 0$
For all the genes with $\bar{c} = 0$, we look at the corresponding $\bar{t}$. If a certain gene with $\bar{c} = 0$ while its $\bar{t}$ is in the highest k% of this kind of genes, we think this gene as significantly expressed, where k can be set by users according to their requirements. Following the same procedure, we process genes with $\bar{t} = 0$. Then, we put the two parts of differentially expressed genes together as significantly expressed genes.

Case 3. For gene expression values where $\bar{c}$ or $\bar{t}$ closes to zero
Either in test group or control group, if there are more than two thirds of the gene expression data is 0, we may consider that the non-zero values may be caused by noise. So, in this case, we need to do some pretreatments, transferring the gene dataset to above two cases. For a certain gene, if there are more than two thirds of the expression data is 0 in test group, then we calculate its $\bar{t}$ After processing all this kind of genes, if the $\bar{t}$ is smaller than the $k\%$ percentile of all $\bar{t}$ s of this kind of genes, we will view that the gene does not express in test group and the small value is drawn from noice. The $k\%$ can be set by users according to their requirements. Following the same procedure, we process the genes with more than two thirds of the expression data is 0 in control group. We will set its expression data in test group all to 0s. For the rest of the genes, when selecting differentially expressed genes, do as the Case 1 and Case 2 do.

The expected results of the differentially expressed genes should scatter uniformly in different expression levels, rather than just concentrate on the areas where the gene expression level is pretty low. To solve this problem, we provide the following methods.

B: A divided section-based method. To identify differentially expressed genes scattered not uniformly in different expression levels, a set of divided sections was given according to the mean expression values. Taking the highest $n\%$ logarithm of the ratio in every divided section as differentially expressed genes, where n can be set by users. Still, we will focus on the gene dataset in normal cases, the procedures are same in previous improved fold-change method. The pretreatment procedure is discussed before.

Step 1. Pretreatments should be done to transfer the gene dataset to normal cases. Identify differentially expressed genes in special situations firstly, and then rank the expression level of genes according to $\bar{t}$.

Step 2. Divide the genes into sections, ensuring that the number of the genes in every section is same (except the last section). It will lead to under-fitting or overfitting if the number of genes is too small or too big. In expected situation, every section includes 0.02%–0.03% of the total genes.

Step 3. In every section, calculate the fold-change score s in (1), taking the genes with highest $n\%$ s score as differentially expressed ones.

This partition method has met the requirement of selecting differentially expressed genes from low to high levels. However, there is still a little problem associating with partition method. We divide the genes into lots of groups, but the boundaries between groups will be very obvious if the number of groups is too small. In turn, it will lead to overfitting, greatly lower the quality of the outcome if the number of groups is too big. Thus, we design the following method to find a smooth curve, which shows that genes above or below are selected to be differentially expressed genes.

C: Fitting Exponential Function Curve. We notice that the exponential function can be a monotonically decreasing function, meeting the expectation that differentially expressed genes are not uniformly distributed in low and high expression level. Here, we just focus on the gene dataset in normal cases, and the pretreatment procedure is discussed in section *A: Improved fold-change method.*

Step 1. Rank the data of the genes according to the mean expression value in test samples $\bar{t}$.

Step 2. Finding out the gene with its mean expression value in test group, if its expression value is in k^{th} percentile of all mean expression value in test group, where k can be set by users according to their requirements, denote this value as x. Similarly, calculate its corresponding mean expression value in control samples, and denote this value as y.

Step 3. According to x and y, we can get a specific exponential function (2).

$$y = a^x \tag{2}$$

However, the exponential function always intersects with y-axis on $(0,1)$. So, the exponential function should be moved rightward. The distance of the right-move is the

k^{th} percentile of all mean expression value in test group of genes, and denote this value as x_0. Then, we will get a new exponential function as follows.

$$y = a^{x-x_0} \qquad (3)$$

Step 4. We use the mean expression value in test group of the genes as abscissa, and absolute value of the logarithm of 2-fold change between mean expression value in two treatments. We choose the corresponding gene as a differentially expressed gene if the point is beyond the exponential function curve.

D: Mean difference-based method. Another alternative method is based on the mean difference. We use the following formula to evaluate the changes of expression level (4) for a certain gene g_i,

$$x_i = |\bar{t}_i - \bar{c}_i| \qquad (4)$$

The genes with the difference in highest $n\%$ of all the differences are selected as differentially expressed genes, where n can be set by users according to their requirements.

E: PCA difference-based method. Similar to the previous method, but the only difference is doing pretreatment before calculating the needed differences.

Step 1. For all the genes, drawing PCA to one dimension on their expression values in test and control samples [10], respectively.

Step 2. For a specific gene, calculate the absolute value of the difference between the principle component in test and control samples.

Step 3. After processing all the genes, choose the genes with the difference in highest $n\%$ as differentially expressed genes.

F: Geometrical distance-based method. In this method, we use the geometrical distance of the raw expression data between two conditions. We view gene values in test and control samples as two points respectively, if the "distance" between the two points is very large, we may say that the expression level of the gene changed greatly in test group and control group. Here, we use 2-norm to describe the "distance" between the two points. Notice that the method is just for the situation where the numbers of samples in test group and in control group are same.

Step 1. For the genes, rank the expression values in test group and control group respectively.

Step 2. Two vectors are constructed by using the sorted values for the test and control groups, respectively.

$$T : \left[t_1' t_2' \dots t_n' \right] \qquad (5)$$

$$C : \left[c_1' c_2' \dots c_m' \right] \tag{6}$$

Step 3. Calculate the 2-norm of the two vectors, just like calculate the distance between two points (7).

$$d = |T - C| \tag{7}$$

Step 4. After processing all the genes, choose the genes with the d in highest $n\%$ of all the ds as differentially expressed genes, where n *can* be set by users according to their requirements.

2.4 Algorithms for Functional Enrichment Analysis

The comparison of individual gene expression values is not sensitive enough to detect the subtle functional changes of biological systems, because cellular changes typically involve in many groups of genes. Multiple genes are linked to a single biological pathway [11, 12], and it is the additive change in expression within gene sets that leads to the difference in phenotypic expression. Thus, researchers will not just focus on the isolated genes, but turn to the pathways and GOs that these differentially expressed genes enriched. The enrichment analysis helps researchers to understand and interpret omics data from the point of view of canonical prior knowledge structured in the forms of pathways and GO diagrams. This allows us finding distinct cellular processes, biological changes or diseases that are statistically associated with selection of differentially expressed genes between two samples [7].

The proposed platform supports four kinds of enrichment analyses, where two of which are proposed to meet the basic requirements for the pathway enrichment and GO enrichment analysis, respectively.

A: **Pathway enrichment.** Traditional enrichment analysis algorithm is based on Fisher's exact test [13]. In addition, the platform also supports additional two more methods for enrichment analysis for different experimental conditions.

Method 1. Fisher's exact test
The detail methods of the fisher's exact test can be found in [14].
Method 2. Using the relationship between gene sets
We know that the membership overlap of two gene sets can be used to describe the relationship about them. The bigger the intersection of the two sets, the closer of their relationships. In this method, we apply this concept to enrichment analysis as follows. For a certain pathway, we can easily know what genes it includes according the information in the database at the backend of the platform. We use a new set U to store the intersection part of the genes included in pathway and the differentially expressed genes found before, and use another set V to store the union part of the genes included in pathway and the differentially expressed genes found before. If the U is not null, we calculate the quotient of the number of genes in U and V. The bigger the value, the more the two sets overlap, which means the more significantly the pathway is enriched.

Method 3. Using the weighted f-measure

We borrow the ideas from the concepts of precision and recall [15] in this method. Let P_i denote the gene set of pathway i, and D the set of differentially expressed genes. The precision is defined as the quotient between the number of genes in the intersection of the two sets and the number of genes in difference, and the recall is defined as the quotient between the number of genes in the intersection of the two sets and the number of genes in interested. Then we calculate the f-measure to describe how the pathway is enriched in given differentially expressed gene set. However, we notice that the number of genes in the intersection of significantly differential is small when doing pathway enrichment, which leads to the fact that precision is small. Thus, the f-measure is largely depended on recall. So, we calculate the weighted f-measure to balance importance of precision and recall, for example, we will make precision more important when we are calculating the f-measure.

B: Pathway Relationship. After finding out the pathways the differentially expressed genes are included, we use the degree of overlap of the pathways to describe the relationship between them.

Step 1. Find out the differentially expressed genes that contained in the pathway.

Step 2. Calculating the number of genes in their intersections and unions for the given pathways.

Step 3. If the numbers of genes in the intersections and unions are not equal to zero, calculate the quotient of the numbers of genes in the intersections and in the unions. Then we use this quotient to describe the relationship between pathways. The bigger the quotient, the closer connection between two pathways.

C: Gene Ontology Enrichment. Gene Ontology (GO) is a major bioinformatics initiative to unify the representation of gene and gene product attributes across all species. The way we take GO enrichment analysis is the same to the previous pathway enrichment. We support traditional Fisher's exact test, as well as use the relationship between precision and recall. The difference is that we use f-measure as a standard when we use precision and recall to describe whether the GO is significantly enriched.

D: GO Relationship. As we do in section *B: Pathway Relationship*, we use the same method to find the relationships between the selected significantly enriched GOs. However, GO is kind of a big term which contains many information. So, we don't pay much attention in this study. The updated version will contain this part.

E: Functional relationship network. A functional relationship network between genes will then be generated by using PPI networks, where each node represents a gene and edge represents the interaction between two genes. This functional relationship network between differentially expressed genes should show potential gene communities related to different biological functions, which is an important supplementary for those canonical prior knowledges structured.

3 Results

3.1 Effects of the Methods for Detecting Differentially Expressed Genes

We compared the performance of the proposed methods in terms of detecting differentially expressed genes and their functional communities, as following figures shows.

The x-axis represents the mean expression value in test group of the genes, and the y-axis is the absolute value of the logarithm of the ratio of the mean expression value between test group and control group (Figs. 3, 4, 5, 6, 7 and 8).

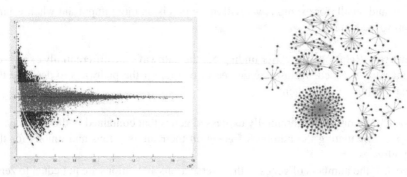

Fig. 3. The performance of *the improved fold-change method* for identifying differentially expressed genes. Red spots represent all raw gene data in the expression data set, the blue spots represent the gene that are differentially expressed. There are 291 nodes and 301 edges in the right functional relationship graph. (Colour figure online)

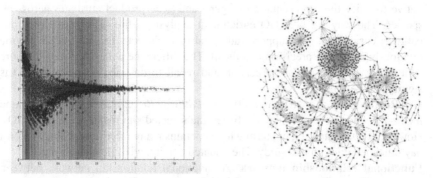

Fig. 4. The performance of *the divided section-based method* for identifying differentially expressed genes. The red spots represent all raw gene data in the expression data set, the blue spots represent the gene that are differentially expressed. There are 955 nodes and 1254 edges in the right functional relationship graph. (Colour figure online)

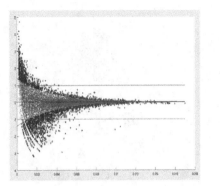
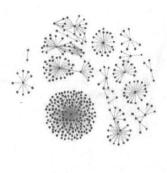

Fig. 5. The. The performance of *the fitting exponential function curve-based method* for identifying differentially expressed genes. The red spots represent all raw gene data in the expression data set, the blue spots represent the gene that are differentially expressed. There are 369 nodes and 404 edges in the right functional relationship graph. (Colour figure online)

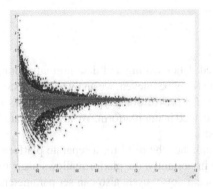
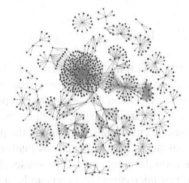

Fig. 6. The. The performance of *the mean difference-based method* for identifying differentially expressed genes. The red spots represent all raw gene data in the expression data set, the blue spots represent the gene that are differentially expressed. There are 750 nodes and 979 edges in the right functional relationship graph. (Colour figure online)

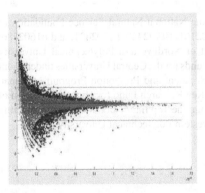
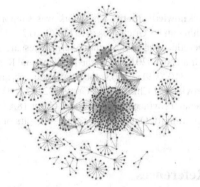

Fig. 7. The performance of *the PCA difference based method* for identifying differentially expressed genes. The red spots represent all raw gene data in the expression data set, the blue spots represent the gene that are differentially expressed. There are 755 nodes and 982 edges in the right functional relationship graph. (Colour figure online)

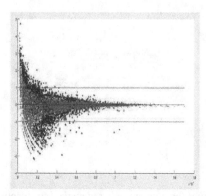 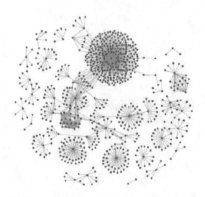

Fig. 8. The performance of *the geometrical difference-based method* for identifying differentially expressed genes. The red spots represent all raw gene data in the expression data set, the blue spots represent the gene that are differentially expressed. There are 729 nodes and 888 edges in the right functional relationship graph. (Colour figure online)

4 Conclusion

We started with microarray data and explained data cleaning and data normalization. Considering that the distribution of data samples may be uneven, we analyzed the distribution of expression data. From the perspective of expressing data types, we proposed six methods for data analyzing, which is greatly reflected in our platform. We also integrated the enrichment analysis function in the platform.

This flexible genetic analysis platform eliminates the need for a separate platform to perform different analysis processes. More importantly, the platform allows users to input raw microarray data and can help filter data noise, users also can set parameters themselves, which greatly improving the shortcomings of traditional platforms. Such an integrated platform returns the final results in text and graphics, facilitating user analysis and visualization of gene expression data. Experiments proved that the proposed platform has good utilization value in both practicability and efficiency.

Acknowledgement. This work was supported by the National Natural Science Foundation of China under Grant Nos. 61972320, 61772426, 61702161, 61702420, 61702421, and 61602386, the education and teaching reform research project of Northwestern Polytechnical University (Grant No 2020JGY23), the Fundamental Research Funds for the Central Universities under Grant No. 3102019DX1003, the Key Research and Development and Promotion Program of Henan Province of China under Grant 182102210213, the Key Research Fund for Higher Education of Henan Province of China under Grant 18A520003, and the Top International University Visiting Program for Outstanding Young Scholars of Northwestern Polytechnical University.

References

1. Tusher, V.G., Tibshirani, R., Chu, G.E.: Significance analysis of microarrays applied to the ionizing radiation response. In: Proceedings of the National Academy of Science of the United

States of America (24 April 2001) 98 (5116–5121)). Proceedings of the National Academy of Sciences of the United States of America, p. 98 (2001)

2. Mutch, D.M., et al.: The limit fold change model: a practical approach for selecting differentially expressed genes from microarray data. BMC Bioinform. **3**(1), 17–20 (2002)

3. Raser, J.M.: Noise in gene expression: origins, consequences, and control. Science (Washington DC), **309**(5743), 2010–2013 (2005)

4. Zhou, Y., et al.: Metascape provides a biologist-oriented resource for the analysis of systems-level datasets. Nature communications (2019)

5. Metascape homepage. http://metascape.org/gp/index.html#/main/step1

6. Dalman, M.R., Deeter, A., Nimishakavi, G., Duan, Z.H: Fold change and p-value cutoffs significantly alter microarray interpretations. BMC Bioinform. **13**, 256–303 (2012)

7. Witten, D.M., Tibshirani, R.A.: comparison of fold-change and the t-statistic for microarray data analysis. Analysis (2007)

8. Robinson, M.D, Smyth, G.K.: Small-sample estimation of negative binomial dispersion, with applications to SAGE data. Biostatistics (2007)

9. Love, M.I., Huber, W., Anders, S.: Moderated estimation of fold change and dispersion for RNA-SEQ data with DESEQ2. Genome Biol. **15**(12), 550 (2014)

10. Mika, S.: Kernel PCA and de-noising in feature spaces. Adv. Neural Inf. Process. Syst. **11**, 65–92 (1999)

11. Hong, M.G., Pawitan, Y., Magnusson, P.K.E., Prince, J.A.: Strategies and issues in the detection of pathway enrichment in genome-wide association studies. Hum. Genet. **126**(2), 289–301 (2009)

12. Gene Ontology Consortium. The Gene Ontology project in 2008. Nucleic acids research, 36(Database issue). D440–D444 (2007). https://doi.org/10.1093/nar/gkm883

13. Huang, D.W., Sherman, B.T., Lempicki, R.A.: Bioinformatics enrichment tools: paths toward the comprehensive functional analysis of large gene lists. Nucleic Acids Res. **37**(1), 1–13 (2009)

14. Fisher, R.A.: On the interpretation of X^2 from contingency tables, and the calculation of P. J. Royal Stat. Soc. **85**(1), 87–94 (1922)

15. Powers, D.: Evaluation: fom precision, recall and f-measure to roc, informedness, markedness and correlation. J. Mach. Learn. Technol. **2**, 37–63 (2007)

Classification of Liver Cancer Images Based on Deep Learning

Hui Ye[1]([⊠]) [iD], Qiaojun Chen[2] [iD], Haimei Wu[1] [iD], and Dong Cao[1] [iD]

[1] School of Medical Information Engineering, Guangzhou University of Chinese Medicine,
Guangzhou, China
yehui@gzucm.edu.cn
[2] School of Software Engineering, South China University of Technology, Guangzhou, China

Abstract. With the rapid development of deep Learning, research into Deep Learning is being increasingly applied to the field of medical imaging. Liver cancer, which has one of the highest rates of morbidity and mortality in the world, is a great threat to people's health. This study aims to apply Convolutional Neural Networks in the grade classification of liver cancer images. DCE- MRI and DWI, two modes of hepatocellular carcinoma images, are originally used separately to grade liver cancers. We combine these two image modes to improve the prediction accuracy. The study finds that the features of the two modes can be complementary, and can improve the grading classification of liver cancer. From comparing the two methods of traditional Machine Learning and Deep Learning, the study demonstrates that the grading accuracy by Machine Learning from the integration of features is 87.8, while the accuracy rate from Deep Learning reaches 90.5. The improvement in grading accuracy is due to Deep Learning can extract the appropriate features. In addition, the presence of micro vascular invasion is an important factor for the recurrence of liver cancer after surgery. The experiment also uses Deep Learning to predict micro vascular invasion. The accuracy of the ADC map prediction reached 69.2, it demonstrates that liver cancer images can also predict micro vascular invasion to a certain extent.

Keywords: Multi-Modality images fusion · Convolutional Neural Network · Liver cancer grade classification · Micro vascular invasion

1 Introduction

Liver cancer has one of the highest rates of morbidity and mortality, and will cause great harm to people's life and health [1]. Early diagnosis of liver cancer is especially important for the effective treatment of liver cancer and the improvement of prognosis. The examination of tumor images is an important reference for doctors to diagnose the degree of tumor malignancy.

Hepatocellular carcinoma (HCC) is the most common primary liver cancer. Histological grading of hepatocellular carcinoma (HCC) is an important marker for evaluating the malignancy of HCC. It can be histologically graded according to the degree of differentiation of cancer cells. Usually grade I is highly differentiated and its malignant degree

© Springer Nature Singapore Pte Ltd. 2020
H. Han et al. (Eds.): IDMB 2019, CCIS 1099, pp. 184–195, 2020.
https://doi.org/10.1007/978-981-15-8760-3_13

is low, grade II is moderately differentiated and its malignant degree is medium; grade III and IV are poorly differentiated and their malignant degree is high. In this experiment, I and II were classified as highly differentiated which is defined to be negative grade, and III and IV were classified as poorly differentiated which is defined to be positive grade.

With the continuous development of Deep Learning, studies have shown that combing DWI, DCE-MRI images for Deep learning has better performance to grade liver cancer than traditional morphological features in traditional method [2].

This study mainly focuses on research into tumor grade classification based on Deep Learning with two images integrating. It demonstrates the advantages of using Deep Learning in the liver cancer grade classification by medical images.

Extraction of low-level features from medical images is the basis for subsequent disease diagnosis and classification of benign and malignant diseases [3]. Considering that medical images are mostly gray-scale images, the extraction methods often used are texture features and local features. However, with the development of Deep Learning and artificial intelligence, it is increasingly common to classify and form a diagnosis from medical images by extracting the features based on Convolutional Neural Networks (CNN).

Dynamic contrast enhanced magnetic resonance imaging (DCE-MRI), a functional imaging method to evaluate the vascular properties of tissue and tumor, uses rapid MRI sequencing to continuously collect MRI data from images during the whole process of contrast media injection.

Diffusion weighted imaging (DWI) is an imaging technique based on the principle design of microscopic motion of water molecules (i.e., diffuse random motion) to quantitatively analyze the random motion of water molecules in tissues [3]. DWI is now widely used in medical imaging and clinical medical research.

The degree of dispersion of water molecule motion in DWI is often described by the apparent diffusion coefficient (ADC). The ADC map can be generated from DWI data [4]. The larger the ADC value, the greater the dispersion of water molecules. According to the dispersion of water molecules, DWI images will produce regions of different signal intensities, and the ADC value will also change and reflect the degree of tumor differentiation, and the degree of micro vascular invasion. Micro vascular invasion (MVI) refers to a cluster of cancer cells in a vascular lumen lined with endothelial cells under a microscope [5], which is common in small branches of veins in adjacent liver tissues. As the tumor enlarges, the proportion of MVI increases, but pathological examination is necessary to determine whether there is MVI. This study conducts research into the use of predictive imagery. In recent years, the academic community has proposed the need for routine diagnosis of MVI since the recurrence rate of liver cancer after surgery is as high as 70–80%. Intrahepatic or extra hepatic tumor cells can migrate along blood vessels. The prediction of micro vascular invasion has a great effect on the postoperative recovery of liver cancer. Because MVI is a mark of invasive biological behavior in liver cancer [6], doctors can understand the risk of liver cancer recurrence by observing MVI.

There are many studies on the prediction of tumor grading at home and abroad [7, 8], but most of them are based on ADC images generated by DCE-MRI images or DWI images alone. for examples, Supu Pu [9] used DCE-MRI to Differ Liver Space-occupying Lesions, Jiawei He [10] used DWI in Differentiating Histopathological

Types of Small Hepatocellular Carcinoma. But there are fewer reteaches on multi-modal fusion images for liver cancer. Wen-Bin [11] showed the diagnostic Value of DWI Combined with DCE-MRI in Focal Nodular Lesions of the Liver. Hoang DA [12] et al. demonstrated the use of two modality images for correlation analysis of prostate cancer, reflecting the advantages of two modal fusion. This study conducts related research for liver cancer. Although there are already related papers using traditional texture features for the prediction of micro vascular invasion [13], it is rare to use Deep Learning to make predictions, so this research helps to promote the application of Deep Learning in the field of micro vascular invasion.

2 The Method

The paper mainly studies two aspects: classifying tumors and predicting micro vascular invasion from medical images. Firstly, the characteristics of image extraction are appropriately selected using traditional statistical methods. Tumor classification is then predicted according to the selected features. The fusion of two image modality is used to observe whether the prediction accuracy can be improved. Finally, Deep Learning is used to train and predict, and to compare the reliability of the results by these two methods.

In addition, because micro vascular invasion is also an important factor affecting the recovery of liver cancer after surgery, the Deep Learning network is also used to study the reliability of images in predicting micro vascular invasion. The main experimental structure of the study is shown in Fig. 1:

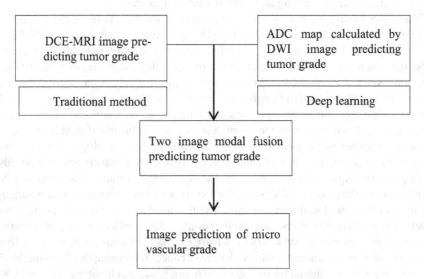

Fig. 1. The experimental structure of the study

2.1 DWI Images Processing

This study included 55 patients with pathologically confirmed tumors from July 2012 to May 2017. The subjects' DCE-MRI images and DWI images were obtained by using a 3T magnetic resonance imaging scanner.

Three b values (0,100,600 s/mm^2) were used for axial observation in DWI. Histological grading of tumors and the invasion of micro vascular are retrieved from clinical histology reports. Some reports lacked of information of histological grading and micro vascular invasion. We made a reasonable selection of data for the experiments. There were 51 males and 4 females, on average. The age was 51.5 years. The region of interest (ROI) of tumors was extracted by an experienced radiologist who has 10years' biological Experience.

The ADC (Apparent diffusion coefficient) map can be generated from DWI data, and the change in ADC value can reflect the degree of tumor differentiation and the degree of micro vascular invasion. The calculation formula of the ADC is as follows:

$$ADC = ln\left(SI_{low}/SI_{high}\right)/\left(b_{high} - b_{low}\right) \tag{1}$$

In the formula, ln is the natural logarithm. b high and b low indicate high b value and low b value respectively; SI low and SI high indicate respectively the signal strength of the low b value and high b value organization on DWI. According to the formula, two different b-value images are needed to calculate the ADC value of the tissue [11]. DWI images are chosen which obtained two b values of 0 and 100.

2.2 Traditional Method and Features Extraction

Logistic Regression (LR) and Generalized Linear Mixed Model (GLMM) are common models of binary classification. The two models were used to classify traditional features.

Based on the acquired region of interest (ROI) of the tumor, the basic texture features of the image are extracted. Each image has its own characteristics that can be distinguished from other images. Some features can be intuitively perceived, such as the brightness, color, texture, etc. of the image, and some can be obtained by recalculation or processing, such as moments, histograms, and so on. Traditional Machine Learning is used to analysis simple regional features such as mean, gray level co-occurrence matrix (GLCM) features, and gray-level run-length matrix features, to compare with Deep Learning results.

We Use Matlab to calculate the characteristic values or use extraction functions to obtain them, such as the calculation of the average intensity values:

$$m = \frac{1}{N} \sum_{(x,y)eR} a(x,y) \tag{2}$$

GLCM can be obtained by texture extraction function. It has four directional features and is based on second order estimation.

Texture analysis method based on combination conditional probability density function. It describes a pair that is spaced at a certain distance in a direction.

The statistical rule of the gray level of the pixels can reflect the comprehensive information of the image in the direction, change range, speed and so on.

In addition, some other features are extracted for feature correlation analysis and feature fusion.

2.3 Deep Learning Images Processing

In general, Deep Learning requires at least thousands of samples as training data. Due to the limited data, a multi-view resampling method was adopted to extract multiple two-dimensional sections of each three-dimensional tumor to increase the training dataset. Here, resampling refers to increasing the number of instances in a sample by copying a few images, thereby increasing the representation of minority classes in the sample. The process of generating a picture is shown in Fig. 2. The dcm slices taken from the hospital were converted into a three-dimensional file format (.mha) using Vol view software, and the three-dimensional volume data was read in Matlab according to the tumor center location (x, y, z) which were already marked by doctors. We use curve-fitting function in Matlab to determine the length, width, height (r, s, t) of the cuboid region surrounding the tumor. After obtaining the tumor body, five further tumor slices were made at intervals of a certain pixel in different direction. Finally, it obtained 5 * 5 * 5 = 125 samples for each tumor. These images are used for subsequent Deep Learning training and testing.

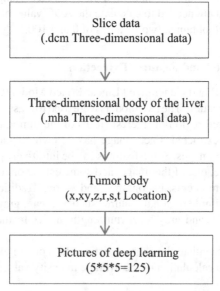

Fig. 2. The process of generating slice data

2.4 Deep Learning Method

The CNN model uses two convolutions, and the convolution kernel size is 5 * 5. The output of the first convolution is 20 feature maps, and the output of the second convolution is 50 feature maps. Each convolution layer is connected to a 2 * 2 maximum pooling layer. There are three fully connected layers, with output neurons of 500, 100 and 2. A dropout layer is added to prevent over fitting. The output of the last softmax layer is used as a classifier. In addition, the Relu activation function is used to introduce nonlinear features. The input image size is 28 * 28, and the output result are 2 types (Fig. 3).

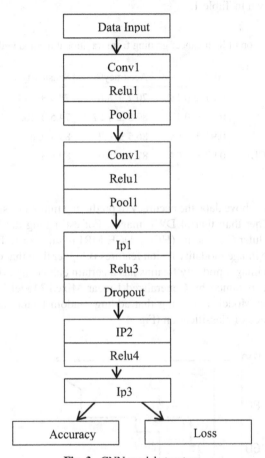

Fig. 3. CNN model structure

2.5 Deep Learning in MVI Prediction

In addition to predicting tumor classification, this study used the same batch of image data to investigate whether there were micro vascular invasions. Using the presence of MVI in the pathological data as the basis for classification, the data was re-divided and approximately 30% of the test set was selected. The two images generated were used as data inputs, and the output results were compared with the presence or absence of the MVI in the actual pathological record to obtain experimental results.

3   Experiment and Results Analysis

3.1   Traditional Machine Learning Analysis

Logistic Regression (LR) and Generalized Linear Mixed Model (GLMM) are common models of binary classification. The two models were used to classify traditional features. The results are shown in Table 1.

Table 1. The prediction of liver cancer grading from traditional method with manual features

	AUC	Accuracy%	Sensitivity%	Specificity%
ADC	0.67 ± 0.18	70.7 ± 8.6	78.9 ± 14.8	59.4 ± 14.5
DCE-MRI	0.85 ± 0.1	80.5 ± 12.7	74.5 ± 10.2	85.2 ± 13.5
LR (ADC + DCE)	0.91 ± 0.05	86.5 ± 4.7	89.4 ± 6.3	83.1 ± 6.6
GLMM (ADC + DCE)	0.92 ± 0.04	87.8 ± 5.6	89.8 ± 5.5	84.2 ± 6.7

According to the above data, the accuracy of predicting tumor classification by DCE-MRI images is higher than that of DWI images. For estimating the histopathological grade of Hepatocellular Carcinoma (HCC), DCE-MRI is superior to DWI. In addition, the accuracy of two image modality fusion features is higher than that of a single image, indicating that two image modality fusions has a certain enhancing effect in improving classification. AUC obtained by Generalized Linear Mixed Model is higher than the Logistic Regression Model, indicating that adding random factors also has a certain effect on the accuracy of classification (Fig. 4).

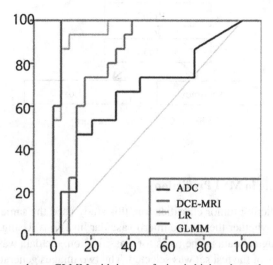

Fig. 4. The AUC map shows GLMM with images fusion is higher than other traditional method

In this project, we use Caffe [14] to process CNN model. Caffe is a fully open-source in-depth learning framework. It is written in C++ language and can be called by Python and MATLAB interface. It contains a series of mature reference models. We divided the images into 10 groups of training and testing pictures for cross-validation. We labialized the pictures by grades one by one, Then the LMDB data is generated with the corresponding label files, and the convolution neural network is trained and tested, and the results are summarized and analyzed. The process is shown in Fig. 5 with manual features

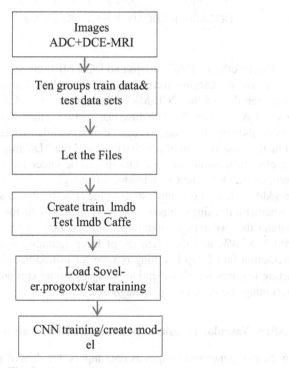

Fig. 5. The deep learning process

3.2 CNN Deep Learning Result Analysis

In this project, we use Caffe [14] to process CNN model. Caffe is a fully open-source in-depth learning framework. It is written in C++ language and can be called by Python and MATLAB interface. It contains a series of mature reference models. We divided the images into 10 groups of training and testing pictures for cross-validation. We labialized the pictures by grades one by one, then the LMDB data is generated with the corresponding label files, and the convolution neural network is trained and tested, and the results are summarized and analyzed. The process is shown in Fig. 5.

The image data were processed and 10 cross-validations were performed. The test set proportion was 25%. Each network operation used 4875 training pictures of 28 * 28 and

1625 test pictures. We ensure the balance of positive and negative sample proportions in the training and test sets. The results are shown in Table 2.

Table 2. Results of image prediction of liver cancer grading with CNN

	Accuracy%	Sensitivity%	Specificity%
ADC	73.7 ± 8.6	77.4 ± 14.6	69.6 ± 13.2
DCE-MRI	83.8 ± 7.7	82.8 ± 11.7	87.1 ± 17.2
DCE-MRI + ADC	90.5 ± 5.4	84.2 ± 8.7	92.8 ± 6.5

Firstly, the effects of ADC maps and DCE-MRI images on the classification of liver cancers in Deep Learning were analyzed. The classification effect of the ADC map is lower than that of the DCE-MRI. The accuracy of DCE-MRI is 83.8%, while the accuracy of ADC is 73.7%. The difference between them is close to 10% points. In the sensitivity and specificity data, it can be observed that DCE-MRI has a slightly better effect on the classification of positive data, and the ADC map has a slightly better effect on the classification of negative data. It can be understood that the characteristics of different images have their own unique advantages.

In addition, Deep Learning tests are more effective than traditional Machine Learning [15], whether it is a single-modality or multi-modality fusion image classification. After combining the two image modalities, the accuracy of traditional Machine Learning GLMM is 87.8%, and the accuracy of Deep Learning reaches 90.5%. This verifies the prediction that Deep Learning is superior to traditional methods in automatically extracting features for classification. Deep Learning can promote the development of medical image classification and diagnosis.

3.3 Micro Vascular Invasion Prediction Results Analysis

Using the two generated images as data inputs, the data of patients without the micro vascular invasiveness flag was removed, and data that has been cross-verified 10 times was generated. 3,750 training images of 28 * 28 and 1375 test images were used, ensuring the balance of positive and negative sample proportions in the training and test sets. After generating 1mdb data for Deep Learning input, the training results were averaged ± standard deviation, and test results of the two images were obtained separately and after the fusion. The results are shown in Table 3.

Table 3. MVI prediction

	Accuracy%	Sensitivity%	Specificity%
ADC	69.2 ± 7.6	78.9 ± 14.9	60.0 ± 15.4
DCE-MRI	66.7 ± 5.8	60.0 ± 11.8	71.4 ± 12.3

According to the test results, the accuracy of the ADC map is 69.2%, and the accuracy of DCE-MRI is 66.7%. DCE-MRI and the calculated ADC map predict that micro vascular invasion is not as high as the predicted tumor classification, but there are certain effects which are also important for understanding the probability of postoperative recurrence of tumors. Two image modal fusions were conducted, but did not lead to an improvement in accuracy. In subsequent research, better fusion methods and better feature extraction techniques should be identified to improve the accuracy of micro vascular invasion prediction.

4 Conclusion

Image slices were converted into three-dimensional bodies from hospitals' medical image data. They were generated more tumor images slices from different angles for Deep Learning by the Matlab image processing program. DWI images were processed to generate a new ADC map for learning and training. The amount of texture features was manually extracted for traditional machine learning method.

Tumor grade classification was studied by texture features, and the accuracy of tumor classification by DCE-MRI images and DWI images was obtained. Multi-modality image fusion features were used in the Logistic Regression Model and the Generalized Linear Mixture Model. The accuracy rate after fusion is 87.8%, which is higher than the 80.5% predicted by the DCE-MRI image alone and the 70.7% predicted by the ADC map. This indicates that multi-modality images fusion features have improved the accuracy in tumor grade classification.

Deep Learning was used in different tests between single-modality and multi- modality fusion. The accuracy of the individual DCE-MRI images was 83.8%, and the accuracy of the ADC map was 73.7%. The accuracy of multi-modality image fusion features was 90.5%. This demonstrates that Deep Learning can extract more effective image information autonomously, thus improving the accuracy of prediction.

The recurrence rate of liver cancer is closely related to micro vascular invasion. Therefore, studies were conducted into whether the image predicts the presence of micro vascular invasion in patients. Results showed that the accuracy of ADC map prediction obtained by the DWI image was 69.2%. The prediction accuracy of DCE- MRI images was 66.7%. This indicates that medical images of liver cancer also have a certain effect on the prediction of micro vascular invasion.

Although this study increased the volume of data by translation or transforming, the available data was still too small to fully reflect the overall situation of hepatocellular carcinoma. In follow-up studies, the volume of data can be increased through closer co-operation with hospitals to further expand the sample sizes. Other more appropriate image processing techniques or new fusion methods could also be used to study the existence of micro vascular invasion. In addition, the use of manual texture features for prediction by Deep Learning may further improve the accuracy of prediction results [16].

All the training data in this study was manually labeled, but on later research, when samples are only partially labeled, we can consider to use the method which is not fully supervised [17]. Luke Metz et al. [18] proposed a deep convolutional generated confrontation network model for unsupervised learning, which could also be one of the directions of future research. In addition, the use of migration learning to study the characteristics of medical images is also an effective method. Many researchers have studied various types of images and constructed many suitable network structures [19], which could be used in medical images of liver cancer to improve prediction accuracy.

Acknowledgement. This study was partially supported by 2017 Chinese National Key Research and Development Program "Cloud Computing and Big Data" Special Program(2017YFB1002302); Youth Innovative Project of High Level Universities in Guangdong Province (2016KQNCX024).

References

1. Lv, G., Chen, L., Wang, H.: Current status and prospects of liver cancer research in China. Department of Life, **27**(03), 237–248 (2015)
2. Wufan, C.: Research progress and future trend of digital medical imaging. China Basic Sci. **16**(05), 21–28 (2014)
3. Xu, Y., Mo, T., Feng, Q., et al.: Deep learning of feature representation with multiple instance learning for medical image analysis. In: IEEE International Conference on Acoustics, Speech and Signal Processing, pp. 1626–1630. IEEE (2014)
4. Liming, X., Jian, S., Rongguo, Z., et al.: Application of in-depth learning technology in the field of mdical imaging. Union Med. J. **1**, 10–14 (2018)
5. Jin, Y., Li, J.: Advances in clinical related factors and molecular markers of microvascular invasion by hepatocellular carcinoma cells. Clin. Hepatobiliary Diseases, **29**(07), 550–553 (2013)
6. Du, M., Chen, L., Zhao, J., et al.: Microvascular invasion (MVI) is a poorer prognostic predictor for small hepaocellular carcinoma. BMC Cancer **14**(1), 1–7 (2014)
7. Zheng, Z., Dong, G., Wang, Q., et al.: Diagnostic vaue of DWI before and after contrast enhancement in grading of brain astrocytic tumors. J. Med. Imag. **21**(6), 797–800 (2011)
8. Peng, Y., Jiang, Y., Yang, C., et al.: Quantitative analysis of multiparametric prostate MR images: differentiation between prostate cancer and normal tissue and correlati on with Gleason score–a computer-aided diagnosis development study. Radiology, **267**(3), 787 (2013)
9. Rouzi, Y., Wang, Y., et al.: Differentiation of hepatic space-occupying lesions by DCE-MRI. Chinese J. Clin. Med. Imag. **26**, 4 (2015)
10. He, J., Xu, Z., Hou, S.: Clinical application of quantitative analysis of diffusion-weighted imaging in differentiating histopathological types of small hepatocellular carcinoma. J. Clin. Radiol. **36**(10), 52 (2017)
11. Wen-Bin, J.: Diagnostic value of DWI combined with DCE-MRI in focal nodular lesions of the liver. Chinese J. CT and MRI (2018)
12. Hoang, D.A., Melodelima, C., Souchon, R., et al.: Quantitative analysis of prostate M ulti-parametric MR images for detection of aggressive prostate cancer in the peri pheral zone: a multiple imager study. Radiology, **280**(1), 151406 (2016)
13. Renzulli, M., Brocchi, S., Cucchetti, A., et al.: Can current preoperative imaging be used to detect microvascular invasion of hepatocellular carcinoma? Radiology, **279**(2), 150998 (2015)
14. Jia, Y., et al.: Convolutional Architecture for Fast Feature Embedding arXiv preprint arXiv: 1408.5093(2014)
15. Sangchixue: Preliminary study on the relationship between MR-DWI imaging manifestations of acute ischemic stroke and TCM syndromes. Guangzhou University of Traditional Chinese Medicine (2007)

16. Wang, Q., Zhang, L., Xie, Y., Zheng, H., Zhou, W.: Malignancy characterization of hepatocellular carcinoma using hybrid texture and deep feature. In: Proceedings 24th IEEE International Conference Image Process, pp. 4162–4166 (2017)
17. Zhou, Z.H.A.: brief introduction to weakly supervised learning. Nat. Sci. Rev. **56**(1), 205 (2018)
18. Salimans, T., Goodfellow, I., Zaremba, W., et al.: Improved Techniques for Training GANs (2016)
19. Pan, S.J., Yang, Q.: A survey on transfer learning. IEEE Trans. Knowl. Data Eng. **22**(10), 1345–1359 (2010)

DSNPCMF: Predicting MiRNA-Disease Associations with Collaborative Matrix Factorization Based on Double Sparse and Nearest Profile

Meng-Meng Yin[1], Zhen Cui[1], Jin-Xing Liu[1], Ying-Lian Gao[2(✉)], and Xiang-Zhen Kong[1]

[1] School of Information Science and Engineering, Qufu Normal University, Rizhao, China
yinmengmeng00@163.com, {cuizhensdws,sdcavell}@126.com,
kongxzhen@163.com
[2] Library of Qufu Normal University, Qufu Normal University, Rizhao, China
yinliangao@126.com

Abstract. Lately, on account of being associated with many human diseases, more and more attentions are paid to microRNAs (miRNAs). Accumulating experimental studies of predicting novel miRNA-disease associations (MDAs) are costly and time-consuming. And there are also many unknown associations between miRNAs and diseases. Therefore, it is a momentous topic to predict possible associations between miRNAs and diseases. Also, it is urgent to increase the accuracy of predictive performance. In this paper, we put forward a computation method of Predicting MiRNA-Disease Associations with Collaborative Matrix Factorization based on Double Sparse and Nearest Profile (DSNPCMF) to estimate underlying miRNA-disease associations. In this model, we integrate Nearest Profile (NP) and Gaussian Interaction Profile (GIP) kernels of miRNAs and diseases to augment information of their neighbors and kernel similarities to improve the predictive ability. In addition, $L_{2,1}$-norm and L_1-norm are introduced into this method to increase the sparseness. Then five-fold cross validation is used for assessing our developed method. At the same time, simulation experiment is used to detect the result of prediction, including both known MDAs and new MDAs that are in descending order. In the end, the results prove that the accuracy of our prediction is better than other previous perfect methods. And our method has the ability to predict latent associations of miRNAs and diseases.

Keywords: miRNA-disease association prediction · Nearest profile · $L_{2,1}$-norm and L_1-norm · Collaborative Matrix Factorization

1 Introduction

MicroRNAs (MiRNAs) are short non-coding RNAs whose length is between 20 and 25 nucleotides. Scientists have tried many various computational methods and experimentally computational models to detect thousands of miRNAs in these recent years since

© Springer Nature Singapore Pte Ltd. 2020
H. Han et al. (Eds.): IDMB 2019, CCIS 1099, pp. 196–208, 2020.
https://doi.org/10.1007/978-981-15-8760-3_14

the first two miRNAs were found [1–4]. Furthermore, many existing evidences have proved that miRNAs play a critical role in important biological processes of cells, such as cell proliferation [5], development [6], and so on. So there is no question that miR-NAs are closely associated with the development, progression, and prognosis of human diseases [7–11]. Hence, it is urgent to develop new computational models to detect novel potential miRNA-disease associations (MDAs).

On the basis of the assumption that miRNAs with similar functions will be related to diseases with similar phenotypes [12–16], plenty of researchers have created many various calculation models to predict novel MDAs. Shi *et al.* proposed a method that used random walk on protein-protein interaction (PPI) to find functional MDAs [17]. Although it is a good model, it also has some non-negligible disadvantages. The method so tightly links with miRNA-target interactions that this would generate a high false positive. In terms of this reason, Chen *et al.* put forward a new method to figure out this problem. Random Walk with Restart for MiRNA-disease Association (RWRMDA) was to make all miRNAs map to a miRNA functional similarity network [18]. Mork *et al.* developed a method based on miRNA-protein-disease (miRPD) associations by considering the information of protein [19]. The method uses protein-disease interactions and protein-miRNA interactions to predict new miRNAs and proteins related to diseases. Based on the assumption that diseases would interact with these miRNAs whose target genes are related to this disease, Xuan *et al.* developed a prediction algorithm based on weighted k most similar neighbors named HDMP which combined miRNA similarity and the distribution of these disease-related miRNAs in neighbors to predict potential MDAs [20]. They integrated disease phenotype similarity, disease semantic similarity and known MDAs in the miRNA functional matrix used in HDMP. It is worth noting that HDMP assigned higher weights to members in the same miRNA cluster. But HDMP cannot be used for the novel disease with unknown associated miRNAs.

In this paper, we develop an improved model named Collaborative Matrix Factorization based on Double Sparse and Nearest Profile (DSNPCMF). This model can predict new MDAs with the help of known MDAs. It is worthy of pointing that it is different from traditional matrix factorization. In this method, we add neighbor information into used networks of miRNAs and diseases for increasing the accuracy of the performance of this model. Therefore, the Nearest Profile (NP) [21] is added to Collaborative Matrix Factorization (CMF) which is beneficial to the experimental results. A lot of missing associations will do harm to the performance of each predictive method. There, the NP will help improve this deficiency and increase the neighbor information of each miRNA and disease. In order to increase the robustness and the sparseness of miRNA functional similarity network and disease semantic similarity matrix, an $L_{2,1}$-norm and an L_1-norm are added into the objective function. There, these norms are used for avoiding overfitting and eliminate the unattached pairs. Also, we augment Gaussian Interaction Profile (GIP) kernel into networks to obtain the network information of miRNA and disease. Coupled with this, Weighted K Nearest Known Neighbors (WKNKN) is used as a pre-processing step [22]. In the end, five-fold cross validation is implemented to evaluate the experimental results of our own method. In addition, we conduct a simulation experiment to predict new MDAs.

Section 2 and Sect. 3 describes materials and methods, including dataset and the description of our method. We produce our results and some analysis in Sect. 4. And Sect. 5 offers conclusions about this paper to readers.

2 Materials

2.1 Human MiRNA-Disease Association

In this paper, the human miRNA-disease association dataset is obtained from HMDD v2.0 [23], which is usually used as gold standard dataset. The dataset contains 5430 miRNA-disease associations, 495 miRNAs and 383 diseases. Adjacency matrix $\mathbf{Y}$ represents known associations between miRNAs and diseases. In this matrix, $\mathbf{Y}(m(i), d(j))$ is equal to 1 if miRNA $m(i)$ is related to disease $d(j)$, otherwise 0. The details of the dataset are listed in Table 1.

Table 1. MiRNAs, Diseases, Associations in Gold Standard Dataset

Datasets	MiRNAs	Diseases	Associations
Gold Standard Dataset	495	383	5430

2.2 MiRNA Functional Similarity

Wang et al. [13] proposed a calculation method of miRNA functional similarity based on the hypothesis that miRNAs that have similar functions will be associated with similar diseases and vice versa [11–14, 20]. The miRNA functional similarity is obtained from http://www.cuilab.cn/files/images/cuilab/misim.zip. We establish a matrix $\mathbf{S}_m$ to represent the miRNA functional similarity network. And in this matrix, $\mathbf{S}_m(m(i), m(j))$ is the functional similarity score between miRNA $m(i)$ and $m(j)$. Furthermore, the self-similarity of a miRNA is 1, so the elements on the diagonal in the matrix $\mathbf{S}_m$ are 1.

2.3 Disease Semantic Similarity

In this paper, Directed Acyclic Graph (DAG) is applied to represent the relationships among various diseases. And $DAG(D_{dis}) = (D_{dis}, T(D_{dis}), E(D_{dis}))$ is used to represent disease D_{dis}, where $T(D_{dis})$ is the node set of all ancestor nodes of D_{dis} and D_{dis} itself, $E(D_{dis})$ represents the corresponding link set including the direct edges from parent nodes to child nodes.

The semantic value of disease D_{dis} can be defined as follows:

$$DSV(D_{dis}) = \sum_{d \in T(D_{dis})} DC_{D_{dis}}(d), \tag{1}$$

$$DC_{D_{dis}}(d) = \begin{cases} 1 & \text{if } d = D_{dis} \\ \max\left\{ \Delta * DC_{D_{dis}}\left(d'\right) \middle| d' \in children\ of\ d \right\} & \text{if } d \neq D_{dis} \end{cases}, \tag{2}$$

where Δ is the semantic contribution factor and $DC_{D_{dis}}(d)$ represents the contribution of disease d in $DAG(D_{dis})$ to the semantic value of D_{dis}. Particularly speaking, the contribution value of disease D_{dis} to the semantic value of itself is 1 and the contribution will decrease as the distance between disease D_{dis} and other diseases increase. Therefore, when the distance between two different diseases and disease D_{dis} is same, the contribution value will be same.

Two diseases will have greater similarity score, which have lager shared part of their own DAGs. So, the formula of the disease semantic similarity score between diseases $d(i)$ and $d(j)$ is described as follows:

$$S_d(d(i), d(j)) = \frac{\sum_{t \in T(d(i)) \cap T(d(j))} \left(DC_{d(i)}(t) + DC_{d(j)}(t) \right)}{DSV(d(i)) + DSV(d(j))}. \tag{3}$$

Similar to $\mathbf{S}_m$, the diagonal elements in the matrix $\mathbf{S}_d$ are 1.

3 Method

3.1 Related Work

Gaussian Interaction Profile Kernel Similarity. Gaussian interaction profile kernel similarity is based on the assumption that similar diseases tend to be related to functional similar miRNAs [11–14]. So Gaussian interaction profile kernel similarity is used for representing the topological structure of known miRNA-disease association network. The network similarity is calculated as follows:

$$GIP_{miRNA}(m(i), m(j)) = \exp\left(-\gamma_m \|\mathbf{Y}(m(i)) - \mathbf{Y}(m(j))\|^2\right), \tag{4}$$

$$GIP_{disease}(d(i), d(j)) = \exp\left(-\gamma_d \|\mathbf{Y}(d(i)) - \mathbf{Y}(d(j))\|^2\right), \tag{5}$$

$$\gamma_m = 1 \left/ \frac{1}{m} \sum_{i=1}^{m} \|\mathbf{Y}(m(i))\|^2, \right. \tag{6}$$

$$\gamma_d = 1 \left/ \frac{1}{d} \sum_{i=1}^{d} \|\mathbf{Y}(d(i))\|^2, \right. \tag{7}$$

where γ_m and γ_d is a parameter that adjusts the kernel bandwidth, $m(i)$ and $m(j)$ are two miRNAs, and $d(i)$ and $d(j)$ are two diseases. In addition, $\mathbf{Y}(m(i))$ and $\mathbf{Y}(m(j))$ represent the interaction profiles of $m(i)$ and $m(j)$, respectively. Similarly, the association profiles of $d(i)$ and $d(j)$ are described as $\mathbf{Y}(d(i))$ and $\mathbf{Y}(d(j))$, respectively. And in Eqs. (6) and (7), m is the number of rows of $\mathbf{Y}$ and d is the number of columns of $\mathbf{Y}$. We integrate $\mathbf{S}_m$ into $\mathbf{K}_m$ to represent the miRNA network similarity matrix. Also, $\mathbf{K}_d$ integrating $\mathbf{S}_d$ is used to represent the disease network similarity matrix. The formulas are as follows:

$$\mathbf{K}_m = \alpha \mathbf{S}_m + (1 - \alpha) GIP_{miRNA}, \tag{8}$$

$$K_d = \alpha S_d + (1 - \alpha) GIP_{disease}, \tag{9}$$

where α is an adjustable parameter which is between 0 and 1. Here, miRNA kernel K_m is the combination of the miRNA functional similarity matrix S_m and the miRNA network similarity matrix GIP_{miRNA}. Disease kernel K_d is similar to K_m.

Nearest Profile. According to the prior study [21], Nearest Profile is an independent prediction model. It can predict latent miRNA-disease associations well. In this paper, based on the assumption that the nearest neighbor information will affect the final results, NP is proposed to process miRNA network similarity matrix K_m and disease network similarity matrix K_d. For a new miRNA $m(i)$ and a new disease $d(j)$, The NP can be calculated as:

$$N_m(m(i)) = K_m(m(i), m(nearest)) \times Y(m(nearest)), \tag{10}$$

$$N_d(d(j)) = K_d(d(j), d(nearest)) \times Y(d(nearest)), \tag{11}$$

where, $m(nearest)$ is the miRNA most similar to $m(i)$, $d(nearest)$ is the disease most similar to $d(j)$, $N_m(m(i))$ is the nearest interaction profile of $m(i)$ and $N_d(d(j))$ is the nearest association profile of $d(j)$.

The process of NP can be performed in four steps. Firstly, the self-similarity in the matrices K_m and K_d is eliminated. Secondly, each miRNA's nearest neighbor and each disease's nearest neighbor will be obtained. Thirdly, all miRNA similarities and disease similarities are reset to 0. Finally, the miRNA nearest neighbor matrix N_m based on K_m and the disease nearest neighbor matrix N_d based on K_d can be obtained.

CMF Algorithm. The traditional CMF is a reliable computational model to predict new MDAs [24]. The objective function of CMF is defined as:

$$\min_{A,B} = \left\| Y - AB^T \right\|_F^2 + \lambda_l \left(\|A\|_F^2 + \|B\|_F^2 \right) + \lambda_m \left\| S_m - AA^T \right\|_F^2 + \lambda_d \left\| S_d - BB^T \right\|_F^2, \tag{12}$$

where $\|\cdot\|_F$ is Frobenius norm, λ_l, λ_m and λ_d are non-negative parameters. And Y is the known association matrix. Here, the adjacency matrix Y is decomposed into two low-rank matrices A and B. And $Y \approx AB^T$. The CMF method uses regularization terms to make sure that the potential feature vectors of similar miRNAs and similar diseases are similar and vice versa [22]. And $S_m \approx AA^T$ and $S_d \approx BB^T$.

3.2 DSNPCMF for MDAs Prediction

In this study, we develop a computational method of DSNPCMF based on CMF. Because CMF ignores the network information of miRNAs and diseases, we combined GIP into CMF. Besides, the aim of using NP is to process miRNA network similarity network K_m and disease network similarity matrix K_d. We will obtain the NP of miRNAs and diseases respectively. Next, we make the NP of miRNAs and diseases integrate with miRNA kernel and disease kernel respectively. Finally, the two matrices $\mathbf{N}_m$ and $\mathbf{N}_d$ are used to replace $\mathbf{S}_m$ and $\mathbf{S}_d$ in the objective function of the CMF method.

Moreover, an $L_{2,1}$-norm and an L_1-norm are added to increase the sparse of matrices $\mathbf{A}$ and $\mathbf{B}$. Therefore, the objective function of DSNPCMF is written as:

$$
\begin{aligned}
\min_{\mathbf{A},\mathbf{B}} = {} & \left\| \mathbf{Y} - \mathbf{A}\mathbf{B}^T \right\|_F^2 + \lambda_l \left(\|\mathbf{A}\|_F^2 + \|\mathbf{B}\|_F^2 \right) + \lambda_l \left(\|\mathbf{A}\|_{2,1} + \|\mathbf{B}\|_1 \right) \\
& + \lambda_m \left\| \mathbf{N}_m - \mathbf{A}\mathbf{A}^T \right\|_F^2 + \lambda_d \left\| \mathbf{N}_d - \mathbf{B}\mathbf{B}^T \right\|_F^2,
\end{aligned}
\tag{13}
$$

where $\|\cdot\|_F$ is Frobenius norm, λ_l, λ_m and λ_d are non-negative parameters. Values of λ_l, λ_m and λ_d are ensured by cross validation and $\lambda_l \in \{2^{-2}, 2^{-1}, 2^0, 2^1\}$, $\lambda_m/\lambda_d \in \{2^{-3}, 2^{-2}, 2^{-1}, 2^0, 2^1, 2^2, 2^3, 2^4, 2^5\}$. The first term is the approximate model of $\mathbf{Y}$. The next term is the Tikhonov regularization term to minimize the norms of $\mathbf{A}$ and $\mathbf{B}$. In the third term, we increase the sparseness of matrix $\mathbf{A}$ and matrix $\mathbf{B}$ by using the $L_{2,1}$-norm and L_1-norm. The last two terms are regularization terms to minimize the squared error between $\mathbf{N}_m(\mathbf{N}_d)$ and $\mathbf{A}\mathbf{A}^T(\mathbf{B}\mathbf{B}^T)$.

3.3 Initialization of A and B

Initializing $\mathbf{A}$ and $\mathbf{B}$ is the first step in the CMF method. Specially, we decompose $\mathbf{Y}$ into $\mathbf{A}$ and $\mathbf{B}$ by adopting the singular value decomposition (SVD) method. The specific formula is as follows:

$$
[\mathbf{U}, \mathbf{S}, \mathbf{V}] = SVD(\mathbf{Y}, k), \mathbf{A} = \mathbf{U}\mathbf{S}_k^{1/2}, \mathbf{B} = \mathbf{V}\mathbf{S}_k^{1/2},
\tag{14}
$$

where $\mathbf{S}_k$ is a diagonal matrix including the k singular values.

3.4 Optimization

In this paper, the least squares method is used for updating $\mathbf{A}$ and $\mathbf{B}$. We update $\mathbf{A}$ and $\mathbf{B}$ until convergence. L is used to represent the Eq. (13). Then set $\partial L / \partial \mathbf{A}$ and $\partial L / \partial \mathbf{B}$ to be 0 to solve the expression of $\mathbf{A}$ and $\mathbf{B}$. In addition, we adopt five-fold cross validation to obtain the optimal values of λ_l, λ_m and λ_d. The update rules are shown below:

$$
\mathbf{A} = (\mathbf{Y}\mathbf{B} + \lambda_m \mathbf{N}_m \mathbf{A}) \left(\mathbf{B}^T \mathbf{B} + \lambda_l \mathbf{I}_k + \lambda_l \mathbf{D}_{21} \mathbf{I}_k + \lambda_m \mathbf{A}^T \mathbf{A} \right)^{-1},
\tag{15}
$$

$$
\mathbf{B} = \left(\mathbf{Y}^T \mathbf{A} + \lambda_d \mathbf{N}_d \mathbf{B} \right) \left(\mathbf{A}^T \mathbf{A} + \lambda_l \mathbf{I}_k + \lambda_l \mathbf{D}_1 \mathbf{I}_k + \lambda_d \mathbf{B}^T \mathbf{B} \right)^{-1}.
\tag{16}
$$

Here, $\mathbf{I}_k$ is $k \times k$ identity matrix, $\mathbf{D}_{21}$ is a diagonal matrix whose i-th diagonal element is $d_{ii} = 1/2 \|(\mathbf{A})^i\|_2$ and $\mathbf{D}_1$ is a diagonal matrix whose i-th diagonal element is $d_{ii} = 1/2 |\mathbf{B}_{ij}|$.

Specially, an iterative experiment is constructed here. The result of this experiment verifies that $\mathbf{A}$ and $\mathbf{B}$ come to converge, when the number of iterations reaches 20.

The appearance is shown in Fig. 1.

Therefore, the algorithm of DSNPCMF is shown as follows:

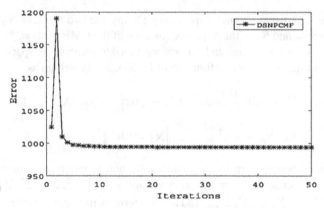

Fig. 1. The Analysis of Convergence

Algorithm 1:DSNPCMF

Input: MDA matrix $\mathbf{Y} \in R^{m \times d}$, miRNA similarity $\mathbf{S}_m$, and disease similarity $\mathbf{S}_d$.

Output: prediction score matrix $\hat{\mathbf{Y}}$

Parameters: $K, p, \lambda_l, \lambda_m, \lambda_d$

Pre-processing: $\mathbf{Y} = WKNKN(\mathbf{Y}, \mathbf{S}_m, \mathbf{S}_d, K, p)$, $\mathbf{S}_m \rightarrow \mathbf{K}_m \rightarrow \mathbf{N}_m$, $\mathbf{S}_d \rightarrow \mathbf{K}_d \rightarrow \mathbf{N}_d$

Initialization: $[\mathbf{U}, \mathbf{S}, \mathbf{V}] = SVD(\mathbf{Y}, k), \mathbf{A} = \mathbf{US}_k^{1/2}, \mathbf{B} = \mathbf{VS}_k^{1/2}$

Repeat

Update $\mathbf{A}$ using Eq. (15).

Update $\mathbf{B}$ using Eq. (16).

Until convergence

$\hat{\mathbf{Y}} = \mathbf{AB}^T$

Return $\mathbf{Y}$

---

4   Results and Discussions

4.1   Five-Fold Cross Validation

Five-fold cross validation is implemented to evaluate the predictive ability of DSNPCMF based on known miRNA-disease associations which are obtained from HMDD v2.0 [23]. Five-fold cross validation tends to randomly divide the MDAs dataset into five subsets, four of which are training sets and the remaining one is testing set. What's more, to eliminate unknown associations, the role of WKNKN lies in preprocessing the matrix $\mathbf{Y}$. At the same time, the NP of miRNAs and diseases help know more specific information to be good for the performance of DSNPCMF. Here, we run 5-fold cross validation for 10 times and adopt the average value as the final result.

In this paper, the Area Under the Curve (AUC) value is used for indicating the performance of this method. The AUC is the area under the Receiver Operating Characteristic (ROC) curve. In this final result, the AUC is selected as the evaluating indicator. The value of this area under ROC curve is no more than 1 and it is between 0 and 1. The predictive performance is low when the value is 0.5.

4.2 Comparison Among These Methods

We compare our method with five state-of-art methods CMF [24], HDMP [25], WBSMDA [26], HAMDA [27], and ELLPMDA [28]. And these results of five methods are obtained from their own studies. Table 2 shows the results. The final result is the average of 10 five-fold cross validations.

Table 2. AUC Results of Cross Validation Experiments

Methods	Gold Standard Dataset
WBSMDA	0.8185(0.0009)
HDMP	0.8342(0.0010)
CMF	0.8697(0.0011)
HAMDA	0.8965(0.0012)
ELLPMDA	0.9193(0.0002)
DSNPCMF	**0.9407(0.0070)**

As listed in Table 2, the average AUCs of WBSMDA, HDMP, CMF, HAMDA, ELLPMDA and DSNPCMF on this gold standard dataset are 0.8185 ± 0.0009, 0.8342 ± 0.0010, 0.8697 ± 0.0011, 0.8965 ± 0.0012, 0.9193 ± 0.0002 and 0.9407 ± 0.0070. The highest value is in bold and the standard deviation is in brackets.

Here, we can see from Table 2 that the AUC of our method DSNPCMF is 12.22% higher than the lowest AUC value of WBAMDA. And our model is 2.14% higher than the state-of-art method called ELLPMDA. Thus it can be seen that our proposed method is better and more reliable than other developed methods.

4.3 Case Studies

Here, an experiment is applied to obtain the final prediction score matrix to make the effect of our method obvious. And three diseases are chosen to describe as detailed below, containing Colon Neoplasms, Brain Neoplasms and Adrenocortical Carcinoma. By seeking from two databases, dbDEMC and miR2Disease, predicted results can be confirmed. And in the final prediction matrix, the corresponding miRNAs will not be selected when this value is less than 0.3. And in the following article, a certain number of miRNAs are selected from the miRNAs with a value greater than 0.3.

The first one is Colon Neoplasms. In the gold standard dataset, there are five known associations between miRNAs and Colon Neoplasms. We choose the first ten here. Four known associations are successfully predicted by our own method. Also, there are six other novel associations predicted. Apart from miR-125b, five other new associations can be found from both dbDEMC and miR2Disease. Thus it can be seen that miR-21, miR-143, miR-106a, miR-20a and miR-155 are strongly interacted with Colon Neoplasms. In addition, although it is only found by dbDEMC, miR-125b also has a close association with Colon Neoplasms. For instance, since no study has evaluated the association between miRNA expression patterns and colon cancer prognosis or therapeutic outcome, Schetter *et al.* constructed an experiment to verify the association between miRNA and colon cancer in 2008 [29]. Then, in 2014, Drusco *et al.* found five differentially expressed microRNAs containing miR-21 and compared to normal, miR-21 was observed to be upregulated in primary tumors [30]. The specific experimental results are listed in Table 3. And the known associations are in bold.

Table 3. Predicted MiRNAs for Colon Neoplasms

Rank	miRNA	Evidence
1	**hsa-mir-1**	Known
2	**hsa-mir-126**	Known
3	**hsa-mir-145**	Known
4	**hsa-mir-17**	Known
5	hsa-mir-21	dbDEMC;miR2Disease
6	hsa-mir-143	dbDEMC;miR2Disease
7	hsa-mir-106a	dbDEMC;miR2Disease
8	hsa-mir-20a	dbDEMC;miR2Disease
9	hsa-mir-125b	dbDEMC
10	hsa-mir-155	dbDEMC;miR2Disease

The second case is Brain Neoplasms. The dataset has eleven known associations and all known associations are predicted. Since there are also many other associations with Brain Neoplasms, we only select the top 20 miRNAs in Table 4 to show our experimental results. Except these known associations, nine new miRNAs are detected and only five of them can be obtained their verified information from dbDEMC. For example, Saydam *et al.* figured out that miR-98 of the overexpressed miRNAs was the most upregulated [31]. For these unconfirmed miRNAs, miR-92a, was confirmed that it was related to Kidney Cancer [32] and Breast Cancer [33]. Although these four miRNAs, including miR-92a, miR-320d and miR-7e, cannot be verified by dbDEMC and miR2Disease, they are tightly related to Brain Neoplasms according to our experimental results. The result is listed as follows in Table 4. The known associations are in bold.

The final term is Adrenocortical Carcinoma. There are 53 miRNAs associated with Adrenocortical Carcinoma from the dataset used in our experiment. We list the top 60 miRNAs on the basis of the association score from the final prediction matrix. In the top 60 miRNAs, all known associations are successfully predicted and it also contains

Table 4. Predicted MiRNAs for Brain Neoplasms

Rank	miRNA	Evidence
1	**hsa-mir-1**	Known
2	**hsa-mir-21**	Known
3	**hsa-mir-9**	Known
4	**hsa-mir-32**	Known
5	**hsa-mir-22**	Known
6	**hsa-mir-34a**	Known
7	**hsa-mir-222**	Known
8	**hsa-mir-92**	known
9	**hsa-mir-129**	Known
10	**hsa-mir-221**	Known
11	**hsa-mir-92b**	Known
12	hsa-mir-92a	Unconfirmed
13	hsa-miR-326	dbDEMC
14	hsa-mir-132	dbDEMC
15	hsa-mir-320d	Unconfirmed
16	hsa-mir-7e	Unconfirmed
17	hsa-mir-98	dbDEMC
18	hsa-mir-373	dbDEMC
19	hsa-mir-20b	dbDEMC
20	hsa-mir-498	Unconfirmed

7 unconfirmed miRNAs, containing miR-221, miR-20a, miR-200a, miR-146b, miR-9, miR-7f and miR-141. Among these unconfirmed miRNAs, miR-221 has the highest predictive score. For miR-221, scientists have verified that it has associations with many diseases, such as Colon Cancer [34], Brain Cancer [35]. Table 5 lists the detailed information. The known associations are in bold.

Table 5. Predicted MiRNAs for Adrenocortical Carcinoma

Rank	miRNA	Evidence	Rank	miRNA	Evidence
1	**hsa-mir-1**	Known	**31**	**hsa-mir-7b**	Known
2	**hsa-mir-203**	Known	**32**	**hsa-mir-335**	Known
3	**hsa-mir-21**	Known	**33**	**hsa-mir-23b**	Known
4	**hsa-mir-22**	Known	**34**	**hsa-mir-17**	Known
5	**hsa-mir-210**	Known	**35**	**hsa-mir-449a**	Known
6	**hsa-mir-200b**	Known	**36**	**hsa-mir-196a**	Known
7	**hsa-mir-7a**	Known	**37**	**hsa-mir-132**	Known
8	**hsa-mir-126**	Known	**38**	**hsa-mir-137**	Known
9	**hsa-mir-145**	Known	**39**	**hsa-mir-135a**	Known
10	**hsa-mir-200**	Known	**40**	**hsa-mir-484**	Known
11	**hsa-mir-143**	Known	**41**	**hsa-mir-449b**	Known
12	**hsa-mir-155**	Known	**42**	**hsa-mir-7d**	Known
13	**hsa-mir-375**	Known	**43**	**hsa-mir-491**	Known
14	**hsa-mir-146a**	Known	**44**	**hsa-mir-133a**	Known
15	**hsa-mir-16**	Known	**45**	**hsa-mir-376c**	Known
16	**hsa-mir-125b**	Known	**46**	**hsa-mir-376a**	Known
17	**hsa-mir-130a**	Known	**47**	**hsa-mir-139**	Known
18	**hsa-mir-7**	Known	**48**	**hsa-mir-194**	Known
19	**hsa-mir-192**	Known	**49**	**hsa-mir-424**	Known
20	**hsa-mir-7c**	Known	**50**	**hsa-mir-301b**	Known
21	**hsa-mir-30a**	Known	**51**	**hsa-mir-376b**	Known
22	**hsa-mir-15b**	Known	**52**	**hsa-mir-483**	Known
23	**hsa-mir-106b**	Known	**53**	**hsa-mir-675**	Known
24	**hsa-mir-100**	Known	54	hsa-mir-221	Unconfirmed
25	**hsa-mir-200c**	Known	55	hsa-mir-20a	Unconfirmed
26	**hsa-mir-301a**	Known	56	hsa-mir-200a	Unconfirmed
27	**hsa-mir-148a**	Known	57	hsa-mir-146b	Unconfirmed
28	**hsa-mir-28**	Known	58	hsa-mir-9	Unconfirmed
29	**hsa-mir-222**	Known	59	hsa-mir-7f	Unconfirmed
30	**hsa-mir-195**	Known	60	hsa-mir-141	Unconfirmed

5   Conclusion

In this paper, a new method based on Nearest Profile and double sparse is proposed to predict novel miRNA-disease associations. The $L_{2,1}$-norm and L_1-norm are introduced into this method to increase sparseness and the nearest neighbors of miRNAs and diseases respectively are described into this method when predicting novel MDAs. In addition, the GIP kernel is applied to this method to obtain the kernel similarity of miRNAs and diseases to be beneficial to the accuracy of this method. Also, WKNKN is used as the preprocessing step to get the known nearest neighbors of miRNAs and diseases respectively. And the AUC is a reliable and significant metric to evaluate the performance of our own method.

In the short run, more and more predictive model will be proposed and more novel miRNAs associated with different diseases will be found out. It is a better expectation that more available datasets can be applied to pre-diction of MDAs. What's more, the most vital affair is that our method DSNPCMF could contribute to prediction of fresh MDAs and its performance is better than other existing models.

Acknowledgment. This work was supported in part by the grants of the National Science Foundation of China, Nos. 61872220, and 61572284.

References

1. Ambros, V.: microRNAs: tiny regulators with great potential. Cell **107**, 823–826 (2001)
2. Bartel, D.P.: MicroRNAs: genomics, biogenesis, mechanism, and function. Cell **116**, 281–297 (2004)
3. Feng, C.M., Xu, Y., Liu, J.X., Gao, Y.L., Zheng, C.H.: Supervised discriminative sparse PCA for corn-characteristic gene selection and tumor classification on multiview biological data. IEEE Trans. Neural Netw. Learn. Syst. **30**, 2926–2937 (2019)
4. Victor, A.: The functions of animal microRNAs. Nature **431**, 350–355 (2004)
5. Angie, M.C., Mike, W.B., Jeffrey, S., Lance P.F.: Antisense inhibition of human miRNAs and indications for an involvement of miRNA in cell growth and apoptosis. Nucleic Acids Res. **33**, 1290–1297 (2005)
6. Karp, X., Ambros, V.: Encountering MicroRNAs in cell fate signaling. Science **310**, 1288 (2005)
7. Ines, A.G., Miska, E.A.: MicroRNA functions in animal development and human disease. Development **132**, 4653–4662 (2005)
8. Meola, N., Gennarino, V.A., Banfi, S.: microRNAs and genetic diseases. Pathogenetics **2**, 1–14 (2009)
9. Lynam-Lennon, N., Maher, S.G., Reynolds, J.V.: The roles of microRNA in cancer and apoptosis. Biological Rev. **84**, 55–71 (2009)
10. Liu, J.X., Feng, C.M., Kong, X.Z., Xu, Y.: Dual graph-laplacian PCA: a closed-form solution for bi-clustering to find checkerboard structures on gene expression data. IEEE Access **7**, 151329–151338 (2019)
11. Lu, M., et al.: An analysis of human microRNA and disease associations. Plos One **3**, 5 (2008)
12. Bandyopadhyay, S., Mitra, R., Maulik, U., Zhang, M.Q.: Development of the human cancer microRNA network. Silence **1**, 6–16 (2010)
13. Wang, D., Wang, J., Lu, M., Song, F., Cui, Q.: Inferring the human microRNA functional similarity and functional network based on microRNA-associated diseases. Bioinformatics **26**, 1644–1650 (2010)
14. Goh, K.I., Cusick, M.E., Valle, D., Childs, B., Vidal, M., Barabasi, A.L.: The human disease network. In: Proceedings of the National Academy of Sciences of the United States of America, vol. 104, pp. 8685–8690 (2007)

15. Pasquier, C., Gardã, S.J.: Prediction of miRNA-disease associations with a vector space model. Sci. Rep. **6**, 27036 (2016)
16. Le, T.D., Zhang, J., Liu, L., Li, J.: Computational methods for identifying miRNA sponge interactions. Brief. Bioinform. **18**, 577 (2016)
17. Shi, Hongbo: Walking the interactome to identify human miRNA-disease associations;through the functional link between miRNA targets and disease genes. BMC Syst. Biol. **7**, 1–12 (2013)
18. Chen, X., Liu, M.X., Yan, G.Y.: RWRMDA: predicting novel human microRNA-disease associations. Molecular Biosyst. **8**, 2792–2798 (2012)
19. Søren, M., Sune, P.-F., Albert, P.C., Jan, G., Lars Juhl, J.: Protein-driven inference of miRNA-disease associations. Bioinformatics **30**, 392 (2014)
20. Xuan, P., et al.: Prediction of microRNAs associated with human diseases based on weighted k most similar neighbors. Plos One **8**, 15 (2013)
21. Yoshihiro, Y., Michihiro, A., Alex, G., Wataru, H., Minoru, K.: Prediction of drug-target interaction networks from the integration of chemical and genomic spaces. Bioinformatics **24**, i232–i240 (2008)
22. Ezzat, A., Zhao, P., Wu, M., Li, X., Kwoh, C.K.: Drug-target interaction prediction with graph regularized matrix factorization. IEEE/ACM Trans. Comput. Biol. Bioinform. **26**, 646–656 (2016)
23. Yang, L., et al.: HMDD v2.0: a database for experimentally supported human microRNA and disease associations. Nucleic Acids Res. **42**, D1070 (2014)
24. Shen, Z., Zhang, Y.H., Han, K., Nandi, A.K., Honig, B., Huang, D.S.: miRNA-disease association prediction with collaborative matrix factorization. Complexity **2017**, 1–9 (2017)
25. Lucherini, O.M., et al.: First report of circulating MicroRNAs in Tumour necrosis factor receptor-associated periodic syndrome (TRAPS). PLoS ONE **8**, e73443 (2013)
26. Chen, X., et al.: WBSMDA: within and between score for MiRNA-disease association prediction. Sci. Rep. **6**, 21106 (2016)
27. Chen, X., Niu, Y.W., Wang, G.H., Yan, G.Y.: HAMDA: hybrid approach for MiRNA-disease association prediction. J. Biomed. Inform. **76**, 50–58 (2017)
28. Chen, X., Zhou, Z., Ao, Y.Z.: ELLPMDA: ensemble learning and link prediction for miRNA-disease association prediction. RNA Biol. **01**, 50 (2018)
29. Schetter, A.J., Leung, S., Sohn, J.J., et al.: Microrna expression profiles associated with prognosis and therapeutic outcome in colon adenocarcinoma. JAMA **299**, 425–436 (2008)
30. Drusco, A., et al.: MicroRNA profiles discriminate among colon cancer metastasis. PLoS ONE **9**, e96670–e96670 (2014)
31. Saydam, O., et al.: Downregulated microRNA-200a in meningiomas promotes tumor growth by reducing E-cadherin and activating the Wnt/beta-catenin signaling pathway. Molecular Cellular Biol. **29**, 5923–5940 (2009)
32. Kort, E.J., Farber, L., Tretiakova, M., Petillo, D., Furge, K.A., Yang, X.J., Cornelius, A., Teh, B.T.: The E2F3-Oncomir-1 axis is activated in Wilms' tumor. Cancer Res. **68**, 4034–4038 (2008)
33. Chan, M., Liaw, C.S., Ji, S.M., Tan, H.H., Wong, C.Y., Thike, A.A., Tan, P.H., Ho, G.H., Lee, A.S.-G.: Identification of circulating MicroRNA signatures for breast cancer detection. Clin. Cancer Res. **19**, 4477 (2013)
34. Zhang, J.-X., et al.: Prognostic and predictive value of a microRNA signature in stage II colon cancer: a microRNA expression analysis. Lancet Oncol. **14**, 1295–1306 (2013)
35. Saydam, O., et al.: miRNA-7 attenuation in Schwannoma tumors stimulates growth by upregulating three oncogenic signaling pathways. Cancer Res. **71**, 852–861 (2011)

Novel Data Science Theory
and Applications

Exploring Blockchain in Speech Recognition

Xuemei Yang[1](✉) and Heming Huang[2]

[1] Xianyang Normal University, Xianyang 712000, China
yangxuemei691226@163.com
[2] Qinghai Normal University, Xining 810008, Qinghai, China
huang-heming@sohu.com

Abstract. Blockchain is changing science and technology in a revolutionary way for its decentralized, incorruptible computing mechanism. This work explores blockchain applications in speech recognition via investigating decentralized deep learning models. The decentralized deep learning models demonstrate a good potential to handle large scale acoustic data by fusing distributed deep learning models to achieve better learning results. To the best of our knowledge, it is a pioneering work to explore blockchain technologies in speech recognition.

Keywords: Blockchain · Speech recognition · Deep learning

1 Introduction

As a new technology that inspired digital currency (e.g., bitcoin), blockchain attracts more and more attentions from almost all fields for its decentralized, incorruptible computing mechanism [1, 2]. For example, IBM blockchain provides decentralized financial services to condense transaction times from hours to seconds [3]. Such decentralized financial services decrease cost by removing intermediaries and make it more affordable for more populations. At the same time, it enhances financial security because of its decentralization and incorruptible computing, because only a customer himself is the only person to know his key to his transaction account. Blockchain technology is transforming the way data is handled and stored in a revolutionary approach. Quite a lot recent work devoted to developing decentralized AI systems by integrating blockchain technologies in machine learning [1]. The integration of blockchain and machine technologies is impacting each filed in science and engineering.

Speech recognition widely employs all kinds of AI techniques (e.g. deep learning) and presents itself as an important independent application field in the whole AI and machine learning domain. Blockchain definitely will play a key role in the future of speech recognition. However, there is no literature available to address it in the existing studies. In this study, we address possible applications of blockchains in speech recognition with the expectation to answer the following questions.

What kinds of advantages can blockchain bring to modern speech recognition? How to realize blockchain in speech recognition? Can the existing decentralized machine learning models be applied to speech recognition? In other words, what theoretical and

H. Han et al. (Eds.): IDMB 2019, CCIS 1099, pp. 211–220, 2020.
https://doi.org/10.1007/978-981-15-8760-3_15

practical challenges we may need to tackle in migrating this exciting technology to modern speech recognition, and how to handle it?

In this work, we aim to answer these questions by introducing blockchain, speech recognition, decentralized machine learning, and discussing possible blockchain models in speech recognition. To the best of our knowledge, it is the first work to explore blockchain technologies in speech recognition. It will inspire future similar work in this fast-growing field, besides providing a pioneering guide.

2   The Principle of Blockchain

Blockchain allows digital information distributed along internet but not allow copies. A blockchain is a decentralized, incorruptible, distributed database without administrators, where each peer has a copy of the whole database. Each block modification or creation must be approved by a majority of peers. There is no an authority to follow in the system. It realizes on consensus protocol and cryptography algorithms besides peers' collaborations to let peers to trust each other.

A block refers to data or transaction records (e.g. contract), which is like a page in a book (blockchain). Each block consists of index and data components at least. Index, also called hash, can be viewed as the fingerprint or id of a block. It is generated by a hash function (e.g. SHA-256). Hash is used to verify the integrity of a block. If a block is modified by some people, its hash (digital signature) will be changed. Geometrically, the blockchain can be viewed as a secured "linked list". Each block is linked to its next one and contains the hash of its parent block besides its own hash.

2.1   How a Block Can Get the Copy of a Whole Blockchain?

The Merkle tree or similar techniques are employed to achieve blockchain. A Merkle tree is a binary tree to record the hashes of all transactions in the blockchain. To get the copy of a whole blockchain, it needs to build a corresponding Merkle tree. Each Merkle tree is constituted by a root node, a group of intermediate nodes, and a group of leaf nodes. The main characteristics of a Merkle tree are that each leaf node contains the stored data or its hash value, and a non-leaf node has the hash value of its two children nodes. Since the Merkle tree records all the hash values of data layer by layer, any changes in the underlying data will be transferred to its parent node, along the path to the root of the tree. Thus, the root actually represents the 'digital summary' of all data at the bottom. If one piece of data in the blockchain is modified, hash will be different.

2.2   How a Block is Created?

Blockchain is a chain of blocks. How a new block (e.g., coin) is created? It is called mining process (e.g., bitcoin mining). Who can create a new block? Only a winner can create a new block instead of everybody. The winner will create the next block: he has right to decide the contents of the next block he wants to create. The winner (miner) also needs to broadcast the new block to all the peers in the whole network so that they can verify the contents of the new block.

Blockchains use a kind of "consensus algorithms" to elect a winner. Each winner has the right to mining. The consensus algorithms create a winner in the network and reach a consensus for blockchain growth. Algorithm results can be easily verified by all peers in the network. The general consensus algorithms include PoW (proof of work), PoS (proof of stake), and DPoS (Delegated proof of Stake). In PoW, when a new block has to be created, a computational problem is sent out to the network. For example, 'find the next number that is divisible by 324349 and divisible by the proof-number of the last block' can be a computational problem in PoW. The miner is able to solve the PoW problem first creates the new block and will be rewarded for creating the new block by a token (e.g. bitcoin).

How to create a new block by a winner? He can start from an 'empty block', which is called a genesis block, somewhere. The genesis block has no data and values for proof and the previous hash is set as a default value. He needs to load data and its hash in the genesis block and set the proof value to complete the block generation.

3 Introduction to Speech Recognition

The task of speech recognition is to convert speech into a sequence of words by a computer program [3]. Figure 1 summarizes a generic speech recognition system. To start a speech recognition procedure, an input audio waveform is first converted into a sequence of fixed size acoustic vectors $X = [x_1, x_2, \cdots , x_T]$. This process is essentially an encoding procedure with feature extraction. The most popular feature vector is Mel-frequency cepstral coefficients (MFCC). Then, a decoder uses both acoustic and language models to find the sequence of words $W = [w_1, w_2, \cdots , w_L]$ with the maximum posterior probability for given input feature vectors X [3]. Mathematically, it is equivalent to the following optimization:

$$\hat{W} = \arg\max_W P(W|X) = \arg\max_W \frac{P(W)P(X|W)}{P(X)} \tag{1}$$

where $P(W)$ and $P(X|W)$ are computed by the language and acoustic modeling respectively. Since a speech signal can be viewed as a piecewise stationary signal or a short-time stationary signal, conventional speech recognition systems utilize Gaussian mixture

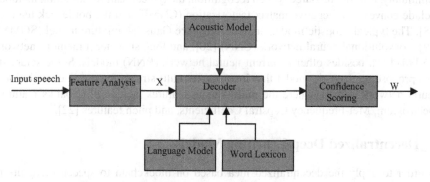

Fig. 1. The block diagram of a speech recognition system

model (GMM) based hidden Markov models (HMMs) to represent the above model. The HMMs speech recognition systems are straightforward and computationally feasible compared to the other peers. However, almost all Gaussian mixture models have limitations in modeling data that lie on a nonlinear manifold in input space [4].

Neural networks trained by back-propagation algorithms or its variants (e.g. Stochastic gradient descent, SGD) have emerged as an attractive approach for speech recognition since 1980s. In contrast to HMMs, neural networks make no assumptions about feature statistical properties. However, in spite of their effectiveness in classifying short-time units such as individual phones and isolated words, neural networks are rarely successful for continuous recognition tasks, largely because of their lack of ability to model temporal dependencies [5].

Deep learning employs different deep neural network models (e.g., DBN) for speech recognition [6–9]. Deep Belief Networks (DBN), which consists of stacked restricted Boltzmann machines (RBM), has achieved very successful applications in some subfield of speech recognition for its powerful feature extraction capabilities [10–12]. To apply DBNs with fixed input and output dimensionality to phone recognition, a context window of n successive frames of feature vectors is used to set the states of the visible units of the lower layer of the DBN that produces a probability distribution over the possible labels of the central frame. To generate speech sequences, a sequence of probability distributions over the possible labels for each frame are fed into a standard Viterbi decoder [13, 14].

Another typical example of using deep learning for speech recognition is convolution neural network (CNN) [15, 16]. CNN is constituted by one or more pairs of convolution layers (C layers), the largest pooling layers (S layers) and fully connected layers. It demonstrates powerful feature extraction capabilities besides mapping. In the C layers, a convolution kernel is used to filter the input signal. Filtering is repeated throughout the input signal space to extract features; The S layers are similar to a maximum filter, in which the convolution result is subsampled to extract more general data characteristics by picking maximum or average values. The input data gets more complicated deep-level feature extraction through convolution and pooling layers before following classification.

The mainstream state-of-the art speech recognition systems rely heavily on the amount of training data, and the recognition performance reduces dramatically as the available data is limited. Therefore, the speech recognition of under-resourced languages is a tough challenge that attracts more attentions for researchers in the speech recognition community. For low-resource speech recognition, the typical feature extraction methods include convex-non-negative matrix factorization (CNMF) and the bottleneck features [18]. The typical acoustic models include subspace Gaussian mixture model (SGMM) [19], convolutional neural network (CNN) [20], and long short-term memory network (LSTM) [21], besides other recurrent neural network (RNN) models. Some scientists also propose an acoustic model that incorporate multi-stream feature in convolutional neural network (CNN), where the multi-stream features include filter banks features, spectrogram, Mel-Frequency Cepstral Coefficients, and pitch features [22].

4 Decentralized Deep Learning Models

In order to apply the decentralized idea based on blockchain to speech recognition, we must first mention distributed deep learning. Distributed deep learning is a machine learning method based on data privacy protection; it is usually used in the field of financial services. As shown in Fig. 2, the distributed deep learning architecture is consisted of several shared models (contributor) and a central controlling agent (fusion of shared deep learning models). Instead of processing data in a single central server, distributed deep learning process data is distributed partially in the individual data providers. When the distributed learning is implemented, each computing contributors independently trains its own deep learning model using local data, and then share the model parameters with a central controlling agent. The central controlling agent fuses the parameters shared by each contributor to form a comprehensive model. Since the deep learning models are trained in distributed locations with smaller sets of data, the computational power required by the individual computing contributors is much lower, compared with that of a central server. However, in this solution, the deep learning architecture is fully controlled by a centralized agent. Therefore, the fusion model is susceptible to a single point of failure [3, 23–26]. In order to overcome the shortcoming, the cooperative and decentralized deep learning architecture is proposed.

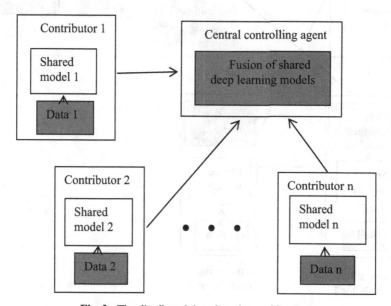

Fig. 2. The distributed deep learning architecture

As shown in Fig. 3, the cooperative and decentralized deep learning architecture is constituted of an application initiators, several computing contributors and verification contributors. In this architecture, each unit has its own decision interface and can make decision independent. Application initiators are in charge of defining the computing tasks, such as the properties of input data and the expected output. They also provide a sample set of data (including the training and verification data) for verification contributor, defining expected accuracy. The computing contributors are charge of building and training deep learning models; each computing contributor can participate or leave the whole computing architecture according to its performance. According to the task given by the initiators, computing contributors will design and train an appropriate machine learning model using local data, and publish it to the verification contributors [27–30]. After receiving the computing models, the verification contributors are responsible for evaluating the performance of the computing contributors, and report the results to the initiator. The initiator decides which computing contributors to fuse and how to fuse [31, 32].

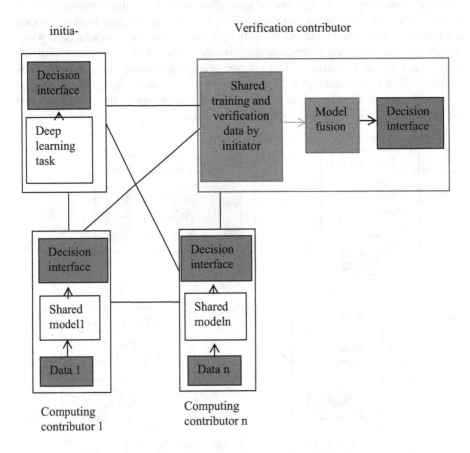

Fig. 3. Cooperative and decentralized deep learning architecture

5   Blockchain in Speech Recognition

The biggest challenge in speech recognition is to process large amounts of data and achieve best accuracy via model identification. The accuracy of recognition depends on the adaptability of model various variations. The cooperative decentralized machine learning based on the blockchain technique can deal with these problems well.

As mentioned in Sect. 3, the system architecture of speech recognition, an important step is to use the acoustic and language models attempting to decode. However, there are many uncertainties exist in acoustic models, associated with speaker characteristics, speech style and rate, noise interference, nonnative accents, microphone and environment variability, gender and dialect, etc. Inspired by the thought of the cooperative and decentralized deep learning method, a fusion model is designed for speech recognition, taking data with different rate, noise, microphone, gender, and dialect as the training data for each shared model.

As shown in Fig. 4, we have five convolutional neural network (CNN) models acting as the computing contributors; they are trained using the data with the characteristic of different rate, noise, microphone, gender, and dialect respectively. After the five shared models are trained sufficiently, five feature vectors f_i ($i = 1, 2, 3, 4, 5$) are obtained. Remove the output layers from the trained models, and fuse the features of each shared models, specifically, it is to concatenate these features. We consider the two-layers-strategy. Each of the layers is achieved by calculating a weighted sum of the corresponding values of the previous layer. Assume f_i is the upper layer feature vector of the i-th model, they are concatenated to form a concatenated feature f_c. The hidden layer h and output layer y are calculated based on the fully connected weights A and B with no constraints. Weight matrices A and B are initiated randomly. The back-propagation algorithm variants (e.g. Adam) are used to calculate the optimum values for the weight matrices A and B [18]. We can also consider the gradual model-fusion strategy. In this strategy, the weight matrix A and B are initiated by a special method, such the correlation between the different computing models are learned, while the uniqueness

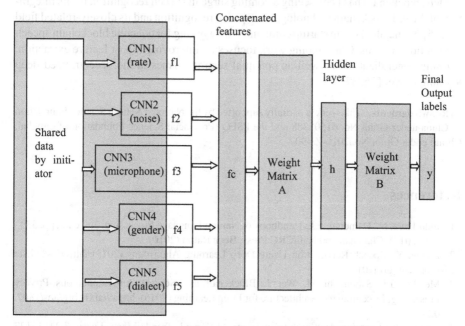

Fig. 4. The fusion model for speech recognition under blockchain

of each computing models is preserved. Because the fusion model takes into account various uncertain factors in the acoustic model, the overall performance of the integrated deep learning model is expected to improve.

6 Conclusion and Discussion

We discuss the application of blockchain in speech recognition by investigating decentralized deep learning models. It explores the powerful large-scale data handling capabilities of deep learning models besides maintaining data privacy [33]. It seems to be suitable to handle complicate speech recognition analysis on blockchain. However, some important issues remain to be solved. The first will be how to guarantee a robust feature extraction under such decentralized learning system via fusing. The weighting matrices in the fusing model may bring blur or shadow some key hidden features extracted from a single deep learning model. The concatenated features may suffer from redundancy because some overlapped or duplicate features extracted in a blockchain node.

Furthermore, there are huge numbers of parameters in a decentralized deep learning system. How to avoid overfitting under the system can be a challenging problem also. It is not sure dropout will still work well under this system because different nodes may have different dropout ratios. In addition, it remains unknown about whether such a system can achieve a real-time speech recognition under such decentralized deep learning systems besides potential security concerns for large input speech datasets under the decentralized deep learning system.

Notwithstanding challenges remains in the discussed decentralized deep learning models, blockchain has been seeing a coming surge in speech recognition for its integration of AI and blockchain technologies in speech recognition and its closely-related field [34, 35]. Some blockchain startup companies are working for concrete blockchain speech recognition systems. Our ongoing work focuses on improving robust feature extraction, learning generalization, as well as potential security issues in the decentralized deep learning model [36, 37].

Acknowledgments. This work is partially supported by the National Natural Science Foundation of China under Grant No. 61501388 and the PSFQ (Provincial Science Foundation of Qinghai, China) under Grant No. 2016-ZJ-904.

References

1. Indurkhya, N., Damerau, F.J.: Handbook of Natural Language Processing, 2nd edn., pp. 339–365 (2010). Chapman and Hall/CRC Press, Boca Raton (2010)
2. Zhang, Y.: Speech Recognition Using Deep Learning Algorithms (2013). http://cs229.stanford.edu/proj2013/
3. Mendis, G.J., Sabounchi, M., Wei, J.: Blockchain as a Service: An Autonomous, Privacy Preserving, Decentralized Architecture for Deep Learning (2018). https://arxiv.org/abs/1807.02515
4. Bengio, Y.: Learning deep architectures for AI. Found. Trends Mach. Learn. **2**(1), 1–127 (2009)

5. Bengio, Y.: Deep learning of representations: looking forward. In: Dediu, A.-H., Martín-Vide, C., Mitkov, R., Truthe, B. (eds.) SLSP 2013. LNCS (LNAI), vol. 7978, pp. 1–37. Springer, Heidelberg (2013). https://doi.org/10.1007/978-3-642-39593-2_1
6. Bengio Y., Courville, A., Vincent, P.: Representation learning: a review and new perspectives. IEEE Trans. PAMI **35**, 1798–1828 (2013)
7. Deng, L.: A tutorial survey of architectures, algorithms, and applications for deep learning. In: APSIPA Transactions on Signal and Information Processing, Cambridge University Press (2014, to appear)
8. Mohamed, A., Dahl, G., Hinton, G.: Deep belief networks for phone recognition. In: Proceedings of the NIPS Workshop Deep Learning for Speech Recognition and Related Applications (2009)
9. Deng, L., Seltzer, M., Yu, D., et al.: Binary coding of speech spectrograms using a deep auto-encoder. In: Interspeech (2010)
10. Dahl, G., Yu, D., Deng, L., Acero, A.: Large vocabulary continuous speech recognition with context-dependent DBN-HMMs. In: ICASSP (2011)
11. Dahl, G., Yu, D., Deng, L., Acero, A.: Context-dependent pre-trained deep neural networks for large vocabulary speech recognition. IEEE Trans. Audio Speech Lang Proc. **20**, 30–42 (2012)
12. Mohamed, A., Dahl, G., Hinton, G.: Acoustic modeling using deep belief networks. IEEE Trans. Audio, Speech Lang. Proc. **20**(1), 14–22 (2012)
13. Mohamed, A., Hinton, G., Penn, G.: Understanding how deep belief networks perform acoustic modelling. In: Proceedings of the ICASSP (2012)
14. Morgan, N.: Deep and wide: multiple layers in automatic speech recognition. IEEE Trans. Audio Speech Lang. Proc. **20**(1), 7–13 (2012)
15. Goodfellow, I., Bengio, Y., Courville, A.: Deep Learning. MIT Press, Cambridge (2016). http://www.deeplearningbook.org
16. Deng, L., Li, J., Huang, J.T., et al.: Recent advances in deep learning for speech research at microsoft. In: Proceedings of IEEE International Conference on Acoustics, Speech, and Signal Processing (ICASSP) (2013)
17. Qin, C.-X., Zhang, L.H.: Deep neural network based feature extraction for low-resource speech recognition. Acta Automatica Sinica **43**(7), 1208–1219 (2017)
18. Wu, W., Cai, M., et al.: Bottleneck features and subspace Gaussian mixture models for low-resource speech recognition. J. Univ. Chin. Acad. Sci. **32**(1), 97–102 (2015)
19. Liu, J., Zhang, W.: Research progress on key technologies of low resource speech recognition. J. Data Acquis. Process. **32**(2), 205–220 (2017)
20. Shu, F., Qu, D., et al.: A speech recognition method using long short-term memory network in low resources. J. Xi'an Jiaotong Univ. **51**(10), 120–127 (2017)
21. Qin, C., Zhang, L.: Acoustic modelling approach of multi-stream feature incorporated convolutional neural network for low-resource speech recognition speech recognition. J. Comput. Appl. **36**(9), 2609–2615 (2016)
22. Graves, A., Mohamed, A.-R., Hinton, G.: Speech recognition with deep recurrent neural networks. In: 2013 IEEE International Conference on Acoustics, Speech and Signal Processing (ICASSP), pp. 6645–6649. IEEE (2013)
23. Zhang, Y., Pezeshki, M., Brakel, P., et al.: Towards end-to-end speech recognition with deep convolutional neural networks. arXiv preprint arXiv:1701.02720 (2017)
24. Abdel-Hamid, O., Mohamed, A.R., Jiang, H., et al.: Convolutional neural networks for speech recognition. IEEE/ACM Trans. Audio Speech Lang. Process. **22**(10), 1533–1545 (2014)
25. Young, T., Hazarika, D., Poria, S., Cambria, E.: Recent trends in deep learning based natural language processing. arXiv preprint arXiv:1708.02709 (2017)
26. Schmidhuber, J.: Deep learning in neural networks: an overview. Neural Netw. **61**, 85–117 (2015)

27. Xu, X., Pautasso, C., Zhu, L., et al.: The blockchain as a software connector. In: 13th Working IEEE/IFIP Conference on Software Architecture (WICSA), pp. 182–191 (2016)

28. Dennis, R., Owen, G.: Rep on the block: a next generation reputation system based on the blockchain. In: International Conference for Internet Technology and Secured Transactions (ICITST), pp. 131–138. IEEE (2015)

29. Watanabe, H., Fujimura, S., Nakadaira, A., et al.: Blockchain contract: a complete consensus using blockchain. In: IEEE 4th Global Conference on Consumer Electronics (GCCE), pp. 577–578 (2015)

30. Shokri, R., Shmatikov, V.: Privacy-preserving deep learning. In: Proceedings of the 22nd ACM SIGSAC Conference on Computer and Communications Security, pp. 1310–1321 (2015)

31. Konecny, J., McMahan, H.B., Ramage, D., Richtárik, P.: Federated optimization: distributed machine learning for on-device intelligence. arXiv preprint arXiv:1610.02527 (2016)

32. Kokkinos, Y., Margaritis, K.G.: Confidence ratio affinity propagation in ensemble selection of neural network classifiers for distributed privacy-preserving data mining. Neurocomputing **150**, 513–528 (2015)

33. Easton, J.: Blockchains: a distributed data ledger for the railway industry. In: Innovative Applications of Big Data in the Railway Industry. IGI Global, Hershey (2018)

34. Ramachandran, S., Krishnmachari, B.: Blockchain for the IoT: opportunities and challenges (2018)

35. Konstantinidis, I., Siaminos, G., Timplalexis, C., Zervas, P., Peristeras, V., Decker, S.: Blockchain for business applications: a systematic literature review. In: Abramowicz, W., Paschke, A. (eds.) BIS 2018. LNBIP, vol. 320, pp. 384–399. Springer, Cham (2018). https://doi.org/10.1007/978-3-319-93931-5_28

36. Kosba, A., et al.: Hawk the blockchain model of cryptography and privacy-preserving smart contracts. In: 2016 IEEE Symposium on Security and Privacy, p. 848 (2016)

37. Theoduli, E., et al.: On the design of a blockchain-based system to facilitate healthcare data sharing. In: 17th IEEE International Conference on Trust Security and Privacy in Computing and Communications/12th IEEE International Conference on Big Data Science and Engineering (2018)

Automatic Segmentation of Mitochondria from EM Images via Hierarchical Context Forest

Jiajin Yi, Zhimin Yuan, and Jialin Peng$^{(\boxtimes)}$

College of Computer Science and Technology, Huaqiao University, Xiamen 361021, China
2004pjl@163.com

Abstract. To solve the problem of automatic mitochondria segmentation from electron microscope (EM) images, a hierarchical context forest (HCF) model using multi-level context features was proposed. Exploring effective contextual information is crucial to address the challenges caused by the varied appearances and shapes of mitochondria, and the complicated image content. To this end, a novel class of features named local patch pattern (LPP) are designed to characterize local contextual information, which are used to resolve the ambiguity caused by similar appearances and intensities of different organelles. Furthermore, to capture long-range contextual information, we also extract LPP features on intermediate probability predictions of the HCF model. Moreover, a multiscale strategy is used to capture different sizes of mitochondria. Solid validations of our method conducted on public dataset demonstrated the effectiveness of both of the proposed LPP features and the proposed model. The result of comparison showed that, the proposed method achieved distinct improvement of results in terms of Precision, Recall and F1-value score.

Keywords: Electron microscopy image · Mitochondria segmentation · Random forest · Contextual information · LPP feature

1 Introduction

An automatic segmentation of mitochondria from electron microscope (EM) images can greatly facilitate the analysis of mitochondria by precisely quantifying their volume, morphology and distribution, which have been directly linked to aging, cancer, and neurodegenerative diseases [1, 2]. High resolution EM is one of the state-of-the-art imaging devices for investigating the ultrastructure of cell, where mitochondria are subcellular organelle. However, with images in the scale of nm^3 and so in huge size, manual segmentation, even with assisted tools [3], is labor-intensive, time-consuming and also subjective. In fact, the very precise annotation could be a challenging task, even for experts, due to the intrinsic ambiguity in the image delineation process (see Fig. 1).

J. Yi and Z. Yuan—These authors contribute equally to this paper.

J. Peng—Supported by National Natural Science Foundation of China (11771160), and Fujian Science and Technology Foundation (2019H0016).

© Springer Nature Singapore Pte Ltd. 2020
H. Han et al. (Eds.): IDMB 2019, CCIS 1099, pp. 221–233, 2020.
https://doi.org/10.1007/978-981-15-8760-3_16

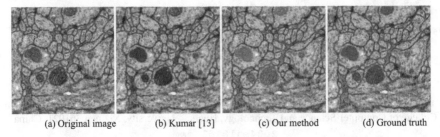

(a) Original image   (b) Kumar [13]   (c) Our method   (d) Ground truth

Fig. 1. Segmentation of mitochondria from EM images, which is a challenge task.

Therefore, automatic methods are highly desirable to process the large volumes of EM images and provide more reproducible results [4]. However, the large variability of mitochondria in density, location, size and shape makes it a challenging task. Moreover, the appearance and content of EM image are rather complex as shown in Fig. 1. For example, image intensities representing mitochondria in EM images show large overlap with intensities of other structures; strong gradients do not necessarily correspond to the semantic boundaries of the target mitochondria. Thus, in order to determine the presence of mitochondria at a given position and delineate their boundaries accurately, high-level contextual information has to be explored, which is the focus of this study.

Recently, the detection and segmentation of mitochondria from EM images have attracted a variety of studies [4–9, 12–15]. For example, Macke et al. [5] introduced a semi-automated method based on level set for tracing of cell membranes, in which segmentation from adjacent slice was utilized as prior constraint. Although the method is much faster than manual tracing, the performance is limited by its simple intensity based features. For mitochondria segmentation, Narasimha et al. [6] explored a filter-bank and textons as texture feature encoders, which were combined with several kinds of popular classifiers. Neila et al. [7] extracted local visual features with both 2D and 3D image filters (e.g., first and second order derivatives) in multiple scales to take into account the varied shape of mitochondria and anisotropic image voxels; conditional random field with an anisotropy-aware regularization is then employed for image segmentation. However, the discriminative ability of these general texture features is still limited.

Rather than texture features, Smith et al. [8] devised a new class of shape features, that considered image characteristics at distant contour points to capture the irregular shape information of mitochondria. In [9], Lucchi et al. combined Ray features with intensity histogram features on supervoxels, and the segmentation was obtained using graph cut with learned potentials. However, this graph learning and partitioning method is costly in memory, which was improved in [10] with max-margin learning and a subgradient descent using working sets. In [11], elliptical descriptor at different scales combined with local Gaussian filters as [7] were investigated. To identify precise boundaries of mitochondria, Giuly et al. [12] performed random-forest-based classification on candidate boundary contours using geometric and intensity based features, while candidate contours were obtained by thresholding prediction results of another random forest classifier; the final fine segmentation is obtained with a geodesic active surface method. To alleviate the negative impact of the complex image content of EM image, Kumar et al. [13] proposed a class of more powerful features, Radon-like features, that allowed for aggregation of spatially distributed image statistics into compact feature descriptors in the EM image. Final segmentation was extracted with thresholdings. Seyedhosseini *et al.* [14] extracted shape and textural features of mitochondria by algebraic curve, using the random forest as the classifier. The mitochondrial boundary curves can be obtained effectively by the method, however, it is easily attracted by the ridge structure of the mitochondria.

In this work, we propose a fully automatic method for mitochondria segmentation by integrating multilevel contextual and appearance features. More specifically, with a class of newly designed features, we propose to extract multilevel context and appearance features with a classifier in cascaded architecture that is similar to the auto-context model in [16, 25]. One contribution of this study is that, we design a novel class of features, named local patch pattern (LPP) to encode local context and appearance from raw images and middle-level context from intermediate predictions. To be specific, under the framework of hierarchical classifier, we extract middle-level features using LPP on prediction results from classifiers in lower layer. In this way, the features extracted include information from a much larger receptive field, the size of which plays a crucial role for the segmentation performance [17]. In this way, the amount of surrounding contextual information is enhanced. To further capture mitochondria of different sizes, we utilize a multi-scale strategy, which improves the efficiency and robustness of the model.

2 Method

In this section, we provide a detailed description of our method. Taking into account anisotropic voxels of our EM images, we perform segmentation in a slice-by-slice strategy using image patches. Specifically, we split each slice of the image into overlapped patches of size $l \times l$, and assign the most probable label to each pixel x of a target image according to the information from the patch I_x, centered on the pixel x. To achieve fine segmentation, we train a hierarchical set of classifier, in which random forest model is employed as the basic classifier due to its efficiency and scalability. The workflow of the proposed framework is shown in Fig. 2.

2.1 Feature Extraction

Local Appearance and Texture Fatures. Gray-scale intensity, gradient magnitude, and local binary pattern (LBP) are employed to capture appearance, edge and texture information. In each patch, we compute 6 statistics (including mean, variance, median, entropy, kurtosis and skewness) of intensities, gradients and local binary pattern (LBP) [18] features in local image patches as local appearance and texture descriptors. Gray-scale intensities and its statistics are the most prominent features of EM images; gradients are the basic features indicating edges in images. LBP is one of the most widely used texture features; it is computed by comparing each pixel in the image with each of its P (such as 8) neighbors, giving an P-digit binary number, which is usually converted to decimal (see Fig. 3 (a)). Specifically, LBP operator is defined as:

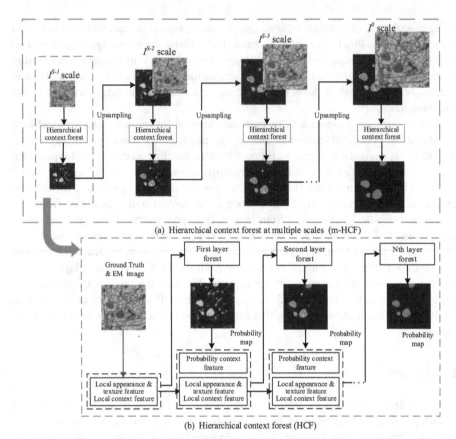

(a) Hierarchical context forest at multiple scales (m-HCF)

(b) Hierarchical context forest (HCF)

Fig. 2. Workflow of the multiscale hierarchical context forest (m-HCF) we proposed. The heat map in the figure is a graphical representation of probability map, in which the colder the color is, the smaller the magnitude of estimated probability is. (Color figure online)

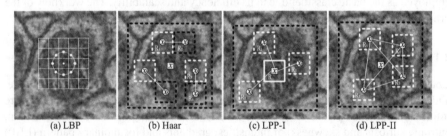

(a) LBP   (b) Haar   (c) LPP-I   (d) LPP-II

Fig. 3. Illustration of LBP feature, Haar feature and the proposed LPP features.

$$f_{\text{LBP}}(x, I) = \sum_{p=1}^{P} 2^{p-1} \delta(x_p - x_c), \tag{1}$$

where

$$\delta(z) = \begin{cases} 1 & z \geq 0, \\ 0 & z < 0. \end{cases}$$

LBP is widely used due to its powerful ability for texture discrimination and simple calculation. However, it only captures the information in a smaller neighborhood, and unable to encode the more discriminative information of context in longer range. Since there are a large number of cell tissues with similar appearances and textures in the EM image, the local features mentioned above is still limited in discrimination.

Local Patch Pattern (LPP) Features. The association between each pixel of the image and its extended neighborhoods, called context information, that provides a way of eliminating the ambiguity of the local appearance and texture features [19, 20]. Inspired by the classical LBP feature and the Haar features [21], we propose a novel class of features (LPP) based on the local patch, which are used to capture local context information effectively. The classical Haar features consider sub-regions of adjacent positions and then calculate the difference of the sum of pixel intensities in these sub-regions, as shown in Fig. 3 (b), it can quickly generate a rich set of features that are robust against the noise. In this paper, we extend the LBP and Haar features into LPP features that capture the context information effectively. The LPP features contain two subtypes, i.e. LPP-I and LPP-II, which are calculated by extracting sparsely distributed sub-regions $\{R_i\}_{i=1}^{L}$ on the image patch I_x.

LPP-I Feature. This feature is a kind of LBP-like Haar feature. For the LPP-I feature extraction, we denote the central sub-region of the patch I_x as R_0, then record the differences between R_0 and other sub-regions R_i, and also the binary code formed by all positive, negative differences. The former is a real type LPP feature (LPP-Ir), the latter is a binary type LPP-Ib feature (actually convert the binary code to decimal).

$$f_{\text{LPP-Ir}}(I_x, R_i) = \frac{1}{|R_i|} \sum_{u \in R_i} I_x(u) - \frac{1}{|R_0|} \sum_{u \in R_0} I_x(v), \tag{2}$$

$$f_{\text{LPP-Ib}}(I_x) = \sum_{i=1}^{L} 2^{i-1} \delta \left(\sum_{u \in R_i} I_x(u) - \sum_{v \in R_0} I_x(v) \right). \tag{3}$$

Specifically, the real valued LPP features $f_{\text{LPP-Ir}}$ for R_i ($i = 1, \cdots, L$) is obtained by calculating the difference of the average pixel intensities of sub-regions $\{R_i\}_{i=1}^{L}$ (as shown in Fig. 3 (c)) and the central one R_0 in I_x. We also record L binarized summaries of $f_{\text{LPP-Ir}}$ features through using different permutations of the $f_{\text{LPP-Ir}}$ values. For each permutation, a $f_{\text{LPP-Ib}}$ feature is calculated by recording the sign of the $f_{\text{LPP-Ir}}$ values, which is converted to decimal for convenience. From these steps, we have achieved the extension of classical LBP feature to the context feature LPP-I.

LPP-II Features. We extend the calculation of the traditional Haar feature to randomly select multiple pairs of sub-region of the same size R_i and R_j from a local image patch (such as R_1 and R_2 that are shown in Fig. 3 (d)). For each pair of sub-regions R_i and

R_j, $f_{\text{LPP-II}}$ is the difference between their average pixel intensities, that is defined as follows,

$$f_{\text{LPP-II}}(I_x, R_i, R_j) = \frac{1}{|R_i|} \sum_{u \in R_i} I_x(u) - \frac{1}{|R_j|} \sum_{v \in R_j} I_x(v). \tag{4}$$

LPP-II features encode the differences between different sub-regions in each local image patch. It is a variant of Haar feature with randomly-extracted non-adjacent sub-regions.

Compared with LBP and Haar features, the LPP-I and LPP-II features reflect contextual information in relatively long-range that the category of a pixel depends on. The contextual information in longer range is extracted by computing LPP features on probabilistic predictions (see Sect. 2.2), which will be used as additional information to refine the segmentation.

2.2 Hierarchical Context Forest (HCF)

We use the classical random forest [22] as the classifier, as it can effectively address the relative large-scale and high-dimensional data. Moreover, it is easily parallelized. Random forest obtains the final classification result by integrating the classification results of an ensemble of separately trained binary decision trees; each tree in the forest is trained only on the subset of the data and feature. The introduction of randomness makes the random forest model relatively robust to noise and mitigate the over-fitting problem. Each random forest can be described by the number of the trees used, the maximum depth and the weak learner used.

Due to the complexity of the EM image content, the context features extracted from raw image may be limited in discrimination. Besides, segmentation by independently inferring the label of each pixel is sub-optimal, as images with semantic content show obvious label dependencies between adjacent pixels. Therefore, it is preferred that the ultimate prediction can be influenced by the model's beliefs about nearby positions.

To solve these limitations, 1) *we use a hierarchical classification model, which cascades multiple random forests to improve the segmentation results iteratively*; 2) *we extract LPP features on the output probabilities of random forest in lower layers as additional features to the random forest in the current layer*. Importantly, the LPP features extracted from the intermediate segmentation probability map of the hierarchical model have a much larger receptive field, since the probability map of each position is influenced by information in the $l \times l$ local patch. Different from the original image, features on the probability map that is produced by the previous classifier can reflect the defect of the previous classifier, which help the latter to pay more attention to the poorly segmentation region in the previous probability map. Therefore, the LPP features on probability maps are expected to be useful to enhance the original context information. The above method follows the philosophy of the auto-context framework proposed in [16]. However, our model differs in that we use the LPP features proposed above to characterize the contextual information in long range.

Specifically, in the training step, a sequence of classifiers $F = \{F_t, t = 1, \cdots, T\}$ are iteratively trained, which with a total of T random forests, as shown in Fig. 2 (b). The first forest is trained by extracting the local image features described in Sect. 2.1 from the training patches; subsequent random forests $F_t(t > 1)$ not only use the original image as training data, but also use the previously-estimated probability map as augmented training data. Importantly, we also extract the LPP features from the probability map as the probability context feature, which used to enhance the discrimination of features extracted on the grayscale images.

In the testing step, each trained classifier F_t is used sequentially to predict the test EM images I_t. In the first iteration, we extract the local image features on I_t described in Sect. 2.1, then predict it with F_1 to obtain initial mitochondrial probability map. In the second iteration, we not only use the features extracted in I_t, but also input the probability context features extracted from the previously-estimated mitochondria probability map as augmented features into F_2 for prediction. Repeat the above steps until all trained classifiers are applied. The output of the final cascaded forest is the segmentation result.

2.3 Hierarchical Context Forest in Multi-scale (M-HCF)

To accurately capture mitochondria of different sizes and improve the validity, robustness of the model, we use the multi-scale strategy to further enhance the performance of the model.

Specifically, at the first, we obtain S images of different resolutions $\{I^0, I^1, \ldots, I^{S-1}\}$ by sequentially downsampling each training image I, where I^0 is the original scale image I and I^j is obtained by downsampling I^{j-1} with 1/2 times. Then, we apply the HCF model on multiscale images (m-HCF), see Fig. 2 (a), in which the input of the model at high resolution is the corresponding scale image and the classification prediction probability map at low resolution. The introduction of multi-scale strategy makes the model utilize context information of different spatial extents. Actually, the multi-scale method has been proved to be effective on both natural image [20] and medical image segmentation [23, 26].

In the model prediction step, the S images of different resolutions are obtained by downsampling the test image I_t at first, and then input the image patches extracted at different resolutions into the m-HCF model, which is already obtained by training. Finally, we can get the result of segmentation from the m-HCF model.

3 Experiments

3.1 Experimental Setting

We evaluated the performance of our method on the Drosophila first instar larva ventral nerve cord (VNC) dataset [3, 24]. DVNC dataset contains 30 Drosophila abdominal nerve images of size 512×512 with a resolution of $5 \times 5 \times 40$ nm^3/voxel, which was acquired using continuous slice transmission electron microscopy (ssTEM). The first 15 images were selected for training and the rest were used for testing. Also we evaluated the robustness of the proposed method with 5 different splits of the dataset.

To evaluate the performance of segmentation, we selected Precision, Recall, F1-value as the measurements, which are defined as follows,

$$\text{Precision} = \frac{TP}{TP + FP}, \tag{5}$$

$$\text{Recall} = \frac{\text{TP}}{\text{TP} + \text{FN}}, \tag{6}$$

$$\text{F1 - value} = \frac{2 \times \text{Precision} \times \text{Recall}}{\text{Precision} + \text{Recall}}, \tag{7}$$

where TP are the true positives, FP are the false positives, and FN are the false negatives. F1-value is a metric that combines precision and recall, equivalent to the Dice similarity coefficient, and it is one of the most commonly used comprehensive metric for evaluating segmentation results. The larger the scores of the three metrics are, the better the segmentation result is.

The model parameters were set as follows: for the multi-scale hierarchical context forest (m-HCF), the scale number S was set as $S = 2$, which included 1/4 original image size (I^1) and the original image size (I^0); for the hierarchical context forest (HCF), we used 2 and 5 layers of random forests at the two scales, respectively; the patch size on the size of 1/4 original image was set as 11×11, and on the original image size, set as 19×19. The discussion about patch size at I^0 scale in Sect. 3.3. For parameters in the random forest, the number of decision trees was 15, the maximum depth of each tree was 50, and the minimum number of samples at the leaf node in each tree was 10. In the experiment, we found that increasing the number of the trees, the depth of the trees and the number of features selected by the non-leaf nodes, can improve the performance of the model. However, this will result in much longer training time for the model and also much larger memory requirement. The parameter setting in our experiments is a balance of efficiency and accuracy.

3.2 Results and Comparisons

We evaluated the segmentation performance of our method and compared it with recent mitochondria segmentation methods on DVNC, which are summarized in Table 1. We compared our method with 1) the method of Kumar et al. [13], which devised a class of powerful feature named Radon-like features for mitochondria segmentation; 2) the novel pipeline of Giuly et al. [12], which combined patch classification, contour pair classification, and automatically seeded level sets; 3) the method of Seyedhosseini et al. [14], which originally extracted the shape and textural features of mitochondria by algebraic curve, using the random forest as the classifier. We also tested the random forest with the proposed features and the proposed HCF method using different scales of images.

As shown in Table 1, our HCF method using 2 scales of images, with a balance of computational efficiency and accuracy, achieved an average F1-value of 80.3%, 84.9% Precision, and 78.5% Recall, indicating that it outperforms other methods in terms of all metrics. In fact, the scores of the HCF model with only 1 scale ($S = 1$) is also higher than most of other methods. Furthermore, using the proposed features, the simple random forest method already achieved promising results, which show the discriminative ability of the proposed features.

Table 1. Performance comparison of different methods with multiple metrics on DVNC.

Methods	F1-value (%)	Precision (%)	Recall (%)
Kumar [13]	56.6	59.3	54.2
Random forest (with proposed features)	62.6	71.4	55.7
Giuly [12]	60.4	64.2	57.0
Seyedhosseini [14]	72.9	78.5	68.0
Proposed method ($S = 1$)	77.8	72.0	**87.5**
Proposed method ($S = 2$)	**80.3**	**84.9**	78.5

3.3 Model Analysis

The Effectiveness of the Proposed LPP Features (see Fig. 4). In order to verify the validity of the LPP features designed in this paper, we tested the multiscale HCF method using different features and different feature combinations. To be specific, we firstly set the baseline features (B in Fig. 4) as the 6 statistics (i.e., mean, variance, median entropy, kurtosis, skewness) and central point pixel intensity from the image patch in the grayscale image, gradient image, and probability map. Then, the baseline features, Radon-like feature [13], LBP feature [18], Haar-like feature, the proposed feature LPP including LPP-I and LPP-II and their combinations are compared.

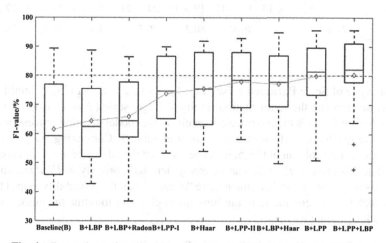

Fig. 4. Comparison of segmentation results with different choices of features.

The results of comparison F1-value metric are demonstrated with a box-plot shown in Fig. 4. It can be observed that LPP features are superior to the classical LBP, Haar features and also the Radon features. The F1-value using baseline features is 61.5%; compared with the LBP feature, the LPP-I feature (i.e., LBP-like feature) was better, and the F1-value was improved from 64.4% to 73.8%; in comparison of LPP-II features with the Haar features, and the Haar + LBP features, we can see that the F1-value increased from 75.5%, 77.2% to 77.9%, so the LPP-II features are not only better than the Haar features, but also better than the combination of Haar and LBP features. In addition, the full LPP features are significantly better than the combination of LBP and Haar features, and the F1-value was increased from 77.2% to 79.8%.

These results proved the advantages of the LPP features designed in this paper. To further increase the robustness of our model, we combined the LPP features with the LBP features in the final model, and the F1-value of the final segmentation was 80.3%. Three reasons may explain the improvements using LPP features: 1) the LPP features contain sufficient image contextual information to resolve the ambiguity of local appearance and texture features; 2) the sub-patch based LPP features are more robust to noise; 3) a large number of features provides guarantee for tree depth, which improves the performance of the model.

The Influence of Patch Size (see Table 2). To verify the influence of different patch sizes at I^0 scale, we compared the segmentation results using different patch sizes at I^0 scale. As shown in Table 1, our model is relatively robust to different patch sizes. Moreover, our model performed best when the patch size was set as 19×19.

Table 2. Segmentation results of different patch sizes at I^0 scale.

Metric	Patch size (pixel)							
	13×13	15×15	17×17	19×19	21×21	23×23	25×25	27×27
F1-value (%)	77.4	79.5	79.8	**80.3**	80.2	79.6	79.5	78.6

The Influence of Scale Parameter (see Fig. 5). To verify the validity of multi-scale strategy, we compare the segmentation results of single-scale ($S = 1$) and multi-scale ($S = 2$). Figure 5 shows the intermediate results in the model iteration process, and the comparative results of performance are shown in Table 1. Comparing the probability maps of the second column of Fig. 5, it can be seen that the left lower non-mitochondrial region detected error at I^0 scale can be easily detected correctly at I^1 scale, and the probability map at the I^1 scale is much more "cleaner" than the probability map at the I^0 scale, which is more conducive to the subsequent classifier to refine the mitochondrial classification probability map.

The Bias of Training Data Selection (see Fig. 6). To verify the robustness of our model, we employed the 5 different splits of the dataset to train model. For each split of the data, we randomly selected 15 images from the dataset as the training set, and the rest as the testing set. Figure 6 shows the results of segmentation accuracy with the 5 different splits. The relatively low standard deviation of the performance show that our method is less affected by the bias of training data selection, and the average value of the results is close to our segmentation result shown in Table 1.

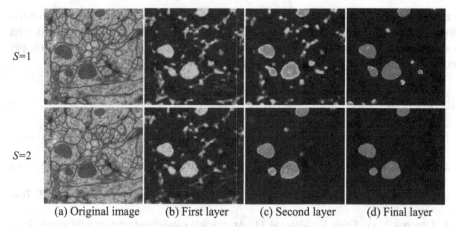

S=1

S=2

(a) Original image   (b) First layer   (c) Second layer   (d) Final layer

Fig. 5. Probability maps at different layers of each scale. (a) Original image & ground truth; (b) probability map of first layer; (c) probability map of second layer; (d) probability map of final layer. The first row (b)–(d) at I^0 scale; the second row (b) at I^1 scale, (c)–(d) at I^0 scale.

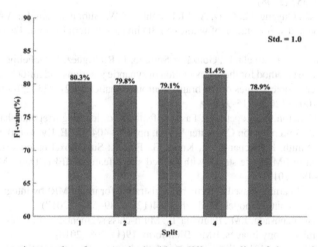

Fig. 6. Segmentation results of our method with 5 different splits of dataset for training and testing. The standard deviation of the 5 segmentation results is 1.0.

4   Conclusion

This paper presented a multiscale HCF model for the automatic segmentation of mitochondria from EM images. To address the challenges of complex image contents, we proposed a novel set of discriminative contextual features, named LPP, to encode multi-level contextual information. Specifically, under the framework of iterative refinement, we extracted features both on raw images and intermediate predictions, which can effectively alleviate the negative impact of complex content of EM images on the mitochondrial segmentation task. Our approach is generic and can be used for similar segmentation tasks. Validation and comparisons on the challenging DNVC dataset indicated that the proposed method is robust and effective for the automatic segmentation of mitochondria form EM images. In the future work, we plan to extend our work to 3D analysis with exploiting the 3D information of the EM image stack. In addition, we intend to apply our model to other objects segmentation task of EM images.

References

1. Campello, S., Scorrano, L.: Mitochondrial shape changes: orchestrating cell pathophysiology. EMBO Rep. **11**(9), 678–684 (2010)
2. Alston, C.L., Rocha, M.C., Lax, N.Z., Turnbull, D.M., Taylor, R.W.: The genetics and pathology of mitochondrial disease. J. Pathol. **241**(2), 236–250 (2017)
3. Cardona, A., et al.: TrakEM2 software for neural circuit reconstruction. PLoS ONE **7**(6), e38011 (2012)
4. Uzunbas, M.G., Chen, C., Metaxas, D.: An efficient conditional random field approach for automatic and interactive neuron segmentation. Med. Image Anal. **27**, 31–44 (2016)
5. Macke, J.H., Maack, N., Gupta, R., Denk, W., Schölkopf, B., Borst, A.: Contour-propagation algorithms for semi-automated reconstruction of neural processes. J. Neurosci. Methods **167**(2), 349–357 (2008)
6. Narasimha, R., Ouyang, H., Gray, A., McLaughlin, S.W., Subramaniam, S.: Automatic joint classification and segmentation of whole cell 3D images. Pattern Recogn. **42**(6), 1067–1079 (2009)
7. Márquez Neila, P., Baumela, L., González-Soriano, J., Rodríguez, J., DeFelipe, J., Merchán-Pérez, Á.: A fast method for the segmentation of synaptic junctions and mitochondria in serial electron microscopic images of the brain. Neuroinformatics **14**(2), 235–250 (2016). https://doi.org/10.1007/s12021-015-9288-z
8. Smith, K., Carleton, A., Lepetit, V.: Fast ray features for learning irregular shapes. In: 12th International Conference on Computer Vision, pp. 397–404. IEEE, Kyoto (2009)
9. Lucchi, A., Smith, K., Achanta, R., Knott, G., Fua, P.: Supervoxel-based segmentation of mitochondria in EM image stacks with learned shape features. IEEE Trans. Med. Imaging **31**(2), 474–486 (2012)
10. Wei, L., et al.: Learning-based deformable registration for infant MRI by integrating random forest with auto-context model. Med. Phys. **44**(12), 6289–6303 (2017)
11. Cetina, K., Buenaposada, J.M., Baumela, L.: Multi-class segmentation of neuronal structures in electron microscopy images. BMC Bioinform. **19**(1), 298 (2018)
12. Giuly, R.J., Martone, M.E., Ellisman, M.H.: Method: automatic segmentation of mitochondria utilizing patch classification, contour pair classification, and automatically seeded level sets. BMC Bioinform. **13**(1), 29 (2012)
13. Kumar, R., Vázquez-Reina, A., Pfister, H.: Radon-like features and their application to connectomics. In: 2010 IEEE Conference on Computer Vision and Pattern Recognition Workshops, pp. 186–193. IEEE, San Francisco (2010)
14. Seyedhosseini, M., Ellisman, M.H., Tasdizen, T.: Segmentation of mitochondria in electron microscopy images using algebraic curves. In: 2013 IEEE 10th International Symposium on Biomedical Imaging, pp. 860–863. IEEE, San Francisco (2013)
15. Li, W., Liu, J., Xiao, C., Deng, H., Xie, Q., Han, H.: A fast forward 3D connection algorithm for mitochondria and synapse segmentations from serial EM images. BioData Mining **11**(1), 24 (2018). https://doi.org/10.1186/s13040-018-0183-7
16. Tu, Z., Bai, X.: Auto-context and its application to high-level vision tasks and 3D brain image segmentation. IEEE Trans. Pattern Anal. Mach. Intell. **32**(10), 1744–1757 (2010)

17. Chen, L.C., Papandreou, G., Kokkinos, I., Murphy, K., Yuille, A.L.: DeepLab: semantic image segmentation with deep convolutional nets, atrous convolution, and fully connected CRFs. IEEE Trans. Pattern Anal. Mach. Intell. **40**(4), 834–848 (2018)
18. Ojala, T., Pietikainen, M., Maenpaa, T.: Multiresolution gray-scale and rotation invariant texture classification with local binary patterns. IEEE Trans. Pattern Anal. Mach. Intell. **24**(7), 971–987 (2002)
19. Zhang, H., et al.: Context encoding for semantic segmentation. In: IEEE Conference on Computer Vision and Pattern Recognition, pp. 7151–7160. IEEE, Salt Lake City (2018)
20. Ding, H., Jiang, X., Shuai, B., Qun Liu, A., Wang, G.: Context contrasted feature and gated multi-scale aggregation for scene segmentation. In: IEEE Conference on Computer Vision and Pattern Recognition, pp. 2393–2402. IEEE, Salt Lake City (2018)
21. Viola, P., Jones, M.: Rapid object detection using a boosted cascade of simple features. In: IEEE Conference on Computer Vision and Pattern Recognition, p. I. IEEE, Kauai (2001)
22. Breiman, L.: Random forests. Mach. Learn. **45**(1), 5–32 (2001). https://doi.org/10.1023/A:1010933404324
23. Kamnitsas, K., et al.: Efficient multi-scale 3D CNN with fully connected CRF for accurate brain lesion segmentation. Med. Image Anal. **36**, 61–78 (2017)
24. Cardona, A., et al.: An integrated micro- and macroarchitectural analysis of the Drosophila brain by computer-assisted serial section electron microscopy. PLoS Biol. **8**(10), e1000502 (2010)
25. Gadde, R., Jampani, V., Marlet, R., Gehler, P.V.: Efficient 2D and 3D facade segmentation using auto-context. IEEE Trans. Pattern Anal. Mach. Intell. **40**(5), 1273–1280 (2017)
26. Roth, H.R., et al.: A multi-scale pyramid of 3D fully convolutional networks for abdominal multi-organ segmentation. In: Frangi, A.F., Schnabel, J.A., Davatzikos, C., Alberola-López, C., Fichtinger, G. (eds.) MICCAI 2018. LNCS, vol. 11073, pp. 417–425. Springer, Cham (2018). https://doi.org/10.1007/978-3-030-00937-3_48

Generating the n-tuples of Natural Numbers by Enzymatic Numerical P System

Zhiqiang Zhang[1], Yuan Kong[2], Zhihua Chen[3], Yunyun Niu[4(✉)], and Mingyuan Ma[1]

[1] School of Electronics Engineering and Computer Science, Peking University,
Beijing 100871, China
[2] College of Mathematics and Systems Science, Shandong University of Science and
Technology, Qingdao 266590, China
[3] Institute of Computing Science and Technology,
Guangzhou University, Guangzhou 510006, Guangdong, China
[4] School of Information Engineering, China University of Geosciences
in Beijing, Beijing 100083, China
yniu@cugb.edu.cn

Abstract. Numerical P systems (NP systems) are a class of computing models inspired by both the cell structure and economics. Enzymatic numerical P systems (ENP systems) are a variant of NP systems, which were successfully applied in autonomous robot control. In this work, we design an algorithm using enzymatic numerical P systems to generate all the n-tuples of natural numbers in a well specified order. Specifically, we improve previously known results on generating all the n-tuples of natural numbers. Based on this method, we prove that a numerical P system with one membrane, and the production function used with polynomials of degree 1 with at most 2 variables can reach universality, which optimizes the previous known results. The results also give a positive answer to a problem formulated in [Fundamenta Informaticae, 2006, 73(1): 213-227].

Keywords: Membrane computing · Numerical P systems · Enzymatic numerical P systems · Turing universality · Recursively enumerable sets

1 Introduction

Membrane computing [6] is a branch of natural computing which aims to abstract computing models from the architecture and the functioning of living cells, as well as from the organization of cells in tissues, organs (brain included) or other higher order structures such as colonies of cells (e.g., bacteria). The computing models in membrane computing are called membrane systems or P systems. The basic P system is to consider a hierarchical arrangement of membrane, like in a cell, delimiting compartments where various chemicals (called objects) evolve according to local reaction rules. Based on the basic model, many variants of P systems have been proposed by considering different strategies of choosing which rules to be applied (such as the maximal parallelism [15, 16], sequential [17], or minimal parallelism [9]), or by introducing new controlling

© Springer Nature Singapore Pte Ltd. 2020
H. Han et al. (Eds.): IDMB 2019, CCIS 1099, pp. 234–244, 2020.
https://doi.org/10.1007/978-981-15-8760-3_17

strategies with biological motivations [20, 21] or with mathematical motivations [18], or with both [23]. Most of these variants lead to computationally universal [14, 19, 22]. P systems have been applied in various fields [5], such as biology, linguistics, theoretical and applied computer science [13, 24, 25]. The recent applications in engineering can be found in [10–12]. For the most up-to-date news and results, one can consult the P systems web page http://ppage.psystems.eu.

Numerical P systems (NP systems) are a variant of P systems, which were first introduced in [7], with inspiration from the economics. The variables evolve in the system are numerical variables, which differs from the symbol variables in the usually P systems. The evolution programs used in the numerical P systems in the sequential way, which means that in each step, each membrane uses only one program, which is another difference from the maximally parallel one (at a given step, one uses a multiset of rules which is maximally in the inclusion sense).

Many variants of numerical P systems have been proposed, such as enzymatic numerical P systems [1], numerical P systems with thresholds [27], numerical P systems with migrating variables [26]. Enzymatic numerical P systems (ENP systems) are introduced to control the application of the programs. In ENP systems, a program is active if and only if the value of enzyme variable is strictly greater than the smallest value of variables involved in the production polynomial. Because of the introduction of the enzyme variables, there can be more than one active production function in each membrane, but the system still is deterministic (the next figuration of the system is determined by the former figuration). This deterministic character makes it a possible use for modeling mobile robot controllers.

The computational power of ENP systems has been investigated both as languages generators [28, 29] and as number generators [8]. The universality of the deterministic ENP systems is proved in [2] by constructing an ENP systems with 254 membranes and the polynomials with 253 variables and $2°$. In paper [3], the indices above are deduced to 4, 1 and 6, respectively. However, in paper [2] and [3], the generating method of the 5-tuples of natural numbers are the same, by which the 5-tuples in base n will be generated again in base n + 1, no considering the efficiency. In this work we use a new method of generating each 5-tuples of the natural numbers only once during the process. By this method, we improve the result by deducing the indices to 1, 1, 2.

2 The Computational Power of Enzymatic Numerical P Systems

2.1 Previous Results

The computing power of ENP systems was first studied in [8], which showed that ENP systems were universal for both non- deterministic and deterministic case. We only recall here the results concerning the deterministic ENP systems in [8].

Theorem 1. $NRE = NP_{254}(poly^{253}(253), enz, allp, det)$.

Theorem 2. $NRE = NP_4(poly^1(6), enz, allp, det)$.

Such universality results are also interesting from the robot control point of view, as they guarantee that any computer program or robot behavior can be implemented using ENP systems. Therefore, it is significant to improve the above universality result by reducing the concerning parameters. In paper [14], the degree of the polynomials of the production functions is reduced from 2 to 1 and the number of membranes from 253 to 4. In this paper, we construct an enzymatic numerical P systems with one membrane and the polynomial used in the system has at most 2 variables. And we proved that this ENP systems are computing universality.

2.2 An Improved Universality Result

Theorem 3. $NRE = NP_{254}(poly^{253}(253), enz, allp, det)$.

Proof. The proof is based on the characterization of recursively enumerable sets of numbers as positive values of polynomials with integer coefficients: for every RE set Q there exists a multivariate polynomial f_Q with integer coefficients such that the set of positive values of f_Q, corresponding to tuples of natural numbers, is Q.

$$\forall Q \subseteq N^+ RE, \exists f_Q \in Z_n[X] \ s.t. Q = \{f_Q(X) | X \in N^n, \ f_Q(x) > 0\}$$

And it is shown in [5] that polynomials of degree at most 5 and with 5 variables are sufficient to generate the elements of *RE* sets. In [13] it is proven that if f_Q is a polynomial of degree 5 with 5 variables, then f_Q can be put in the following form:

$$f_Q(x_1, \cdots, x_5) = \sum_{i=1}^{252} \beta_i \cdot (a_{1,i}x_1 + \cdots + a_{5,i}x_5 + a_{6,i})^5$$

For an arbitrary set $Q \in N^+ RE$ and $f_Q(x_1, \cdots, \xi_5)$ as above, we construct the following ENP system which can enumerate all 5-tuples of natural numbers and compute the corresponding values of f_Q: $\prod = (1, H, \mu, (V, P, V_0), enum)$

$$H = 1,$$

$$\mu = [_1]_1,$$

$$V = \{x_i, e_i, x_i^c, 1 \le i \le 5, t, enum, e_f, g, g_c, x_{11}, f_Q\} \cup \{e_s, n_2, u\},$$

$$P = \{1 + x_2|_{e_1} \rightarrow 1|ez_2, g + x_1|_{e_1} \rightarrow 1|x_{11}, e_1 \rightarrow 1|e_2, g + e_i|_{ez_i} \rightarrow 1|e_{i+1},$$

$$1 + x_{i+1}|_{ez_i} \rightarrow 1|ez_{i+1}, ez_i + 1|_{e_i} \rightarrow 1|e_s, ez_i - 1|_{e_i} \rightarrow 1|x_i,$$

$$x_{11} + ez_i + 1|_{e_i} \rightarrow 1|x_1, x_{11} - 2|_{ez_i} \rightarrow 1|x_1, e_i + ez_i \rightarrow 1|g_c : 2 \le i \le 4$$

$$ez_5 - 1|_{e_5} \rightarrow 1|x_5, x_{11} + ez_5 + 1|_{e_5} \rightarrow 1|x_4, x_{11} - 2|_{ez_5} \rightarrow 1|x_4,$$

$$1 + g + x_{11}|_{e_{z5}} \to 1|x_5, g + 1|_{e_5} \to 1|_{e_s}, e_5 + e_{z5} \to 1|_{g_c},$$

$$g + 2x_j|_{e_s} \to 1|x_j + 1|x_j^c, g + 2|_{e_s} \to 1|_{e_i^c}, e_s \to 1|_{g_c} : 1 \leq j \leq 5\} \cup$$
$$\{-1 + u + 2a_{k,l}x_k^c|_{e_i^c} \to 1|c + 1|d, e_i^c \to e_{i,1}^c, -1 + u + 2(a_{5,l}x_5^c + a_{6,l})|_{e_i^c}$$

$$\to 1|c + 1|d, u|_{e_i^c} \to 1|u : 1 \leq k \leq 4, 1 \leq l \leq 252\} \cup$$
$$\{-3c - 2d|_g \to 1|e_1^- + 1|a + 1|b + 1|s_1 + 1|s_2, g + 2c|_d \to 1|a + 1|b,$$

$$g + 2d|_c \to 1|s_1 + 1|s_2, g + 2a|_b \to 1|a + 1|s_1, b - 1 \to 1|b, b + 1|_u \to 1|b,$$

$$b + a|_u \to 1|b, b + s_1|_u \to 1|a, b - s_2|_u \to 1|a, b + 2s_2|_u \to 1|s_1 + 1|s_2,$$

$$b - 1|_u \to 1|n_1, a + 1|_u \to 1|n_1, n_1 - a + 3|_{n_2} \to 1|a, n_1 - b + 4|_{n_2} \to 1|b,$$

$$n_1 - a + 3|_{n_2} \to 1|s_1, n_1 + 3 - s_2|_{n_2} \to 1|s_1, n_1 + 3 + e_{i,1}^c|_{n_2} \to 1|e_{i+1}^c,$$

$$g - 2\beta_l t|_{e_{i+1}^c} \to 1|aux + 1|e_f : 1 \leq l \leq 252, n_1 + 3 + a|_{n_2} \to 1|t,$$

$$n_1 + 3 + e_1^-|_{n_2} \to 1|e_2^-, g - t|_{e_2^-} \to 1|t, e_2^- \to 1|_{g_c}, 2e_{253}^c \to 1|e_{254}^c + 1|e_f,$$

$$aux|_{e_f} \to 1|f_Q, -f_Q|_g \to 1|enum, -(f_Q + e_{254}^c) \to 1|e_f, enum + f_Q \to 1|_{g_c}, e_{254}^c \to 1|e_1\}$$

$V_0 = \{0, \cdots, 0\} \cup \{0, \cdots, 0\} \cup \{1, -2, 1\}.$

The ENP system constructed above is deterministic, and mainly includes two parts: the generation of 5-tuples and the computation of the polynomial's value.

In [15], Gheorghe Păun arose a question that whether or not the set of all vectors (a_1, ..., a_n) of natural numbers can be generated in a deterministic way (e.g., in a well specified order). In the following we show that this question can be affirmatively answered, by using enzymatic variables to control the application of evolution programs.

For every 5-tuple (x_5, x_4, x_3, x_2, x_1), we attach a label to it by denoting $a_{i,j}$, where $i = \sum_{k=1}^{5} x_k, j = \sum_{l=1}^{5} x_l \cdot 10^l$. We generate all the 5-tuples in the following order: first, we generate all the 5-tuple with label $i = 0$, which has only one tuple (0, 0, 0, 0, 0), then the tuples with label $i = 1$, the number of which is 5, then the tuples with label $i = 2$, $i = 3, \cdots$ and so on.

The generation order of the 5-tuples with the same label $i = n$, is determined by the index j, namely that if $j_1 < j_2$, then a_{n,j_1} is generated after a_{n,j_2}. For example, the ordering of 5-tuples with the same label $i = 3$ (which means that the tuple with the component sum is 3) is the following: the initial tuple is (3, 0, 0, 0, 0) (which is the biggest vale of index $j = 3 \cdot 10^5$, the next generated tuples are (2, 1, 0, 0, 0), (2, 0, 1, 0, 0), (2, 0, 0, 1, 0), (2, 0, 0, 0, 1), and then (1, 2, 0, 0, 0), $\cdots$, (1, 0, 0, 0, 2), (0, 3, 0, 0, 0), (0, 2, 1, 0, 0), $\cdots$, and so on until $(0, 0, 0, 0, 3)$, the tuple with the smallest value of index $j = 3$, is generated, then all the 5-tuples with index $i = 3$ are generated.

The algorithm of generating all 5-tuples with the same index i can be programmed as follows:

- Step 1. $x_1 \rightarrow 1|x_{11}$: Clear the value of x_1, and save its value to a new variable x_{11}.
- Step 2. Test if x_2 is zero. If $x_2 \neq 0$, decrements the vale of x_1 and add the value of $x_{11} + 1$ to x_1 simultaneously, then go to step 6. If $x_2 = 0$, go to step 3.
- Step 3. Test if x_3 is zero. If $x_3 \neq 0$, decrement the value of x_3 and add the value of $x_{11} + 1$ to x_2 simultaneously. Then go to step 6. If $x_3 = 0$, go to step 4.
- Step 4. Test if x_4 is zero. If $x_4 \neq 0$, decrement the value of x_4 and add the value of $x_{11} + 1$ to x_3 simultaneously. Then go to step 6. If $x_4 = 0$, go to step 5.
- Step 5. Test if x_5 is zero.

If $x_5 \neq 0$, decrement the value of x_5 and add the value of $x_{11} + 1$ to x_4 simultaneously. Then go to step 6. If $x_5 = 0$, the index of i increase 1, then go to step 6.

Step6. Transfer the values of $x_i (1 \leq i \leq 5)$ to the polynomial $f_Q(x_1, \cdots, x_5) = \sum_{i=1}^{252} \beta_i \cdot (a_{1,i}x_1 + \cdots + a_{5,i}x_5 + a_{6,i})^5$. After the computation, go to step 1. The programs used to generate all the 5-tuples are as follows:

$$\{1 + x_2|_{e_1} \rightarrow 1|ez_2, g + x_1|_{e_1} \rightarrow 1|x_{11}, e_1 \rightarrow 1|e_2,$$

$$g + e_i|_{ez_i} \rightarrow 1|e_{i+1}, 1 + x_{i+1}|_{ez_i} \rightarrow 1|ez_{i+1}, ez_i + 1|_{e_i} \rightarrow 1|e_s, ez_i \rightarrow 1|e_i \rightarrow 1|x_i,$$

$$x_{11} + ez_i + 1|_{e_i} \rightarrow 1|x_1, x_{11} - 2|_{ez_i} \rightarrow 1|x_1, e_i + ez_i \rightarrow 1|gc : 2 \leq i \leq 4$$

$$ez_5 - 1|_{e_5} \rightarrow 1|x_5, x_{11} + ez_5 + 1|_{e_5} \rightarrow 1|x_4, x_{11} - 2|_{ez_5} \rightarrow 1|x_4,$$

$$1 + g + x_{11}|_{ez_5} \rightarrow 1|x_5, g + 1|_{e_5} \rightarrow 1|e_s, e_5 + ez_5 \rightarrow 1|gc,$$

$$g + 2x_j|_{e_s} \rightarrow 1|x_j + 1|x_j^c, g + 2|_{e_s} \rightarrow 1|e_1^c, e_s \rightarrow 1|gc : 1 \leq j \leq 5\}$$

The generation of all the 5-tuples in the increasing order of the the sum of the tuple's components, namely that starting with 0, successively generate the 5-tuples with the sum of $1, 2, \cdots$. For the tuples with the component's sum of n, the first one generated is $(n, 0, 0, 0, 0)$, and the last one generated is $(0, 0, 0, 0, n)$, which signals that all the sum of n has been generated and the next step will generate the 5-tuples with the component's sum of $n + 1$.

The generation of the 5-tuples with the component's sum n complies with the following 2 rules:

Rule 1: If $x_1 = 0$, and the first digit not equal to 0 from the right is x_{i_0}, which means all the digits to its right are 0, then except decrementing x_{i_0}, and adding 1 to $x_{i_0?1}$, others have no change.

Rule 2: If $x_1?0$, first save the value of x_1 to an auxiliary x_{11}, then reset x_1 to zero. If the first digit is not equal to 0 from the right is x_{i_0}, then subtracts 1 from x_{i_0}, and add $x_{1,1}+1$ to $x_{i_0}?1$. Others keep the value unchanged. Before expatiating how the programs run, we give some explanations of the variables involved.

- $x_i[0]$, $1 \leq i \leq 5$ represent the components of 5-tuples from left to right.
- $e_1[1]$, enzymatic variable to start all the programs.
- $e_i[0]$, $2 \leq i \leq 5$ are the enzymatic variables used to control the execution flow of the system.
- $ez_i[0]$, used to test whether the corresponding variable x_i is zero.
- $e_s[0]$, the variables to control terminating the generation of the 5-tuples.

The process of generating a 5-tuple takes 6 time steps. In the following we will present the computation in each time step.

- Step 1:$e_1 = 1$

The active programs are:

(1) $g + x_1|_{e_1} \to 1|x_{11}$, saves the value of x_1 to the auxiliary x_{11}, and resets the value of x_1 to zero.
(2) $1 + x_2|_{e_1} \to 1|ez_2$, is used to test whether the value of x_2 is zero. Because e_1 is 1, the program activates only when x_2 is zero.
(3) $e_1 \to 1|e_2$, transfers the value of e_1 to e_2. Note that e_1 is consumed in this program and is not produced by any other one. Thus the value of e_1 in the next time step is zero.

  - Step 2 :$e_2 = 1$

The active programs are:

(1) $g + e_2|_{ez_2} \to 1|e_3$ is active if $ez_2 = 1$, which means $x_2 = 0$. When $x_2 \neq 0$, thus $ez_2 = 0$, this program will not be active.
(2) $1 + x_3|_{ez_2} \to 1|ez_3$ are active only when $x_3 = 0$ and simultaneously $ez_2 = 1$.
(3) $1 + ez_2|_{e_2} \to 1|e_s$ transfers 1 to the variable e_s, which activates the computation of the polynomials and terminates the program of the 5-tuple generation. The condition of this program active is the variable $ez_2 = 0$.
(4) $ez_2 - 1|_{e_2} \to 1|x_2$ decrements x_2 if $ez_2 = 0$, corresponding that $x_2 \neq 0$.
(5) $1+x_{11}+ez_2|_{e_2} \to 1|x_1$ adds the value of $x_{11}+1$ to the variable of x_1. This program is supposed to activate only when the condition $ez_2 = 0$ is satisfied. But we can see that when $ez_2?0$ and $x_{11} = 0$, it can activate undesirably. In order to compensate this condition, we design the program (6).
(6) $x_{11} - 2|_{ez_2} \to 1|x_1$ offsets the superfluous value brought to x_1 by program (5).
(7) $e_2 + ez_2 \to 1|gc$ rests the value of e_2 and ez_2 to zero.

  - Step 3 and step 4 are similar to step 2. All the programs in step 2 can translate to step 3 and step 4. We just skip to step 5.

- Step 5: $e_5 = 1$

The active programs are:

(1) $ez_5 - 1|_{e_5} \rightarrow 1|x_5$ decrements x_5 if $ez_5 = 0$, corresponding that $x_5 \neq 0$.
(2) $x_{11} + ez_5 + 1|_{e_5} \rightarrow 1|x_4$ adds the value of $x_{11} + 1$ to the variable of x_4. This program is supposed to activate only when the condition $ez_5 = 0$ is satisfied. But when $ez_5 \neq 0$ and $x_{11} = 0$, this program activates undesirably. In order to compensate this condition, we design the program (3).
(3) $x_{11} - 2|_{ez_5} \rightarrow 1|x_4$ offsets the superfluous value brought to x_4 by program (2).
(4) $1 + g + x_{11}|_{ez_5} \rightarrow 1|x_5$ adds the value of $x_{11} + 1$ to the leftmost digit x_5. This program activates when the variable $ez_5 = 1$, which implies that both the variable x_5 and the variables x_i, $2 \leq i \leq 4$ are equal to zero. This corresponds to the 5-tuple in the form $(0, 0, 0, 0, n)$, which is the last 5-tuple with all the components' sum n. The next generation should transfer to the tuples with the components' sum $n + 1$. This program realizes the transformation above-mentioned.
(5) $g + 1|_{e_5} \rightarrow 1|e_s$ goes to the next step.
(6) $e_5 + ez_5 \rightarrow 1|g_c$ rests the value of e_5 and ez_5 to zero.

- Step 6: $e_s = 1$

The active programs are:

(1) $g + 2x_j|_{e_s} \rightarrow 1|x_j + 1|x_j^c : 1 \leq j \leq 5\}$ copy the generated 5-tuples for processing.
(2) $g + 2|_{e_s} \rightarrow 1|e_1^c$ transfers the control to the next group grograms.
(3) $e_s \rightarrow 1|g_c$ resets the variable e_s to zero.

After the generation of one 5-tuple, the value of enzymatic variable e_s is transferred to e_1^c, which initiate the computation of the polynomial $f_Q(x_1, \cdots, x_5) = \sum_{i=1}^{252} \beta_i \cdot (a_{1,i}x_1 + \cdots + a_{5,i}x_5 + a_{6,i})^5$. The process of the computation is composed by 3 steps. The first step is to compute the value of $a_{1,i}x_1 + \cdots + a_{5,i}x_5 + a_{6,i}$. Then raise the value of $a_{1,i}x_1 + \cdots + a_{5,i}x_5 + al6, i$ to its fifth power. The last step is to sum up the values of $(a_{1,i}x_1 + \cdots + a_{5,i}x_5 + a_{6,i})^5$, $1 \leq i \leq 252$.

When $e_1^c = 2$, the programs activated are

$$-1 + u + 2a_{1,1}x_1^c|_{e_1^c} \rightarrow 1|c + 1|d,$$

$$-1 + u + 2a_{2,1}x_2^c|_{e_1^c} \rightarrow 1|c + 1|d,$$

$$-1 + u + 2a_{3,1}x_3^c|_{e_1^c} \rightarrow 1|c + 1|d,$$

$$-1 + u + 2a_{4,1}x_4^c|_{e_1^c} \rightarrow 1|c + 1|d,$$

$$-1 + u + 2(a_{5,1}x_5^c + a_{6,1})|_{e_1^c} \rightarrow 1|c + 1|d,$$

$$u|e_1^c \;\rightarrow\; 1|u,$$

$$e_1^c \;\rightarrow\; e_{1,1}^c.$$

After these 7 programs are implemented, the variables c and d will get the value of $a_{1,1}x_1 + \cdots + a_{5,1}x_5 + a_{6,1}$. For the coefficient is integer, so the value of $a_{1,1}x_1 + \cdots + a_{5,1}x_5 + a_{6,1}$, namely c and d is possible negative. In order to compute the 5 power of c and d, we first distinguish the value of c and d between positive and negative. If the values of c and d are negative, the condition $c < g$ and $d < g$ will be satisfied, which can activate the program $-3c - 2d|g \rightarrow 1|e_1^- + 1|a + 1|b + 1|s_1 + 1|s_2$, and prohibit the execution of programs $g + 2c|d \rightarrow 1|a + 1|b, g + 2d|c \rightarrow 1|s_1 + 1|s_2,$. While the values of c and d are positive, the condition $c > g$ and $d > g$ satisfied, the programs $g + 2c|d \rightarrow 1|a + 1|b, g + 2d|c \rightarrow 1|s_1 + 1|s_2$, activate, and the program $-3c - 2d|g \rightarrow 1|e_1^- + 1|a + 1|b + 1|s_1 + 1|s_2$ will be prohibited from execution. The variable e_1^- is used to record the negative sign of c and d. When the value of c and d are negative, e_1^- will equal to the absolute value of c or d. When the value of c and d are positive, e_1^- will be zero unchanged. Each the variable a, b, s_1 and s_2 will get the absolute value of c or d.

Computing the power of a number can be done by using only multiplication; computing the 5-th power of n can be done by first computing $a = n \cdot n = n^2$, then $b = n \cdot a = n \cdot n^2 = n^3$ and finally $c = a \cdot b = n^2 \cdot n^3 = n^5$. This process can be programmed as: $a = s_1, b = a, s_1 = a \cdot b$. The program starts with $a = b = s_1 = n$ and the variable s_1 will be n^5 after running 3 steps. For n is a positive integer, multiplication can be realized by repeated addition, $a \cdot b = a + \ldots + a$.

The process of the 5-power computation is controlled by the state of b.

The multiplication $a \cdot b$ is executed by the programs $g + 2a|b \rightarrow 1|a + 1|s_1$ and $b - 1 \rightarrow 1|b$. For every time step, the program $g + 2a|b \rightarrow 1|a + 1|s_1$ accumulates one a to s_1 and the program $b - 1 \rightarrow 1|b$ decrements one from b. When $b=0$, the program $g + 2a|b \rightarrow 1|a + 1|s_1$ terminates and s_1 receives $a \cdot b$. Note that the variable s_1 has the initial value a, and by now it is not consumed, so the current value of s_1 is $n + n^2$. The condition $b = 0$ is not only the sign of the multiplication $a \cdot b$ is finished, but also the sign of the following programs start: $b + 1|u \rightarrow 1|b, b + a|u \rightarrow 1|b, b + s_1|u \rightarrow 1|a, b - s_2|u \rightarrow 1|a, b + 2s_2|u \rightarrow 1|s_1 + 1|s_2, b - 1|u \rightarrow 1|n_1$.

Program $b + 1|u \rightarrow 1|b$ is an auxiliary program, which is used to compensate the value of b generated by the program $b - 1 \rightarrow 1|b$ when $b = 0$. Program $b + a|u \rightarrow 1|b$ is used to assign the value of a to b. Similarly, program $b + s_1|u \rightarrow 1|a$ together with $b - s_2|u \rightarrow 1|a$ is used tfo replace the value of a with a^2. Program $b + 2s_2|u \rightarrow 1|s_1 + 1|s_2$ keeps the variables s_1 and s_2 equal to the value of a. Program $b - 1|u \rightarrow 1|n_1$ saves the times of $b = 0$ in the variable n_1.

For $a = n^2$, $b = n$, repeating the process as mentioned above, we can get $s_1 = n^3 + n$ and $n_1 = -2$. Applying program $b + a|u \rightarrow 1|b$ to reset the value of b, programs $b + s_1|u \rightarrow 1|a$ and $b - s_2|u \rightarrow 1|a$ to reset the value of a, we can obtain $a = n^3, b = n^2$. After implementing the multiplication programs $g + 2a|b \rightarrow 1|a + 1|s_1$ and $b - 1 \rightarrow 1|b$ for the third time, the value of variable n_1 is changed to -3, which activates the program with enzymatic variable n_2 for the reason that $n_1 = -3 < n_2 = -2$.

Programs $n_1 - a + 3|_{n_2} \rightarrow 1|a$, $n_1 - b + 4|_{n_2} \rightarrow 1|b$, $n_1 - a + 3|_{n_2} \rightarrow 1|s_1$, $n_1 + 3 - s_2|_{n_2} \rightarrow 1|s_1$ are used to set the values of a, b, s_1, s_2 to the initial ones. The value of a, which is equal to n^5, is transferred to the variable t by program $n_1 + 3 + a|_{n_2} \rightarrow 1|t$. If the initial value $a_{1,1}x_1 + \cdots + \alpha_{5,1}\xi_5 + \alpha_{6,1}$ is negative, programs $n_1 + 3 + e_1^-|_{n_2} \rightarrow 1|e_2^-$ and $g - t|_{e_2^-} \rightarrow 1|t$ convert the value of t to its contrary number. Program $n_1 + 3 + e_{1,1}^c|_{n_2} \rightarrow 1|e_2^c$ passes the value of $e_{1,1}$ to e_2, which activates the computation of the next term of the polynomial. The process goes on until the last term of the polynomial is computed. After all the $m = 252$ terms are computed, they are summed up by the programs $g - 2\beta_l t|_{e_{l+1}^c} \rightarrow 1|aux + 1|e_f$, $1 \leq l \leq 252$. Note that the number stored in the variable aux and e_f is the contrary value of the polynomial f_Q, the aim of which is to choose the positive value of f_Q as the output of Π in the next steps. Enzymatic variable e_{253}^c activates the program $g - 2\beta_{252}t|_{e_{253}^c} \rightarrow 1|aux + 1|e_f$, which adds the last term to the variables e_f and aux, and simultaneously confers its value to the variables e_{254}^c and e_f by the program $2e_{253}^c \rightarrow 1|e_{254}^c + 1|e_f$. Program $aux|_{e_f} \rightarrow 1|f_Q$ is used to transfer the value of aux (which contains the value of the polynomial) to f_Q, and to test the value of the polynomial against zero. Because e_f and aux are equal, this program is not active until the program $2e_{253}^c \rightarrow 1|e_{254}^c + 1|e_f$ (all terms were computed) breaks the balance. Program $-f_Q|_g \rightarrow 1|enum$ ensures that the value saved in $enum$, where the ENP system's results are collected, is positive. The value of variables f_Q, e_{254}^c, $enum$ are cleaned up by programs $-(f_Q + e_{254}^c) \rightarrow 1|e_f$ and $enum + f_Q \rightarrow 1|g_c$. Finally, program $e_{254}^c \rightarrow 1|e_1$ transfers the enzymatic variable e_{254}^c to e_1, which starts generating the next 5-tuple.

3 Conclusions and Discussions

In this work, we design an algorithm using enzymatic numerical P systems to generate all the n-tuples of natural numbers in a well specified order. Specifically, we improve previously known results on generating all the n-tuples of natural numbers. Based on this method, we prove that a numerical P system with one membrane, and the production function used with polynomials of degree 1 with at most 2 variables can reach universality, which optimizes the previous known results. The results also give a positive answer to a problem formulated in [MCE].

It remains open whether the indices in the systems constructed in our proofs can be decreased. In particular, it is of interest to investigate whether the system can reach universality by using less programs with no increment of other indices.

Generating n-tuples by numerical P systems is first proposed as an challenge by Gheorghe Păun in [MCE]. In this work we generate n-tuples by enzymatic numerical P systems. It is also interesting to investigate whether other numerical P systems, such as numerical P systems with thresholds, numerical P systems with migrating variables, can also generate n-tuples in a deterministic way.

Acknowledgments. This work is supported by the National Natural Science Foundation of China (61872325, 61876047, 61872399); the Fundamental Research Funds for the Central Universities (2652019028) and Youth Program of National Natural Science Foundation of China (61802009).

References

1. Pavel, A.B., Arsene, O., Buiu, C.: Enzymatic numerical P systems–a new class of membrane computing systems. In: IEEE Fifth International Conference on Bio-Inspired Computing: Theories and Applications (BIC-TA), Liverpool, pp. 1331–1336 (2010)
2. Pavel, A.B., Buiu, C.: Using enzymatic numerical P systems for modeling mobile robot controllers. Nat. Comput. **11**(3), 387–393 (2012)
3. Pavel, A.B., Vasile, C.I., Dumitrache, I.: Robot localization implemented with enzymatic numerical P systems. In: Prescott, T.J., Lepora, N.F., Mura, A., Verschure, P.F.M.J. (eds.) Living Machines 2012. LNCS (LNAI), vol. 7375, pp. 204–215. Springer, Heidelberg (2012). https://doi.org/10.1007/978-3-642-31525-1_18
4. Păun, G., Rozenberg, G., Salomaa, A. (eds.): The Oxford Handbook of Membrane Computing. Oxford University Press, New York (2010)
5. Ciobanu, G., Păun, G., Pérez-Jiménez, M.J. (eds.): Applications of Membrane Computing. Springer, Heidelberg (2006)
6. Păun, G.: Computing with membranes. J. Comput. Syst. Sci. **61**(1), 108–143 (2000)
7. Păun, G., Păun, R.: Membrane computing and economics: numerical P systems. Fundamenta Informaticae **73**(1), 213–227 (2006)
8. Vasile, C.I., Pavel, A.B., Dumitrache, I., Păun, G.: On the power of enzymatic numerical P systems. Acta Informatica **49**(6), 395–412 (2012)
9. Ciobanu, G., Pan, L., Păun, G., Pérez-Jiménez, M.J.: P systems with minimal parallelism. Theoret. Comput. Sci. **378**(1), 117–130 (2007)
10. Wang, J., Shi, P., Peng, H.: Membrane computing model for IIR filter design. Inf. Sci. **329**, 164–176 (2016)
11. Wang, T., Zhang, G., Zhao, J., He, Z., Wang, J., Pérez-Jiménez, M.J.: Fault diagnosis of electric power systems based on fuzzy reasoning spiking neural p systems. IEEE Trans. Power Syst. **30**(3), 1182–1194 (2015)
12. Peng, H., Wang, J., Pérez-Jiménez, M.J., Wang, H., Shao, J., Wang, T.: Fuzzy reasoning spiking neural P system for fault diagnosis. Inf. Sci. **235**, 106–116 (2013)
13. Liu, X., Li, Z., Liu, J., Liu, L., Zeng, X.: Implementation of arithmetic operations with time-free spiking neural P systems. IEEE Trans. Nano Biosci. **14**(6), 617–624 (2015)
14. Păun, G.: Membrane Computing–An Introduction. Springer, Berlin (2002). https://doi.org/10.1007/978-3-642-56196-2
15. Song, T., Pan, L.: Spiking neural P systems with rules on synapses working in maximum spikes consumption strategy. IEEE Trans. Nanobiosci. **14**(1), 37–43 (2015)
16. Song, T., Pan, L.: Spiking neural P systems with rules on synapses working in maximum spiking strategy. IEEE Trans. Nanobiosci. **14**(4), 465–477 (2015)
17. Song, T., Pan, L., Jiang, K., Song, B., Chen, W.: Normal forms for some classes of sequential spiking neural P systems. IEEE Trans. Nanobiosci. **12**(3), 255–264 (2013)
18. Song, T., et al.: Normal forms of spiking neural P systems with anti-spikes. IEEE Trans. Nanobiosci. **11**(4), 352–359 (2012)
19. Song, T., Xu, J., Pan, L.: On the universality and non-universality of spiking neural P systems with Rules on Synapses. IEEE Trans. Nanobiosci. **14**(8), 960–966 (2015)
20. Zeng, X., Zhang, X., Song, T., Pan, L.: Homogeneous spiking neural P systems. Fundamenta Informaticae **97**(1), 275–294 (2009)
21. Zeng, X., Zhang, X., Song, T., Pan, L.: Spiking neural P systems with thresholds. Neural Comput. **26**(7), 1340–1361 (2014)
22. Zhang, X., Pan, L., Păun, A.: On universality of axon P systems. IEEE Trans. Neural Networks Learn. Syst. **26**(11), 2816–2829 (2015)

23. Zhang, X., Zeng, X., Pan, L.: Weighted spiking neural P systems with rules on synapses. Fundamenta Informaticae **134**(1), 201–218 (2014)
24. Zeng, X., Song, T., Pan, L., Zhang, X.: Performing four basic arithmetic operations by spiking neural P systems. IEEE Trans. Nanobiosci. **11**(4), 366–374 (2012)
25. Zhang, G., Rong, H., Ou, Z., Pérez-Jiménez, M.J., Gheorghe, M.: Automatic design of deterministic and non-halting membrane systems by tuning syntactical ingredients. IEEE Trans. Nanobiosci. **13**(3), 363–371 (2014)
26. Zhang, Z., Wu, T., Păun, A., Pan, L.: Numerical P systems with migrating variables. Theor. Comput. Sci. **641**, 85–108 (2016)
27. Zhang, Z., Pan, L.: Numerical P systems with thresholds. Int. J. Comput. Commun. Control **11**(2), 292–304 (2016)
28. Zhang, Z., Wu, T., Pan, L., Păun, G.: On string languages generated by numerical P systems, Romanian. J. Inf. Sci. Technol. **18**(3), 273–295 (2015)
29. Zhang, Z., Wu, T., Pan, L., Păun, G.: On string languages generated by sequential numerical P systems. Fundamenta Informaticae **145**(4), 485–509 (2016)

DNA Self-assembly Computing Model for the Course Timetabling Problem

Zheng Kou[1], Zhibao Xing[2], Wenfei Lan[2], and Xiaoli Qiang[1]([⊠])

[1] Institute of Computing Science and Technology,
Guangzhou University, Guangzhou 510006, Guangdong, China
qiangxl@gzhu.edu.cn
[2] College of Computer Science, South-Central University for Nationalities,
Wuhan 430074, Hubei, China

Abstract. The course timetabling problem is a highly-constrained combination problem, which cannot be solved in polynomial time by a deterministic algorithm. In this paper, a DNA self-assembly computing model was presented to solve a course timetabling problem which can meet all students' course-choosing requests using the least number of classes. There are three subsystems in this DNA self-assembly computing model, which are initial solution space generation system, detection system, and time slots counting system. The results demonstrated that the university timetable self-assembly system can obtain the solution using $O(n^2)$ tiles.

Keywords: DNA computing · Self-assembly · Course timetabling problem

1 Introduction

In 1994, Adleman [1] explored the feasibility of computing directly with DNA molecules, which was the first work to apply DNA computing in vitro. Since then, DNA molecule has been used as a computing substrate, because of its advantages of stability and specificity. DNA molecules have been used for molecular computing [2, 3], construction of nano-scale structures [4, 5], logic gates [6, 7], information storage [8], and even nano-machines [9] through the programmed hybridization of complementary strands.

The course timetabling problem is often encountered in colleges and universities, and university course timetabling problem have been proven to be an NP problem [10]. The complexity and practicality of the problem has attracted many scholars. Currently, the algorithms for solving the timetabling problem are mainly divided into several categories. Such as, the operational research algorithm, which can be divided into the graph-based coloring algorithm and linear programming [11], or the integer programming algorithm and constraint-based solution algorithm and the meta-heuristic algorithm, which includes the genetic algorithm, ant colony optimization algorithm and mimetic algorithm [12] et al.. In 2007, Zhou et al. [13] proposed a DNA computing model by using closed circle DNA molecules for course timetabling problem. In 2014, Shan et al. [14] propose

© Springer Nature Singapore Pte Ltd. 2020
H. Han et al. (Eds.): IDMB 2019, CCIS 1099, pp. 245–253, 2020.
https://doi.org/10.1007/978-981-15-8760-3_18

a surface DNA computing model based on micro-spotting technology to solve course timetabling problem.

Here, a DNA self-as computing model was presented for the course timetabling problem by using DNA tiles. This algorithm takes advantage of DNA self-assembly calculations and uses $O(n^2)$ tiles to solve the course timetabling problem.

2 Course Timetabling Problem

The course timetabling problem is one of the complex problems that is composed of a set of instructors teaching specific courses, a set of courses offered during the academic semester, a list of students registered to a number of courses, and a fixed number of room location. The problem aims to find a possible scheduling meeting some certain constraints.

Here we consider the course timetabling problem as follows: given a set of course, each student can be registered some of them to study with the constraint that different courses chosen by a student should not be arranged in a same time slot. Bipartite graph $G = (V_c, V_s, E)$ could be used to show the requirements of timetabling problem. $V_c = \{k_1, k_2, ..., k_n\}$ represents the course set, $V_s = \{s_1, s_2, ..., s_m\}$ represents the student set and edge set E represents the selection requirements. $(k_i, s_j) \in E$ means the requirement that student s_j choose course k_i. An example (see Fig. 1) proposed in reference [15] is used here to show how to get a timetable according to the students' requirement by using DNA self-assembly computing model.

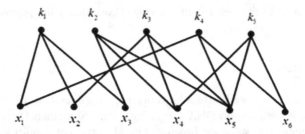

Fig. 1. A bipartite graph to show the student selection requirements.

Assuming that there are 2 time slots in both the morning and afternoon, and 1 slot in the evening. If classes can be scheduled from Monday to Friday, the number of available time slots should be $5 \times 5 = 25$ different time slots.

3 DNA Self-assembly Computing Model for Course Timetabling Problem

3.1 Designing DNA Self-assembly System

The abstract tile assembly model, which provides a rigorous framework for analyzing algorithmic self-assembly, was originally proposed by Winfree [16]. It extends the theoretical model of tiling proposed by Wang to include a mechanism for growth based on the physics of molecular self-assembly.

To solve the course timetabling problem, three sub-systems (initial solution space generation system S_{INIT}, detection system S_{INIT}^{T}, and time slots counting system S_{CNT}) are constructed to implement the following algorithm.

Step 1: Given five classes, randomly assign courses to form initial course solution space E_{init}.

Step 2: According to students' requirements to form the candidate solution space note as E'_{init}.

Step 3: According to the course information to delete the non-feasible solutions from E'init.

Step 4: Use statistics tiles to count the number of time slots used in a feasible solution.

Step 5: Read all solutions to the optimal solution space.

3.2   Initial Solution Space Generation Subsystem

(1)   DNA self-assembly model of the initial solution space generation system

Let $S_{INIT} = <T_{INIT}; g_{INIT}; \tau_{INIT}>$ be the initial solution space generation system, where T_{INIT} is the set of all tiles required for the initial solution space generation system. The set of binding domains for T_{INIT} is $\Sigma_{INIT} = \{I, II, -, \alpha, null\}, \alpha \in \{k_1, k_2,\ldots, k_5\}$ and g_{INIT} is the function of the binding domain. The definition is as follows:

$\forall \sigma \in \Sigma_{INIT}, g_{INIT}(\sigma, null) = g_{INIT}(null, \sigma) = 0;$
$\forall \sigma \in \Sigma_{INIT} \setminus \{null\}, g_{INIT}(\sigma, \sigma) = 1;$
$\forall \sigma, \sigma' \in \Sigma_{INIT}, if \ \sigma \neq \sigma', then \ g_{INIT}(\sigma, \sigma') = 0.$

The binding strength between any binding domain and null is 0. The binding strength of any binding domain other than null to its own binding domain is 1. The system thermodynamic temperature is set to $\tau_{INIT} = 3$, which means that only when the sum of the binding domains strength is greater than or equal to 3 can it be stably integrated into the existing configuration.

(2)   DNA tiles in the initial course solution spatial generation system

The DNA tiles used in the initial course solution space system S_{INI} are shown in Fig. 2. There are 4 binding domains for each DNA tile, binding domain bd_E and bd_S are the input edges of information, binding domain bd_W and bd_N are the output edges of information. Binding domain I means no course is arranged, and the tile with binding domains $bd_E = bd_S = I$ will transport the information from right to left, and $bd_W = bd_N = I$ can transport the information from the bottom to the top. The tile with binding domains $bd_S = I$ and $bd_N = \alpha$ is computing tile, and the result is arranging course α in a time slot.

The seed configuration S'_{INIT} are made up of gray tiles and notice that the M tile are used here as a separator between time slots. The position of the rightmost tile (S tile) in row 0 of S'_{INIT} is (x_0, y_0). The tiles shown in Fig. 2a (1)–(3) can be assembled in seed configuration S'_{INIT} during the self-assembly process according to the complementary binding domain. For example, the computing tile with binding domains $bd_S(t) = bd_E(t) = I$ can be assembled at the position (x_{0-1}, y_{0+1}). The self-assembly operation is performed under the

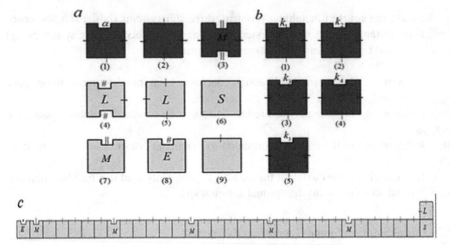

Fig. 2. DNA tiles used for S_{INI}. a Two types of tile. Gray tiles are used for the seed configuration, and their binding domain bd_S can be combined with the binding domain bd_N of blue tiles. b The basic computing tiles for S_{INI}, there $\alpha \in \{k_1, k_2, ..., k_5\}$. c The seed configuration denoted as S'_{INIT}. (Color figure online)

control of combination domain function and system temperature τ_{INIT}. The self-assembly process is shown in Fig. 3.

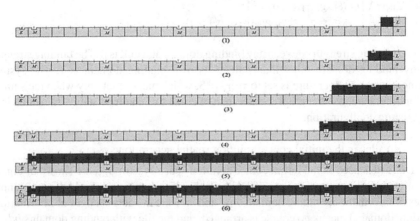

Fig. 3. The computation process Initial course solution space system self-assembly S_{INIT}.

By using S_{INIT} the courses will be arranged in a random order, and we need to choose and detect the feasible solutions by using detection system S_{INIT}^T according to the constraint that all the requirement of students will be satisfied without conflict.

3.3 Detection Subsystem

(1) The DNA self-assembly model of detection subsystem

Let $S_{INIT}^T = <T_{INIT}^T, g_{INIT}^T, \tau_{INIT}^T>$ be the DNA self-assembly model of detection system, where T_{INIT}^T is the set of all tiles designed for S_{INIT}^T. The set of binding domains for tile in T_{INIT}^T is $\Sigma_{INIT}^T = \{v, r, *, +, =, \alpha, \|, |, \#, \text{null}, 0\}$, $\alpha \in \{k_1, k_2, ..., k_5\}$, and g_{INIT}^T is the function of the binding domain. The definition is as follows:

$$\forall \sigma \in \sum_{INIT}^T, g_{INIT}^T(\sigma, null) = g_{INIT}^T(null, \sigma) = 0;$$

$$\forall \sigma \in \sum_{INIT}^T \setminus \{null\}, g_{INIT}^T(\sigma, \sigma) = 1;$$

$$\forall \sigma, \sigma' \in \sum_{INIT}^T, \text{ if } \sigma \neq \sigma', \text{ then } g_{INIT}^T(\sigma, \sigma\prime) = 0.$$

The value of τ_{INIT}^T are set to be the same as τ_{INIT} in S_{INIT}.

(2) DNA tiles in detection system

The DNA tiles in S_{INT} are known as double-crossover (DX) tiles with four binding domains. In detection system, more binding domain should be used and designed for marking the state of student or the course. Here the tiles with 6 binding domains are designed (see Fig. 4).

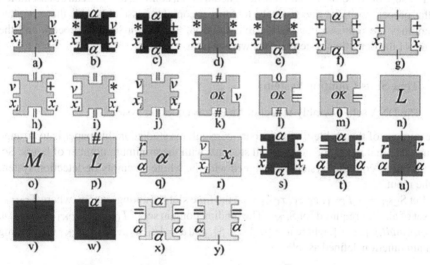

Fig. 4. DNA tiles designed for S_{INT}^T.

Binding domains designed here are all satisfies the requirements shown in Fig. 1 without conflict. Binding domain vx_i means that the timetable detected is no conflict for student x_i. $*x_i$ means student x_i have a course in this time slot. $+xi$ shows that no course chosen by student xi arranged in the previous time slot.

The DNA tiles to detect the state of course are also designed using binding domain $r\alpha$ and $= \alpha$. $r\alpha$ means no course α detected, and $= \alpha$ means course α detected.

To detect the feasible solution, the tiles q and r shown in Fig. 4 should be assembled to S_{INT} as boundary tile firstly (See Fig. 5) to form the seed configuration of S_{INT}^T.

Fig. 5. The seed configuration of S_{INT}^T

As you can see from Fig. 5, the self-assembly operation process of the detection system can be summarized as follows: Firstly, student x_i chooses the course according to his requirements. Secondly, detecting the courses to ensure each of them is arranged correctly.

Because the binding domains are designed without conflict, the tiles can be assembled to the proper positions successively to form a complete assembly. All the complete assemblies are representing the feasible solution. The self-assembly process and the results of the detection system are shown in Fig. 6.

3.4   Time Slots Counting System

(1) The DNA self-assembly model of time slots counting system

The purpose of the problem of university curriculums studied in this paper is to arrange a timetable that meets the needs of all students with the minimum number of hours. So, the time slots counting system is designed with the whole assembly of detection system as the input.

Let $S_{CNT} = <T_{CNT}, g_{CNT}, \tau_{CNT}>$ be the time slots counting system, where T_{CNT} is the set of all tiles required for S_{CNT}. The binding domain set of T_{CNT} is $\Sigma_{CNT} = \{\times, v, n, n', \alpha, |, \|, null, 0\}, z, n'\}$, where $n = \{0, 1, ..., 5\}$, and $\alpha = \{k_1, k_2, ..., k_5\}$. g_{CNT} is a binding domain function defined as follows:

$\forall \sigma \in \Sigma_{CNT}, g_{CNT}(\sigma; null) = g_{CNT}(null; \sigma) = 0;$
$\forall \sigma \in \Sigma_{CNT} \setminus \{null, n'\}, g_{CNT}(\sigma; \sigma) = 1;$
$\forall \sigma, \sigma' \in \Sigma_{CNT}, \text{if } \sigma \neq \sigma', \text{then } g_{CNT}(\sigma, \sigma') = 0;$
$g_{CNT}(n', n') = 4$

The value of τ_{INIT}^T are set to be the same as τ_{INIT} in S_{INIT}.

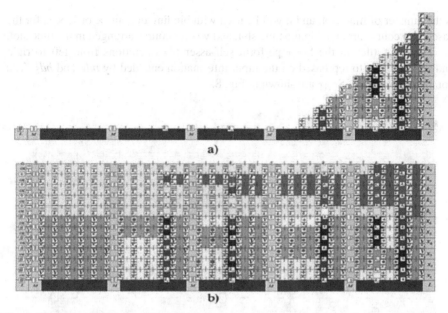

Fig. 6. The self-assembly process and the results of the detection system. (a) The self-assembly process of S_{INT}^T. (b) A complete assembly representing a feasible solution.

(2)   DNA tiles in time slots counting system

The system S_{CNT} is designed for counting the time slots for each possible solution. If there is no course arranged in a time slot, the counting number should be 0, otherwise, the counting number should be added 1. The DNA tiles designed for SCNT are shown in Fig. 7.

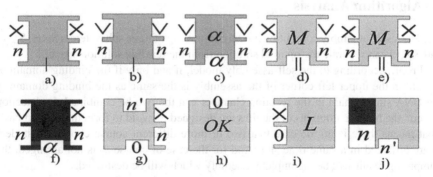

Fig. 7. DNA tiles designed for S_{CNT}.

As shown in Fig. 7, the binding domains denoted as α, I, and II represent that course α was arranged in the time slot, no course was arranged in the time slots, and the special binding domain for M tile (the separator for time slots) respectively. Binding domain n

is the number of time slot, and it will be used with binding domain × or v. × is for the case of no course arranged in the time slot, and v is for course arranged in the time slot.

The DNA tiles in the S_{CNT} perform self-assembly operations from left to right and from bottom to top based on the input information encoded by bd_E and bd_S. The computing results of S_{CNT} are shown in Fig. 8.

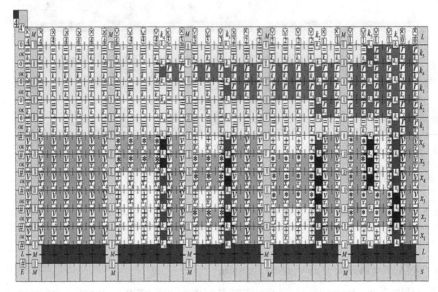

Fig. 8. The computing result of SCNT.

4 Algorithm Analysis

The DNA self-assembly computing model composting of three subsystems (S_{INIT}, S_{INIT}^T, and S_{CNT}) can separate the optimal solution from the feasible solution space.

Proof: According to the self-assembly model, if and only if the binding domain of the tile in the upper left corner of the assembly is the same as the binding domain in the DNA single chain in the reaction chamber can the two be combined. In detection system, the binding domains of the tiles are designed to avoid to form the conflict case. That means there is no the case that two or more different course chosen by student x_i are arranged in a same time slot. And for the case of course k_j is not arranged, the computing result will be incomplete assembly which will be destroyed.

According to the time slot counting system, the value of the time slots number is set to be 0 at the beginning of the self-assembly calculation. During the self-assembly calculation process, each time slot will be scanned in order. The total number only increases by 1 if there are one or more courses in the time slot. Hence, it is possible to accurately count the number of class hours that can be arranged in a feasible solution.

5   Conclusion

To solve the course timetabling problem, three sub-systems (initial solution space generation system S_{INIT}, detection system S_{INIT}^{T}, and time slots counting system S_{CNT}) are constructed to implement the following algorithm. The model can use 38 tile type $O(n^2)$ tiles to obtain the optimal solution of the university curriculum in linear time.

It is well known that the course timetabling problem is one of the complex problems and it should be satisfied with a number of hard and soft constraints. How to construct the DNA computing model to solve course time tabling problem with more constraints need to further study.

References

1. Adleman, L.M.: Molecular computation of solutions to combinatorial problems. Science **266**, 1021–1024 (1994)
2. Xu, J.: Probe machine. IEEE Trans. Neural Networks Learn. Syst. **27**(7), 1405–1416 (2016)
3. Lan, W.F., Xing, Z.B., Huang, J., Qiang, X.L.: The DNA self-assembly computing model for solving perfect matching problem. J. Comput. Res. Dev. **53**(11), 2583–2593 (2016)
4. Ke, Y., Ong, L.L., Shih, W.M., et al.: Three-dimensional structures self-assembled from DNA Bricks. Science **6111**(338), 1177–1183 (2012)
5. Praetorius, F., Kick, B., Behler, K.L., et al.: Biotechnological mass production of DNA origami. Nature **552**, 84–87 (2017)
6. Yang, J., Wu, R., Li, Y., et al.: Entropy-driven DNA logic circuit regulated by DNAzyme. Nucleic Acids Res. **46**(16), 8532–8541 (2018)
7. Qian, L., Winfree, E.: Scaling up digital circuit computation with DNA strand displacement cascades. Science **6034**(332), 1196–1201 (2011)
8. Goldman, N., Bertone, P., Chen, S., et al.: Towards practical, highcapacity, low-maintenance information storage in synthesized DNA. Nature **494**(7435), 77–80 (2013)
9. Thubager, A.J., Li, W., Johnson, R.F., et al.: A cargo-sorting DNA robot. Science **357**(6356), 1095–1096 (2017)
10. Fong, C.W., Asmuni, H., Lam, W. S., McCollum, B., McMullam, P.: A novel hybrid swarm based approach for curriculum based course timetabling problem. In: 2014 IEEE Congress on Evolutionary Computation, pp. 544–550. IEEE Press, Beijing (2013)
11. Qu, R., Burke, E.K.: Hybridizations within a graph-based hyper-heuristic framework for university timetabling problems. J. Oper. Res. Soc. **60**, 1273–1285 (2008)
12. Abdullah, S., Turabieh, H.: On the use of multi neighborhood structures within a tabu-based memetic approach to university timetabling problems. Inf. Sci. **191**, 146–168 (2012)
13. Zhou, K., Tong, X., Liu, W.: Algorithm of closed DNA computing model for time-table problem. Comput. Appl. **27**(4), 991–993 (2007)
14. Shan, J., Yin, Z.: The solution to the time-table problem based on surface DNA computing model. J. Anhui Univ. Sci. Technol. (Nat. Sci.) **34**(1), 11–14 (2014)
15. Wu, J.: On time-table problem for arranging course in University. Oper. Res. Manag. Sci. **11**(6), 66–71 (2002)
16. Winfree, E.: Algorithmic self-assembly of DNA [dissertation]. California Institute of Technology, Pasadena (1998)

Optimization of GenoCAD Design Based on AMMAS

Yingjie Wang[1] and Yafei Dong[1,2(✉)]

[1] School of Computer Science, Shaanxi Normal University, Xian 710119, China
dongyf@snnu.edu.cn
[2] College of Life Science, Shaanxi Normal University, Xian 710119, China

Abstract. By successively click on prescribed grammar captured in successful genetic designs to assemble a range of genetic parts for example BioBricks, large and complicated genetic systems composed of substantial functional blocks can be constructed. As the number of genetic parts increases, each category of genetic parts includes so many parts that the process of assembling a great deal of genetic parts is costly, time-consuming and error-prone. GenoCAD is a web-based application for synthetic biology to guide users through the design of artificial gene networks, protein expression vectors and other complex genetic constructs by continuously click on predefined grammars according to the notion of genetic parts. However, at the last step of a design in GenoCAD, it's difficult for users to determine which basic part will be taken in every category. On the basis of statistical language model, a probability distribution over a string S reflecting how frequently a string S will occur and a mathematical model to select basic genetic parts to form a genetic construct can be determined. After converting the parts assembly process into a mathematical model, adaptive maximum-minimum ant system (AMMAS) proposed in this paper can be applied to the mathematical model to figure out an optimal combination of parts of a design with maximum probability automatically within seconds. The adaptive maximum-minimum ant system (AMMAS) can not only optimize the parts selection process of a design but also can devise particular projects performing specific functions based on former successful parts assembly experience. Consequently, redundant operations can be reduced and cost as well as time spent in biological experiments can be minimized drastically.

Keywords: GenoCAD · AMMAS · Statistical language model · Grammars

1 Introduction

The rapid development of synthetic biology makes it essential to develop methodologies to streamline the design of custom genetic systems [1]. Gene expression network, metabolic engineering and protein expression vector are some of applications in this field [2, 3]. GenoCAD, a web-based application for synthetic biology, can satisfy the needs of scientific studies in synthetic biology and allow users to quickly devise genetic constructs based upon the notion of genetic parts [4]. GenoCAD is built upon a solid computational linguistic foundation and can guide users through genetic designs by successively click

© Springer Nature Singapore Pte Ltd. 2020
H. Han et al. (Eds.): IDMB 2019, CCIS 1099, pp. 254–271, 2020.
https://doi.org/10.1007/978-981-15-8760-3_19

on prescribed grammars capturing design strategies of specific applications [4]. Users, who elect to create a personal account, can log in the system to engineer project-specific parts libraries, upload new parts into their workspace and save designs for later use [5]. Designers usually decompose large biological sequences into functional blocks as genetic parts to expand parts database including promoter, terminator, plasmid backbone, gene and ribosome binding site (RBS) which are necessary for designing genetic constructs [6]. The compelling vision of libraries of biological parts enabling a fast and cheap assembly of large biological systems is one of the foundations of synthetic biology [7, 8]. There are several assembly standards to follow when assembling a set of genetic parts into genetic constructs and the BioBrick Foundation (BBF) has been favorable for promoting the BioBrick standard. A BioBrick standard compliant part is a DNA fragment flanked by a prefix and a suffix segment having particular restriction sites [9, 10]. In addition, two BioBrick compliant parts can be assembled together using multiple specific ligations and restriction digestions independent of both parts sequences, which indicates that any number of parts compliant with a same assembly standard can be assembled into a new complex genetic construct by means of specific restriction digestions and ligations.

When assembling series of genetic parts into intricate genetic systems, users commonly are unsure of selecting an appropriate basic part in a part category. In Consequence, it's always costly, error-prone and time-consuming for biologists to determine a combination of parts of a genetic construct with biological experiments. For the sake of minimizing time and cost spent in the parts assembly process, researchers have developed robotic platforms to automate the parts assembly process which can be used to devise genetic systems by continuous click on the pre-defined grammars to convert the structure of genetic designs. Ultimately users will choose a genetic part from every parts category to fulfill their designs [11]. However, with the development of synthetic biology, an increasing number of genetic parts are developed, which makes users confused to select a suitable part from every parts category at the last step of a design (Fig. 1). Therefore, this study is carried out to settle the problem of selecting a reasonable genetic part from every parts category to build a project-specific genetic constructs. Above all, statistical language model (SLM) is introduced to facilitate the parts assembly process by transforming the parts assembly process into a mathematical model according to interaction between parts. Applications of statistical language model are as diverse as speech recognition, machine translation, word segmentation, part of speech tagging and other natural language applications. The established mathematical model in this paper can be solved by statistical parameters extracted from BioBrick standard parts downloaded from iGEM website and the proposed AMMAS to work out an optimal combination of parts with maximum probability to accomplish a genetic construct. Taking former successful iGEM part assemblies and resulted statistical parameters into account, our algorithm can be used to minimize cost and time spent in the parts assembly process. Our suggested scheme can not only select an optimized combination of parts at the last step of a genetic design in robotic platforms for example GenoCAD, but also can devise new projects performing specific functions based on former successful experience.

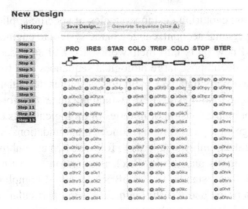

Fig. 1. Every icon has so many options.

2 Materials and Methods

A static snapshot of iGEM Registry content in June 2014 containing 7242 parts is available at http://parts.igem.org/das/parts/entry_points/ and Perl script is employed to parse out and analyze the content of each part at the link http://parts.igem.org/das/parts/features/?segment=part. Decomposed into a structured and unified data format, these parts can be imported into a relational database MySQL. Using SQL sentences, finally 75744 features sequences are acquired. Genetic parts comprise both basic parts which are unable to further divided into subparts such as promoter, terminator and ribosome binding site (RBS) [12, 13] and composed parts consisting of multiple basic parts such as system and device. Employing SQL sentences, basic parts and their usage frequencies in composed parts can be counted as well as composite parts and the usage frequencies of two adjacent basic parts (parts pair) in composed parts can be uniformly calculated. By querying the MySQL database, we extracted 1682 basic parts including 405 promoters, 57 terminators, 42 RBSs and 1178 genes, which can be used to build complicated genetic constructs. The experience of previous successful assembly of genetic constructs is utilized to guide us into the parts selection so that the resulted genetic constructs according to our algorithm will be reliable and scientific.

3 Mathematical Model

At the last step of a design in robotic platforms for example GenoCAD, there are so many choices in every parts category to complete the design making it a hard nut to crack to select the most suitable part from every part category (Fig. 1). It seems impossible and impractical to carry out exhaustive search method by testing all possible combinations of parts with wet biological experiments. And representing the structure of a genetic construct by using a reasonable mathematical model can simplify the process of choosing a suitable combination of parts [14]. To meet our expectations, statistical language model (SLM) is suggested to convert the parts assembly process into a mathematical model which takes the probability of occurrence of a sentence as a proof of its reasonableness.

A sentence S, denoted as a genetic construct, is composed of a strand of words which can be regarded as basic parts and the probability of a genetic construct can be evaluated accordingly.

$$S = part_1, part_2, \ldots, part_n \tag{1}$$

$$P(S) = P(part_1, part_2, \ldots, part_n) \tag{2}$$

In accordance with conditional probability formula, following formula (3) can be derived.

$$
\begin{aligned}
P(S) &= P(part_1, part_2, \ldots, part_n) \\
&= P(part_1) \cdot P(part_2|part_1). \\
&\quad P(part_3|part_1, part_2). \\
&\quad \ldots \cdot P(part_n|part_{n-1}, part_{n-2}, \ldots part_1)
\end{aligned}
\tag{3}
$$

In above formula (3), $P(part_1)$ is the probability $part_1$ occurs in a design and $P(part_2|part_1)$ means the probability that $part_2$ appears with $part_1$ prior to it. Moreover, the probability $part_n$ occurs hinges on all the parts prior to it making it difficult to calculate $P(part_n|part_{n-1}, part_{n-2}, \ldots, part_1)$ compared with computing $P(part_1)$ and $P(part_2|part_1)$ owing to so many variables involved in it distinctly. According to Markov Hypothesis, the probability a part will occur is merely related to one or more parts prior to it. Hence formula (3) can be described in a simplified way as follows:

$$
\begin{aligned}
P(S) &= P(part_1, part_2, \ldots, part_n) \\
&= P(part_1) \cdot P(part_2|part_1) \cdot \\
&\quad \ldots \cdot P(part_n|part_{n-1})
\end{aligned}
\tag{4}
$$

The formula (4) presented above is bi-gram of statistical language model which implies whether a part will appear in a design is simply concerned with one part prior to it. In line with the notion of conditional probability formula, conditional probability formulas involved in formula (4) can be deduced.

$$P(part_i|part_{i-1}) = \frac{P(part_{i-1}, part_i)}{P(part_{i-1})} \tag{5}$$

Utilizing usage frequencies of basic parts and parts pairs extracted from the downloaded feature sequences, we can estimate $P(part_i)$ and $P(part_i, part_{i-1})$ respectively.

$$P(part_i, part_{i-1}) \approx \frac{count(part_i, part_{i-1})}{count(all\_parts)} \tag{6}$$

$$P(part_i) \approx \frac{count(part_i)}{count(all\_parts)} \tag{7}$$

By means of above formulas (5–7), we can calculate conditional probability formulas involved in formula (4).

$$P(part_i|part_{i-1}) = \frac{count(part_{i-1}, part_i)}{count(part_{i-1})} \tag{8}$$

In this way, all components in formula (4) can be estimated meaning the combination of parts with maximum probability can be figured out. In accordance with the statistical language model (SLM), the combination of parts with maximum probability is the most meaningful and reasonable. There are too many candidate paths to accomplish a design and a path will lead to an S (a path $=$ an $S = part_1, part_2, \ldots, part_n$). The optimal path among all candidate paths can be represented by $PATH$.

$$
\begin{aligned}
PATH &= \arg \max_{all\_S}(P(S)) \\
&= \arg \max_{all\_S}((P(part_1) * \prod_{i=2}^{n} P(part_i|part_{i-1}))
\end{aligned}
\tag{9}
$$

Since the database we use is extracted from a relatively sparse corpus, zero-frequency problem is inevitable when parts pairs never occur in the corpus meaning their corresponding *count* will be zero. This circumstance causes the deviation in calculating $P(S)$ and $PATH$. To overcome this difficulty, Add-k data smoothing technology is applied to settle the zero-frequency problem [15], and the corresponding conditional probability formulas involved in formula (4) can be gauged as follows.

$$
P(part_i|part_{i-1}) = \frac{count(part_{i-1}, part_i) + k}{count(part_{i-1}) + k|W|}
\tag{10}
$$

Furthermore, the target function in this paper is represented by $f(S)$.

$$
\begin{aligned}
f(S) &= P(S) \\
&= P(part_1) * \prod_{i=2}^{n} P(part_i|part_{i-1})
\end{aligned}
\tag{11}
$$

W is the total number of parts pairs and formula (10) is employed to replace the conditional probability formula (8). Hence all components in formula (4) can be gauged and the resulted $PATH$ is regarded as the optimal path (S) with maximum probability of occurrence among all candidate paths. After transforming the parts assembly process into a mathematical model, adaptive maximum-minimum ant system (AMMAS) can be exploited to automate this parts selection process efficiently.

4 Algorithms

The next step is to use the proposed algorithm to figure out a path, composed of a sequence of basic parts with the largest probability, in this lattice. This algorithm can direct us through the process of solving the target function $f(S)$.

The ant system algorithm, inspired by the observation of ant colonies in real world, was first proposed by Dorigo et al. in 1991 as a population-based approach to settle difficult combinatorial optimization problems [16–18]. Furthermore, an interesting and important behavior of ants is how to find a shortest path between their nest and food sources. While walking from the nest to food sources and vice versa, ants deposits a substance called pheromone on the ground, forming in this way a pheromone trail [19]. When deciding the direction to go, they choose paths which are marked by stronger pheromone centration with higher probability. Ant's tendency to determine a specific path is positively correlated with the intensity of a found trail. The basic behavior is a foundation for a cooperative interaction relationship that results in the emergency of the shortest paths. The ant system algorithm has been applied to plenty of difficult combinatorial optimization problems such as the quadratic assignment problem [20, 21] and traveling salesman problem (TSP). The performance of ant system algorithm can be enhanced by introducing maximum and minimum trail strengths on arcs, named maximum and minimum ant system, to alleviate the problem concerning early stagnation. However, long runs of the maximum and minimum ant system (MMAS) still show stagnation behavior, despite of using minimum and maximum trail limits. Therefore, we raised three main changes to further improve its performance.

4.1 Simulated Annealing Mechanism

Beginning by randomly generating an initial solution, simulated annealing is a neighborhood search technique to resolve combinatorial problems. At each stage, a new solution taken from the neighborhood of the current solution will be accepted as a new current solution if it has a lower or equal cost; If it has a higher cost it will be accepted with a probability which decreases as the difference in cost between two solutions increases and as the temperature of the method decreases [22]. This temperature, simply a positive number, is reduced periodically according to the following formula, so that it can gradually from a relatively high value to near zero as the algorithm progresses. At the beginning of simulated annealing, most worsening moves are accepted, however, only improving ones are more likely to be accepted in the later stage of the algorithm. Furthermore, to enhance the intensity and diversity of simulated annealing searching procedure, when a solution doesn't show better performance within a prescribed number of continuous cooling process, the restarting solution mechanism, called tempering mechanism, is designed to generate new solutions for the further solution improving and the maximum number of tempering is $H_{\max}$.

$$T(N + 1) = a \times T(N) \tag{12}$$

In addition, in order to avert premature convergence of the algorithm, the simulated annealing algorithm presented by the following formula (13) is carried out.

$$p = \begin{cases} \exp(-\frac{f(S_{global})-f(S)}{T(t)}) f(S) \leq f(S_{global}) \\ 1 \qquad\qquad\qquad f(S) > f(S_{global}) \end{cases} \tag{13}$$

Where S_{global} is the global optimal route and S is a collection of paths resulted in this round. To make the performance of our proposed algorithm more robust, comparing the calculated p by taking out solutions from S one by one with a random number γ within the interval [0, 1] is necessary. It is noted that a good quality solution can be confirmed in the case of $p = 1$ or $p > \gamma$; Otherwise a path defined as a bad solution will be determined. Moreover, the temperature reduction factor a of 0.9 is chosen, which has been indicated to be satisfactory in the gradual temperature reduction process [23, 24].

4.2 Adaptive Pheromone Concentration Updating Mechanism

The pheromone updating mechanism is designed to allocate a great amount of pheromone concentration to short tours, in a sense, which is similar to the reinforcement learning schema. It is widely recognized that better solutions will get a higher reinforcement. The pheromone updating formula was intended to simulate the change of the amount of pheromone in virtue of both the addition of new pheromone deposited by ants on the visited edges and of pheromone evaporation. Ants have memory ability, however, as time goes on, information is lost. In order to prevent the algorithm from getting into local optimum due to large differences of pheromone density between the worst path and the best path, pheromone updating formula is designed to dynamically adjust the volatilization coefficient of pheromone as follows.

$$\rho(N+1) = 1/\log_2(1+N_c) \tag{14}$$

In this research, the volatilization coefficient of pheromone ρ reduces gradually up to $1/\log_2(1+N_c) < \rho_{min}$.

4.3 Adaptive Change of Weight Coefficient

As previously noted, to intensify and diversify the searching procedure and to make the solution found more robust, a dynamic change mechanism of weight coefficient α and β is designed to fulfill the purpose when the current global optimal path does not change within 50 rounds. The idea is to maintain a high ability to search for new solutions and prevent algorithm from getting into local optimum not only by reducing the relative influence of pheromone, but also by increasing the relative influence of the heuristic information, presented as follows, thus the goals of intensification and diversification of the algorithm can be achieved.

$$\beta = sl/(3 \cdot sum\_column) \tag{15}$$

Where sum_column is the total number of columns of the built lattice. In the above formula, sl reflects the total number of ants which go through all edges of the optimal solution of one iteration.

4.4 State Transition Rules

To achieve better balance between using prior knowledge and exploring new paths, the pseudo-random rate rule is selected when ant k chooses next node j from node i, specifically described as follows.

$$j = \begin{cases} \arg\max_{u \in allowed_k} \left\{ [\tau_{iu}(t)]^\alpha * [\eta_{iu}(t)]^\beta \right\}, & if \ q \le q_0 \\ p_{ij}^k(t), & else \end{cases} \tag{16}$$

$$p_{ij}^k(t) = \begin{cases} \dfrac{(\tau_{ij}(t))^\alpha (\eta_{ij}(t))^\beta}{\sum\limits_{l \in Allowed_k(i)} (\tau_{il}(t))^\alpha (\eta_{il}(t))^\beta} & j \in Allowed_k(i) \\ 0 & else \end{cases} \tag{17}$$

where q is a random number with uniform distribution in [0, 1] and variable q_0, defined as below, determines the relative importance degree between using prior information and exploring new paths. α and β emphasize the importance degree of pheromone concentration and heuristic information respectively while η_j represents the heuristic information guiding the selection of the next node.

$$q_0 = sum\_column / sl \qquad (18)$$

$$\eta_j = P(part_j | part_i)$$
$$= \frac{P(part_i, part_j)}{P(part_i)} \qquad (19)$$

4.5 The AMMAS Algorithm

Then the improved ant colony system is designed to solve the problem of biological parts selection and it consists of two steps as below.

First of all, a lattice is created. Each column, expressed by one icon, corresponds to one parts category and every node in each column refers to as a basic part. Part name is unique in the workspace.

Afterwards, the improved algorithm, composed of five steps, can be applied to solve the built statistical language models and the algorithm flow chart is as follows.

Step 1: The initial pheromone information $\tau_{ij} = \tau_{max}$ (a constant), the iteration counter of the algorithm N_c, an adjustable parameter q_0 involved in the state transition rule, the maximum iteration N_{max}, the maximum number of nodes N_m an ant passes in each round are all to be initialized. The variable n is the number of nodes one ant goes through and it is set to $n = 1$. To ensure that the pheromone distribution varies along the paths at the beginning of the algorithm, the threshold U ($U \leq 10$) can be employed to determine whether ants choose next node according to formula (17) or not in the early stage of the algorithm. m ants are randomly assigned to all nodes of the first column of the established lattice.

Step 2: If $N_c \leq U$, ant k chooses next node in accordance with formula (17) and makes a step forward according to the selected node as well as places that node into its taboo list $Tabu_k$. If $N_c > U$, ants will choose appropriate node selection formula based on the comparison between q_0 and q.

Step 3: If the number of nodes ant k passed in this round has not reached the given number N_m, the algorithm will go to Step 2 to decide next node, on the contrary go to Step 6.

Step 4: When all ants complete this iteration, the route with maximum probability among all candidate paths in this iteration can be gained. And volatilization factor of pheromone ρ can be changed based on formula (14). In addition, pheromone information on all paths can be updated globally, however, updating global pheromone information in this way can't well guide ants towards the global optimum. In view of this situation, the introduction of reward and punishment mechanism regarding better and worse solutions respectively is necessary. Different edges have diverse impacts on guiding ants towards global optimal solutions. Considering this characteristic, we are intended to allocate more pheromone on good paths and less pheromone on bad paths as below. Updating pheromone concentration in this way can accelerate the convergence speed.

$$\tau_{ij}(t+1) = (1-\rho)\tau_{ij}(t) + \sum_{k=1}^{m} \Delta\tau_{ij}^{good} \qquad (20)$$

$$\tau_{ij}(t+1) = (1-\rho)\tau_{ij}(t) - \sum_{k=1}^{m} \Delta\tau_{ij}^{bad} \qquad (21)$$

Where

$$\Delta\tau_{ij}^{good} = w \cdot f(s^{best}) \qquad (22)$$

$$\Delta\tau_{ij}^{bad} = w \cdot f(s^{worst}) \qquad (23)$$

Where $w(0 \leq w \leq m)$ expresses the total number of edge (i, j) appearing in all candidate paths. Moreover, $f(s^{best})$ and $f(s^{worst})$ depict best and worst value of the target function in this round respectively.

Step 5: Compared with global optimal path, if the optimal path obtained in this iteration is better than the current global optimal one, the global optimum will be replaced. In order to jump out of the local optimal and expand the search range, β can be further dynamically changed if global optimal path remains unchanged within fifty rounds.

The temperature T should be updated according to $T \leftarrow aT, t \leftarrow t+1$. If $T \geq T_{min}$, the proposed algorithm goes to Step 2 to start a new iteration. If $T < T_{min}$ and $H = H_{max}$, the algorithm outputs the saved global optimized path. If $T < T_{min}$ and $H < H_{max}$, then $H \leftarrow H+1$, $T \leftarrow T_{max}$ and it goes to Step 2.

Ultimately, the optimal combination of parts to form a genetic construct can be worked out by the algorithm in seconds after entering the parts category sequences, which is also an optimal solution with maximum probability to the target function $f(S)$. In comparing the proposed algorithm with exhaustive algorithm, one must bear in mind that the running time of the former is on the order of seconds, which is acceptable in synthetic biology (Fig. 2).

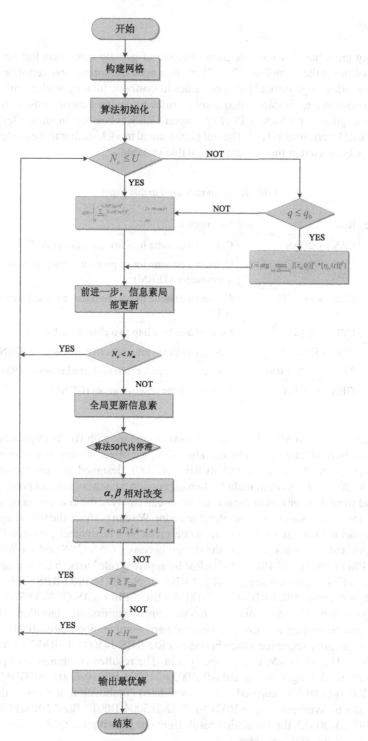

Fig. 2. The algorithm and flow chart.

5   Result

Developing grammars for modeling the structure of genetic constructs has become a routine tool in synthetic biology [25]. There is a need for simple and versatile design strategies to allow high throughput approaches in synthetic biology studies (rule-based design). Therefore, we implemented a set of rules to design genetic constructs based on the basic grammar which is PRO (promoter)-RBS (ribosome binding sites)-GEN (genes)-TERM (terminator) [26]. The full grammatical model, similar to the context-free grammar (CFG), used in this system is available in Table 1.

Table 1. Grammars used in this paper

Rule code	Rule	Description
1	CAS → 2CAS	Convert a cassette into two cassettes (CAS)
2	CAS → Pro CIS TERM	Convert a cassette into a promoter (Pro), a cistron (CIS), a terminator (TERM)
3	CAS → Pro CIS	Convert a cassette into a promoter (Pro) and a cistron (CIS)
4	CIS → 2CIS	Convert a cistron into two cistrons (CIS)
5	CIS → RBS GEN	Convert a cistron into a rbs (RBS) and gene (GEN)
6	TERM → 2TERM	Convert a terminator into two terminators (TERM)
7	GEN → 2GEN	Convert a gene into two genes (GEN)

To describe how to assemble series of parts compliant with BioBrick standard into a functional biosynthetic system by our algorithm, we select wintergreen odor biosynthetic system (http://parts.igem.org/Part:BBa_J45700), designed and implemented by MIT iGEM 2006. This system includes two expression cassettes: one can produce salicylate acid from cellular metabolic and the other can catalyze the conversion of the salicylate acid to methyl salicylate or wintergreen odor. We can perform the following grammatical model to direct users through the wintergreen odor biosynthetic system. Starting with a CAS and by means of rule1, the design becomes CAS-CAS and the following design is PRO-CIS-TERM-PRO-CIS-TERM by applying rule2 to both CAS. Employing rule5 to both CIS, the design turns into PRO-RBS-GEN-TERM-PRO-RBS-GEN-TERM and finally it becomes PRO-RBS-GEN-TERM-TERM-PRO-RBS-GEN-TERM-TERM according to rule6. After determining genes we want to express, our algorithm in Python language can be applied to choose an optimal parts combination automatically for the input parts category sequence which becomes PRO-RBS-J45004-TERM-TERM-PRO-RBS-J45017-TERM-TERM to form the system. The resulted combination of parts by our bi-gram model algorithm is R0040-B0032-J45004-B0010-B0012-R0010-B0032-J45017-B0010-B0012. Compared with the validated combination of parts of the wintergreen odor biosynthetic system R0040-B0032-J45004-B0010-B0012-R0011-B0032-J45017-B0010-B0012, the simulation result from our algorithm (Fig. 3) is very similar to that verified one of this system.

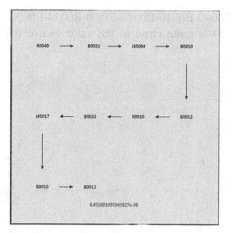

Fig. 3. The simulation results of the first system.

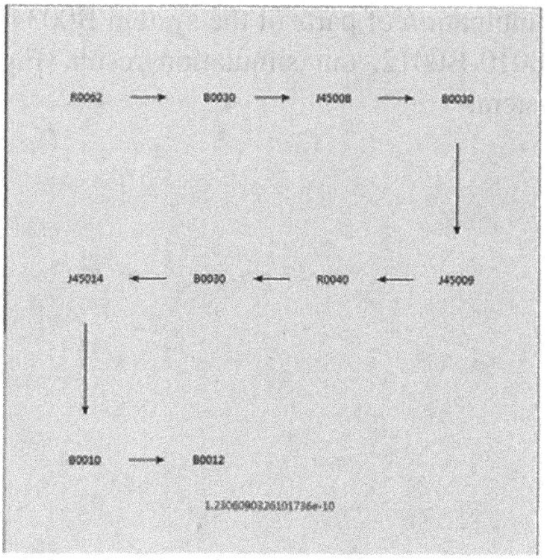

Fig. 4. The simulation results of the second system.

Moreover, for another example, we elect the banana odor biosynthetic system (http://parts.igem.org/Part:BBa_45900), implemented and designed by teams enrolling in iGEM in 2006. This system includes two expression cassettes: one can produce isoamyl alcohol with *BAT*2 and *THI*3 and the other can catalyze the conversion of the cellular metabolite leucine to isoamyl acetate or banana odor. We can implement the following grammatical model to direct users through the banana odor biosynthetic system. Starting with a CAS and by means of rule1, the design becomes CAS-CAS and the following design is PRO-CIS-PRO-CIS-TERM by applying rule3 to the first CAS and rule2 to the second CAS. Employing rule4 to first CIS and rule5 to the second CAS, the design turns into PRO-CIS-CIS-PRO-RBS-GEN-TERM and it becomes PRO-RBS-GEN-RBS-GEN-PRO-RBS-GEN-TERM according to rule5. Finally the design is PRO-RBS-GEN-RBS-GEN-PRO-RBS-GEN-TERM-TERM according to rule6. After determining genes to be expressed in this design, our algorithm in Python language can be utilized to pick out an optimal parts combination automatically for the input parts category sequence which becomes PRO-RBS-J45008-RBS-J45009-PRO-RBS-J45014-TERM-TERM to form the system. The resulted parts series of this design by our bi-gram model algorithm is R0010-B0030-J45008-B0030-J45009-R0040-B0030-J45014-B0010-B0012. Compared with the validated combination of parts of the banana odor biosynthetic system R0011-B0030-J45008-B0030-J45009-R0040-B0030-J45014-B0010-B0012, the simulation result from our algorithm (Fig. 4) is very close to that verified one of this system.

In addition, we select the design RBS.GFP + PBad CFP (http://parts.igem.org/Part: BBa_I13404), also designed and implemented by the team participating in iGEM in 2006, as another example to illustrate efficiency of our algorithm. Under the input design RBS-E0040-TERM-TERM-PRO-RBS-E0020–TERM-TERM, the bi-gram model algorithm proposed in this paper recommends the parts series B0034-E0040-B0010-B0012-R0010-B0034-E0020-B0010-B0012. As can be seen during comparison with the actual combination of parts of the system B0034-E0040-B0010-B0012-I0500-B0034-E0020-B0010-B0012, our simulation result (Fig. 5) is quite close to the valid one of the system.

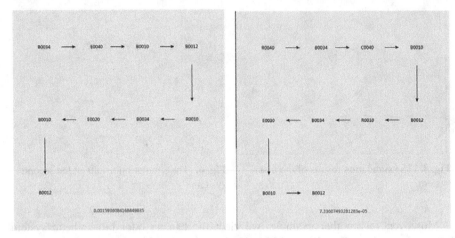

Fig. 5. The simulation results of the third system.

Fig. 6. The simulation results of the fourth system.

For the fourth example, we select the example I0500.Q04400.E0430 (http://parts.igem.org/Part:BBa_E0611) which was designed and implemented by the group taking part in iGEM in 2004. Under the input part category sequence PRO-RBS-C0040-TERM-TERM-PRO-RBS-E0030-TERM-TERM for this genetic construct, our bi-gram model algorithm recommends a combination of parts R0040-B0034-C0040-B0010-B0012-R0010-B0034-E0030-B0010-B0012. According to the comparison between simulation results and real results I0500-B0034-C0040-B0010-B0012-R0040-B0034-E0030-B0010-B0012, there are two basic parts that differ from the real parts series in simulation ones (Fig. 6).

Furthermore, we select a design from the link http://parts.igem.org/Part:BBa_S01664, designed and implemented by the team participating in iGEM in 2004, as the fifth example. Under the input design PRO-RBS-C0051-TERM-TERM-PRO-RBS-C0012-TERM-TERM-PRO, the bi-gram model algorithm suggests the parts sequence R0040-B0034-C0051-B0010-B0012-R0010-B0034-C0012-B0010-B0012-R0010. As is illustrated during comparison with the actual combination of parts of the system R0040-B0034-C0051-B0010-B0012-R0051-B0034-C0012-B0010-B0012-R0011, our simulation result (Fig. 7) is similar to the verified one of the system.

The example [TetR][rbs][LuxR][dblTerm][LuxPR] + [rbs][LacI][dblTerm] "AHL-dependent inverter" (http://parts.igem.org/Part:BBa_J23040), designed and implemented by the group participating in iGEM in 2006, is chosen as the sixth example. Based on the input parts category sequence of the design PRO-RBS-C0062-TERM-TERM-PRO-RBS-C0012-TERM-TERM, a set of parts can be figured out (R0040-B0034-C0062-B0010-B0012-R0010-B0034-C0012-B0010-B0012) by our algorithm according to the specifications required for this design. In comparison with the real parts

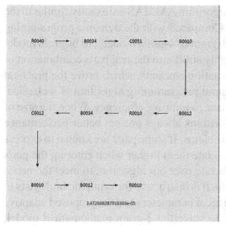

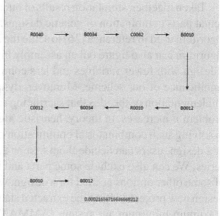

Fig. 7. The simulation results of the fifth system.

Fig. 8. The simulation results of the sixth system.

combination of the design R0040-B0034-C0062-B0010-B0012-R0062-B0034-C0012-B0010-B0012, our simulated result (Fig. 8) has only one different basic part from the real one.

We consider the example QPI Test Construct Intermediate (Q04121.E0430) (http://parts.igem.org/Part:BBa_I13021), designed and implemented by the group joining in iGEM in 2004, as the seventh example. To meet specific needs of the design, our algorithm presents a genetic construct B0034-C0012-B0010-B0012-R0010-B0034-E0030-B0010-B0012 within seconds on the basis of the input parts series RBS-C0012-TERM-TERM-PRO-RBS-E0030-TERM-TERM. It is obvious that our simulation result (Fig. 9) is pretty similar to the valid series of parts of the design B0034-C0012-B0010-B0012-R0011-B0034-E0030-B0010-B0012.

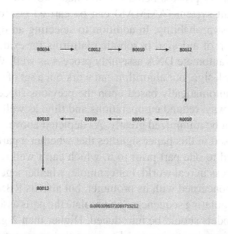

Fig. 9. The simulation results of the seventh system.

Taken together, simulation results of our algorithm AMMAS are exactly similar to the actual parts combination of genetic designs. Compared with the dynamic programming algorithm used in reference [26] to settle the 3-gram or 4-gram model, our bi-gram model algorithm can also figure out an assembly highly similar to the real parts combination of a design with fewer variables and less computation amounts, which prove the practical significance of our scheme. Moreover, dynamic programming algorithm is a classical implementation of the idea that sacrificing space to improve efficiency. When the size of problem n increases, in theory, heuristic algorithms always present better performance in solving such combinatorial optimization problems. If some parts are known to express in a design, users can decide them first or evaluate them higher when entering the parts series. We can also exclude some parts and iterate over our algorithm to meet the needs of some other options needed for a design, which is instructive for synthetic biologists to design new projects. Using the extracted statistical parameters and the proposed adaptive maximum-minimum ant system (AMMAS) to solve the 2-gram mathematical model, the resulted optimal combination of parts is scientific and reliable.

In above cases, R0010 always appears at the same time as B0012. One of the reasons for this is that the database we use is sparse, and the other reason is that the bi-gram statistical language model can't well reveal the interaction relationships between parts categories. To address this issue, it is essential to adopt higher-gram model to depict the interaction relationships between parts categories.

6 Discussion

This paper presents an efficient algorithm AMMAS to guide users through the design of genetic constructs performing specific function by selecting an optimal parts combination for a genetic construct at the last step of a design in GenoCAD and to devise projects meeting specific requirements. Utilizing the concept of statistical language model and conditional probability, the parts assembly process can be converted into a mathematical model. The parts assembly process being transformed into a bi-gram model, adaptive maximum-minimum ant system (AMMAS) can be carried out to choose an optimal solution with the largest probability. In addition to selecting an optimized parts combination at the final step of a design in robotic platforms for example GenoCAD, this method can be used to automate DNA assembly process as well. We entering the parts category sequence of a design, our algorithm can work out a set of suitable parts to form the genetic construct automatically based upon the previous successful assemblies on iGEM website. In this way, redundant operations and time as well as cost spent in biological experiments can be minimized greatly. As depicted above, bi-gram of statistical language model proposed in this paper signifies that whether a part can be enabled in a design is simply related to one part prior to it, which can't well reveal the mechanism of DNA assembly process in real world. For example, whether a gene will be expressed efficiently is not only concerned with its promoter, but also its RBS and plasmids backbone as well as other regulating sequences. To simulate the parts assembly process in real world, higher-gram models should be introduced. Higher than 2-gram models indicate that one part involved in a design is related to more than one part prior to it. However, there are so many variables involved in these higher-gram models making calculating conditional probability formulas a hard nut to crack. When developing higher-gram models, computers of high performance are necessary [27] though the accuracy of the results has improved greatly.

Since the dataset we used was extracted from a relatively sparse corpus, zero-frequency issue is inevitable when some parts pairs never appear. When calculating conditional probability formulas involved in the mathematical model, we employed Add-k smoothing technology to address the zero-frequency issue. However, it's not a good idea to utilize Add-k smoothing due to its disadvantages such as considerable amount of the probability space allocated to unseen events. It is used just for simplicity. Therefore, other smoothing techniques will be considered to improve the accuracy of the results such as Katz Backoff smoothing, Good-Turing smoothing, Witten-Bell and so on [28]. Some parts are likely to appear in any analysis so that the simulation results of the algorithm have a certain deviation from the real values. The reason may be that the existence of noisy data in the dataset results in the deviation. We intend to adopt the commonly used basic parts and parts pairs with high usage frequency (more than three times for basic parts and parts pair) in the corpus next. The commonly used basic parts and parts pairs with high usage frequency (more than three times for basic parts and parts pair) are regarded as successful words while the others are referred to as noisy data. In addition to improving data smoothing techniques, it is also of great importance to expand the database. However, expanding the corpus needs more operations to represent the notion and definition of parts and features in a unified format. That is, we should eliminate inconsistencies between features and redundant data in the corpus. The problem can be resolved by developing the ontology giving the community a controlled vocabulary to depict parts and features in a uniform format. And developing the synthetic biology open language (SBOL) will accelerate this process remarkably.

Based upon statistical language model, we present an efficient computational supplement for designs in robotic platforms of synthetic biology for example GenoCAD. In synthetic biology, it's an important question that too many choices are offered at the last step of a design. It's a matter of considerable interest to take the previous successful assemblies into account when we develop new projects. For those who don't have the expertise in synthetic biology, it's fairly difficult to elect a suitable part in a particular category. Users can choose a train of suitable parts to form a design according to a body of existing experiences by our algorithm. Our newly proposed method will facilitate the popularity of synthetic biology to a wider community and can help to eliminate inconsistencies in this field. In the future, further successful assemblies will be considered and we can devote ourselves to developing efficient algorithm to guarantee the reliability of results.

Funding. Natural Science Foundation of China [61572302].

Conflict of Interest Statement. None declared.

References

1. Goler, J.A., Bramlett, B.W., Peccoud, J.: Genetic design: rising above the sequence. Trends Biotechnol. **26**(10), 538–544 (2008)

2. Graslund, S., Nordlund, P., Weigelt, J., et al.: Protein production and purification. Nat. Methods **5**(2), 135–146 (2008)
3. Ghaemmaghami, S., Huh, W.K., Bower, K., et al.: Global analysis of protein expression in yeast. Nature **425**(6959), 737–741 (2003)
4. Czar, M.J., Cai, Y., Peccoud, J.: Writing DNA with genoCAD. Nucleic Acids Res. **37**, W40–W47 (2009)
5. Cai, Y., Wilson, M.L., Peccoud, J.: GenoCAD for iGEM: a grammatical approach to the design of standard-compliant constructs. Nucleic Acids Res. **38**(8), 2637–2644 (2010)
6. Isaacs, F.J., Dwyer, D.J., Ding, C., et al.: Engineered riboregulators enable post-transcriptional control of gene expression. Nat. Biotechnol. **22**(7), 817–841 (2004)
7. Endy, D.: Foundations for engineering biology. Nature **438**(7067), 449 (2005)
8. Baker, D., Church, G., Collins, J., et al.: Engineering life: building a FAB for biology. Sci. Am. **294**(6), 44–51 (2006)
9. Arkin, A.: Setting the standard in synthetic biology. Nat. Biotechnol. **26**(7), 771–774 (2008)
10. Canton, B., Labno, A., Endy, D.: Refinement and standardization of synthetic biological parts and devices. Nat. Biotechnol. **26**(7), 787–793 (2008)
11. Coll, A., Wilson, M.L., Gruden, K., Peccoud, J.: Rule-based design of plant expression vectors using GenoCAD. PLoS ONE **10**(7), e0132502 (2015)
12. Gardner, T.S., Cantor, C.R., Collins, J.J.: Construction of a genetic toggle switch in Escherichia coli. Nature **403**(6767), 339 (2000)
13. Atkinson, M.R., Savageau, M.A., Myers, J.T., et al.: Development of genetic circuitry exhibiting toggle switch or oscillatory behavior in Escherichia coli. Cell **113**(5), 597–607 (2003)
14. Ellis, T., Wang, X., Collins, J.J.: Diversity-based, model-guided construction of synthetic gene networks with predicted functions. Nat. Biotechnol. **27**(5), 465–471 (2009)
15. Bahaaddini, M., Hagan, P.C., Mitra, R., Hebblewhite, B.K.: Parametric study of smooth joint parameters on the shear behaviour of rock joints. Rock Mech. Rock Eng. **48**(3), 923–940 (2015)
16. Dorigo, M., Gambardella, L.M.: Ant colony system: a cooperative learning approach to the traveling salesman problem. IEEE Trans. EC **1**(1), 53–66 (1997)
17. Dorigo, M., Gambardella, L.M.: Ant colonies for the travelling salesman problem. Bio Syst. **43**(2), 73 (1997)
18. Dorigo, M., Maniezzo, V., Colorni, A.: Ant system: optimization by a colony of cooperating agents. IEEE Trans. Syst. Man Cybern. Part B Cybern. **26**(1), 29 (2002). A Publication of the IEEE Systems Man & Cybernetics Society
19. Watanabe, I., Matsui, S.: Improving the performance of ACO algorithms by adaptive control of candidate set. In: Congress on Evolutionary Computation, pp. 1355–1362. IEEE (2003)
20. Maniezzo, V., Colorni, A., Dorigo, M.: The ant system applied to the quadratic assignment. IEEE Trans. Knowl. Data Eng. **11**(5), 769–778 (1994)
21. Gambardella, L.M., Taillard, E., Dorigo, M.: Ant colonies for the quadratic assignment problem. J. Oper. Res. Soc. **50**(2), 167–176 (1999)
22. Low, C., Yeh, J.Y., Huang, K.I.: A robust simulated annealing heuristic for flow shop scheduling problems. Int. J. Adv. Manuf. Technol. **23**(9–10), 762–767 (2004)
23. Van Laarhoven, P.J.M., Aarts, E.H.L., Lenstra, J.K.: Job shop scheduling by simulated annealing. Oper. Res. **40**(1), 113–125 (2007)
24. Low, C., Wu, T.H.: Mathematical modelling and heuristic approaches to operation scheduling problems in an FMS environment. Int. J. Prod. Res. **39**(4), 689–708 (2010)
25. Cai, Y., Hartnett, B., Gustafsson, C., et al.: A syntactic model to design and verify synthetic genetic constructs derived from standard biological parts. Bioinformatics **23**(20), 2760–2767 (2007)

26. Fang, G., Zhang, S., Dong, Y.: Optimizing DNA assembly based on statistical language modeling. Nucleic Acids Res. **45**(22), e182 (2017)
27. Jelinek, F.: Statistical Methods for Speech Recognition. MIT Press, Cambridge (1997)
28. Katz, S.M.: Estimation of probabilities from sparse data for the language model component of a speech recogniser. IEEE Trans. Signal Process. **35**(3), 400–401 (1987)

Node Isolated Strategy Based on Network Performance Gain Function: Security Defense Trade-Off Strategy Between Information Transmission and Information Security

Gang Wang[1] , Shiwei Lu[1] , Yun Feng[1] , Wenbin Liu[2(✉)] , and Runnian Ma[1]

[1] Information and Navigation Institute, Air Force Engineering University, Xi'an 710077, China
[2] Institute of Advanced Computational Science and Technology, Guangzhou University, Guangzhou 510006, Guangdong, China
wbliu6910@126.com

Abstract. The development of information transmission has greatly accelerated the progress of society, and along with this, the security of information brings hidden threats to individual in the network. In this paper, a novel security defense trade-off strategy between information transmission and information security is proposed based on network performance gain function. First, the network performance gain function is defined on network average information intensity and network security index, and dedicated to make a trade-off between information transmission and information security of the network. Based on the gain function, node isolated strategies are provided, and corresponding algorithms named maximum degree isolation algorithm (MDIA) and maximum performance gain isolation algorithm (MPDIA) are devised respectively to isolate network node and improve network security. The results in simulation demonstrate that MDIA can quickly improve network security performance and MPDIA can acquire the larger gain within network robustness.

Keywords: Information transmission · Information security · Node isolated strategy · Network performance gain

1 Introduction

With the rapid development of information theory and technology, human life and social progress depends more and more on information transmission. In reality, numerous infrastructures for information transmission constitute various complex network, such as social network, computer network, wireless communication network and so on. Although the general networks operation and management can ensure information transmission normally, they are still confronted with the threat of malicious information intrusion.

For malicious information, many defensive models and strategies are employed in complex network to prevent its damage and diffusion. On microscopic level, Adewole et al. [1–4] proposed taxonomy of methods for identifying and detecting malicious accounts in

© Springer Nature Singapore Pte Ltd. 2020
H. Han et al. (Eds.): IDMB 2019, CCIS 1099, pp. 272–286, 2020.
https://doi.org/10.1007/978-981-15-8760-3_20

social networks. Although these fruitful works have greatly facilitated the network protection against malicious activities in the cognitive domain, there are still multiple novel malicious information intruding networks. Therefore, some scholars attempted to investigate the defensive method of malicious information from the macro perspective. Cheng et al. [5] proposed novel information dissemination models to segregate between useful and malicious types of information, which could eventually aid in the design of more efficient wireless communication networks. Zhao et al. [6] employed a propagation coupling model to investigate how the structure is coupled with the diffusion in online social networks from the view of weak ties, and proved that in practical application, weak ties can be used to control virus transmission and private information transmission. Zhu et al. [7] proposed a modified susceptible-infected-removed (SIR) model to explore rumor diffusion on complex social networks. In addition, many researchers addressed questions linked to virus propagation properties, spreading models, epidemic dynamics, tipping points, and control strategies [8–13]. These achievements provide references for the construction of secure network architecture, and for more solve some reality security problems of information diffusion in the network. However, the proposed methods and strategies mainly concentrate on the network security unilaterally, neglecting their influences on network information transmission capacity. Thus, the improvement of network security may lead to a significant decline in network information transmission capacity and affect the basic network services bearing capacity. In addition, the employed security solutions are lack a quantitative decision framework, causing the incapability of quantifying the benefits of strategies to the network performance. Thus, it is of great significance to explore a way to balance network security and network services bearing capacity, and to measure network benefits quantitatively.

In wireless sensor network, Soini et al. [14] explored the trade-offs between added security of sensor and sensor network performance, and the results showed the energy consumption increase caused by added security is tolerable and doesn't substantially affect sensor lifetime. Lin et al. [15] established the relationship between the worst-case security risk and packet delivery ratio and gave the theoretical limit on the security performance tradeoff of node-disjoint multipath routing in multi-hop wireless networks. The results of the above researches demonstrate that in a reasonable scope, even if network security measures weaken the ability of network information transmission, it will not affect the network own service bearing capability (the main reason is that the network has certain robustness). Therefore, by reasonably improving the network security index, the basic information transmission capability of the network can still be maintained, and the balance between network security and network service bearing capacity can be achieved. Certainly, since different types of networks have different requirements for security and information transmission, we should also consider the influence of network types on rational decision.

In this paper, we first propose two indicators, network average information intensity and network security index, according to network structure and susceptible-infected-removed-susceptible (SIRS) information diffusion paradigm. Afterwards, the network performance gain function, related to the two indicators and network types, is devised to describe network performance changes in the process of network adjustment quantitatively. Two strategies to improve network security are introduced based on the gain function to adjust network under different requirements of time and defensive effect.

The main contributions of this paper are summarized as follows:

(i) Network performance gain function is defined on network average information intensity and network relative security index, to measure network performance gain quantitatively. By calculating network performance gain, appropriate methods can be found to balance network information transmission capability and information security.

(ii) Based on the gain function and trade-off strategies of node isolated, two algorithms, maximum degree isolation algorithm (MDIA) and maximum performance gain isolation algorithm (MPGIA) are devised to isolate network nodes properly, where MDIA has a better time complexity than MPGIA. Within the scope of network robustness, the network security index and network performance gain are calculated by searching eligible nodes gradually.

(iii) The two algorithms are compared with random isolation algorithm (RIA) in simulation. The results demonstrate that regardless of network information transmission, MDIA make network healthy with the fastest speed and the least number of nodes isolated. In consideration of network communication robustness, MPGIA can get the largest network performance and make a trade-off between network information transmission capability and network information security.

2 Network Performance Gain Model

2.1 Network Average Information Intensity

A complex network can be represented as an undirected graph $G = (V, E)$. Supposing that the total number of network vertex or node is N, the vertex or node set can be denoted as $V = \{v_1, v_2, \ldots, v_N\}$. Each edge in the graph represents the information transmission channel between two adjacent nodes. If v_i is adjacent to node v_j, there is an edge e_{ij}. Because it is undirected graph, the edge e_{ij} can be described by $e_{ij} = (v_i, v_j) = (v_j, v_i) = e_{ji}$, where $e_{ij}, e_{ji} \in E$. According to the network structure, the network adjacent matrix $A = (a_{ij})_{N \times N}$ can be obtained as follow

$$a_{ij} = \begin{cases} 1, \text{ if } (V_i, V_j) \in E, \\ 0, \text{ if } (V_i, V_j) \notin E. \end{cases} \tag{1}$$

Specifically, given a pair of nodes (v_i, v_j), if a path connects the two users, then the distance d_{ij} is defined as the length of the shortest path between v_i and v_j [16]. If there is no path from v_i to v_j, then $d_{ij} = \infty$. Let $D = (d_{ij})_{N \times N}$ denote shortest path matrix of network node. Figure 1 gives a simple representation of network graph. For example, the distance between A and F is 4.

The information from node i to node j is directed and can be denoted by I_{ij}. Suppose the information transmits between any two nodes along with the shortest path in the network. Generally, in a period of time, information transmission between two nodes is limited. The shorter the distance d_{ij} is, the more frequently information transmits between v_i and v_j. Assume that the information transmission time between nodes with the same distance in the network is the same. Here we define the time of transmitting single message between nodes with a distance of 1 as unit time t_U and the single transmission time between other nodes is an integral multiple of the unit time. The single transmission time between any two nodes t_S^{ij} can be computed by

$$t_S^{ij} = d_{ij} \cdot t_U. \tag{2}$$

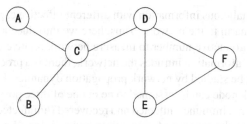

Fig. 1. A simple graph representation of the network

To investigate the network average information intensity, we replace single information transmission time t_S^{ij} with the information transmission probability p_{ij}, where p_{ij} represents the possibility of v_i sending information to v_j in unit time. The information transmission probability p_{ij} can be denoted as

$$p_{ij} = \frac{1}{t_S^{ij}} = \frac{1}{d_{ij} \cdot t_U}. \tag{3}$$

The information transmission probability matrix is defined as $\mathbf{P}_I = (p_{ij})_{N \times N}$ and it determines how much information will be transmitted in the network per unit time. For the capability of network service bearing, it is not only related to the number of information transmitted, but also associated with the importance of the information transmitted. In other words, loss of important information has a greater impact on network service bearing capability. Let $\mathbf{IM}(t) = (im_{ij}(t))_{N \times N}$ denote the information importance matrix at time t, where im_{ij} is usually decided by specific network and corresponding nodes and it is not our major concern. Here we select the random number matrix with same dimension as information importance matrix at time t and it changes with time dynamically. Afterward, we define the network average information intensity at time t as $\mu(t)$ and it can be computed by

$$\mu(t) = \frac{\sum_{i=1}^{N} \sum_{j=1}^{N} p_{ij} \cdot im_{ij}(t)}{N^2}, \tag{4}$$

where N is the total number of network node. To compare the network average information intensity with network security index mentioned later, $\mu(t)$ is set in the range [0, 1] (i.e. $\mu(t) \in [0, 1]$). Thus, for a specific network, when the number of shortest paths is large and the transmission of information is important, the average network information intensity is high.

2.2 Network Relative Security Index

From the perspective of network security, if there is no artificial adjustment for network, the network security index is determined by network invulnerability [17, 18]. However, the network invulnerability is just related to network structure in these researches, which doesn't consider the actual infective capability of malicious information intrusion. For a specific network, malicious information with different infective capabilities will cause different degrees of harm to the network. Thus, here we introduce a new method based on network basic reproductive number to measure network relative security index.

When malicious information intrudes the network, there is a process for information diffusion, which can be studied by network propagation dynamics. From Refs. [19, 20], we know the network node can be assumed to be in one of three possible states: susceptible (uninfected but not immune), infected and recovered (uninfected and immune). Let $S(t)$, $I(t)$, $R(t)$ denote the number of susceptible nodes, infected nodes and recovered nodes at time t. N is the total number of the network node, where $N = S(t) + I(t) + R(t)$. We can define the network relative security index at time t as $\delta(t)$ and let NS denote whether the final network is infected. If the infected node persists in the network, then $NS = 1$. Otherwise, $NS = 0$. Thus, $\delta(t)$ can be computed by

$$\delta(t) = \begin{cases} 1, & \text{if } NS = 0, \\ 1 - \frac{I(t \to +\infty)}{N}, & \text{if } NS = 1, \end{cases} \tag{5}$$

where $I(t \to +\infty)$ represents the finial number of infected nodes in the network and it is related to the network basic reproductive number at time t. From Eq. (5), we know the fewer infected nodes in the network, the higher the relative security index of the network.

Generally, malicious information diffusion has the equilibrium or stability points and we can replace $I(t \to +\infty)$ with the number of infected nodes at stable state. To further explore network relative security index, the SIRS model is employed to calculate the equilibrium points of the network. Before constructing SIRS model, there are some hypotheses should be addressed:

(H1) Malicious information infects single susceptible node with probability $\beta > 0$. According mean-field theory, every susceptible node in the network is infected by infected nodes with probability $\langle k \rangle \beta I(t)/N$, where $\langle k \rangle$ is the average degree of network nodes.

(H2) Due to network self-defense capability, each infected node recovers with probability ε.

(H3) Due to the appearance of new vaccine, susceptible nodes can obtain temporary immunity with probability φ.

(H4) Every recovered node in the network loss immunity with probability γ.

Let $TP = \{\beta, \varepsilon, \varphi, \gamma\}$ denote the transition probability parameter set. After malicious information intrusion, the state transition diagram of the network nodes with SIRS paradigm is shown in Fig. 2.

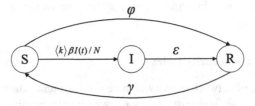

Fig. 2. The state transition diagram of SIRS paradigm

The corresponding dynamic model is established as

$$\begin{cases} \frac{dS(t)}{dt} = \gamma R(t) - \langle k \rangle \beta S(t)I(t)/N - \varphi S(t), \\ \frac{dI(t)}{dt} = \langle k \rangle \beta S(t)I(t)/N - \varepsilon I(t), \\ \frac{dR(t)}{dt} = \varepsilon I(t) - \gamma R(t) + \varphi S(t), \end{cases} \quad (6)$$

The two equilibriums, virus-free equilibrium and viral equilibrium, can be obtained as

$$\begin{aligned} P^0(S^0, I^0, R^0) &= \left(\frac{\gamma N}{(\gamma + \varphi)}, 0, \frac{\varphi N}{(\gamma + \varphi)} \right), \\ P^1(S^1, I^1, R^1) \\ &= \left(\frac{\varepsilon}{\langle k \rangle \beta} N, \frac{\langle k \rangle \beta \gamma - (\gamma + \varphi)\varepsilon}{\langle k \rangle \beta (\varepsilon + \gamma)} N, \frac{\varepsilon(\langle k \rangle \beta + \varphi - \varepsilon)}{\langle k \rangle \beta (\varepsilon + \gamma)} N \right). \end{aligned} \quad (7)$$

The process of detailed solution can refer to in Refs. [19, 20]. Let Rv denote basic reproductive number of the model (6), which indicates the number of susceptible nodes infected by an infected node per unit time [21, 22]. According to I^1 of viral equilibrium P^1, Rv can be represented by

$$Rv = \frac{\langle k \rangle \beta \gamma}{(\varphi + \gamma)\varepsilon}. \quad (8)$$

If $Rv \leq 1$, the network can suppress malicious information diffusion by self-defense system and there is no infected node in the network at stable state. Otherwise, if $Rv > 1$, the infected nodes will persist in the network (Detailed proof can refer to Refs. [19, 20]). Generally, if the relevant parameters don't change, the network basic reproductive number remains unchanged. Therefore, network relative security index in Eq. (5) can be rewritten as

$$\delta(t) = \begin{cases} 1, & \text{if } Rv \leq 1, \\ 1 - \frac{I^1}{N}, & \text{if } Rv > 1. \end{cases} \quad (9)$$

2.3 Network Performance Gain Function

The two indicators for network information transmission and network information security has been introduced in Subsect. 2.1 and Subsect. 2.2. To unilaterally measure the network information transmission capability and information security quantitatively, we can calculate the value of relevant indicator.

For network security practitioner, node isolated method is usually employed to reduce the network basic reproductive number and suppress malicious information diffusion. However, Node isolated will lead to the loss of network information transition capability. Too many nodes isolated will make network information transmission capability decline rapidly and cause losses to the interests of network operators. Thus, we propose the network performance gain function $Q(t)$ based on $\mu(t)$ and $\delta(t)$ to measure the impact of node isolated operation on network performance. Suppose that a node isolated operation occurs at time t and it is instantaneous. Let $\Delta\mu(t)$ and $\Delta\delta(t)$ denote the difference between $\mu(t+1)$ and $\mu(t)$, $\delta(t+1)$ and $\delta(t)$ respectively, which can be represented by

$$\begin{cases} \Delta\mu(t) = \mu(t+1) - \mu(t), \\ \Delta\delta(t) = \delta(t+1) - \delta(t). \end{cases} \quad (10)$$

Let λ denote the cumulative information loss proportion caused by multiple node isolated operation. For a specific network, suppose that without affecting normal communication, the maximum proportion of information loss tolerable by the network is λ_{max}, which is determined by network robustness. In this paper, we regard the network performance is only affected by network information transmission capability and information security. Thus, we define the network performance gain function $Q(t)$ as follow

$$Q(t) = \begin{cases} a_1 \cdot \Delta\mu(t) + b_1 \cdot \Delta\delta(t), & \text{if } \lambda \leq \lambda_{max}, \\ a_2 \cdot \Delta\mu(t) + b_2 \cdot \Delta\delta(t), & \text{if } \lambda > \lambda_{max}, \end{cases} \tag{11}$$

where $a_1 + b_1 = 1$ and $a_2 + b_2 = 1$. Specifically, $a_1 > b_1$ and $a_2 < b_2$ because fewer nodes isolated have relatively less impact on network information transmission capability, but excessive nodes isolated have a greater impact. Let $GP = \{a_1, b_1, a_2, b_2, \lambda_{max}\}$ denote the parameter set of gain function $Q(t)$. According to Eq. (11), if we can isolate node properly within the permissible range of network robustness, network performance gain will increase and the overall performance of the network will be improved.

3 Node Isolated Strategies Based on Network Performance Gain Function

In this section, we propose two strategies based on network performance gain function to obtain maximum network performance gain in different network conditions. Each strategy has respective advantage or disadvantage in time complexity and isolation effect.

Assuming that information transmission probability matrix doesn't change with time, the matrix $IM(t)$ can be represented as IM. In this assumption, if there is no artificial adjustment for the network, network information transmission capability doesn't change with time and $Q(t)$ is unchangeable as well. When network basic reproductive number $Rv \leq 1$, because the network is in a secure state and the node isolated strategy has no influence on network performance gain $Q(t)$, we don't need to adopt node isolated strategy. However, when the network basic reproductive number $Rv > 1$, we propose the following strategies to improve network security based on network performance gain function.

3.1 Node Isolated Strategy for Local Maximum Network Performance Gain

The main idea of this strategy is to isolate one node at a time, which will bring the greatest gain to the network performance, so as to enhance network security and ensure network performance gain increase.

Let $MP = \{TP, GP\}$ denote the merged parameter set. Algorithm 1 is designed for the strategy to search a node with maximum performance gain after isolation in the current network, called maximum performance gain isolation algorithm (MPGIA).

Algorithm 1 MPGIA

Input: current network graph G , current nodes total number N, information importance matrix $IM(t)$, merged parameter set MP initial network average information intensity μ_0

Output: The node v with maximum gain and its gain Q

1: Counting the degree of each node in the network k=count_deg(G)

2: Calculate shortest path matrix with Dijkstra algorithm D=Dijkstra(G)

3: Initialize current network average information intensity $\mu \leftarrow 0$ and network average degree $k \leftarrow 0$

4: **for** $i = 1$ to N **do**

5:   $k = k + k[i]$

6:   **for** $j = 1$ to N **do**

7:     $\mu \leftarrow \mu + IM[i,j] / D[i,j]$

8:   **end for**

9: **end for**

10:   $\mu \leftarrow \mu / N^2$, $k = k / N$

11: Initialize transition probability β=MP[1] , ε=MP[2] , φ=MP[3] , γ=MP[4]

12: Calculate the number of infected nodes at stable state $I = N\left[k\beta\gamma - (\gamma + \varphi)\varepsilon\right]/k\beta(\varepsilon + \gamma)$

13: Current network security index $\delta \leftarrow 1 - I/N$

14: **for** $i = 1$ to N **do**

15: Try to isolate node i from graph G and get the new graph G_N=Remove(G, i)

16: Calculate the new shortest path matrix D_N=Dijkstra(G_N)

17: Calculate the new network average information intensity $\mu 1 \leftarrow 0$

18:   **for** j=1 to N-1 **do**

19:     **for** k=1 to N-1 **do**

20:       $\mu 1 \leftarrow \mu 1 + IM[j][k] / D\_N[j][k]$

21:     **end for**

22:   **end for**

23:   $\mu 1 \leftarrow \mu 1 / N^2$

24: Calculate the new network average degree

      $k1 \leftarrow k * N - k[i]$

25:   $I\_N = N\left[k1\Box\beta\gamma - (\gamma + \varphi)\varepsilon\right]/k1\Box\beta(\varepsilon + \gamma)$

26: The new network security index $\delta 1 \leftarrow 1 - I\_N/N$

27:   $\Delta\mu \leftarrow \mu 1 - \mu$, $\Delta\delta \leftarrow \delta 1 - \delta$

28: Cumulative information loss proportion caused by isolating node $\lambda = (\mu 0 - \mu 1)/\mu 0$

28: Initialize transition probability $a1$=MP[5] , $b1$=MP[6] , $a2$=MP[7] , $b2$=MP[8] ,

      $\lambda\_\max = MP$[9]

29: Calculate the gain of isolating node i

      if $\lambda < \lambda\_\max$ **then**

30:   $Q[i] \leftarrow a1 \cdot \Delta\mu + b1 \cdot \Delta\delta$

31: **else**

32:   $Q[i] \leftarrow a2 \cdot \Delta\mu + b2 \cdot \Delta\delta$

33: **end if**

34: **end for**

35: The node v = find_max_index(Q) and its gain $Q = Q[v]$

Every time algorithm 1 is executed, we can get a node that can maximize benefits by isolation. We record these nodes and their gain in turn. Generally, when $\lambda > \lambda\_\max$, node isolated operation will bring negative gains to the network, but it is not absolute.

Thus, we create a queue Q for the nodes to be isolated, where the elements in Q are arranged in descending order of gain. If we find a node i by Algorithm 1 whose gain is less than 0, the program will terminate. Before termination, suppose the number of nodes to be isolated is n, and $Q = \{v_{i1}, v_{i2}, \ldots, v_{in}\}$ is the isolation queue.

The time complexity of the Dijkstra algorithm is $O(VN + N^2 \lg N)$, where N and E are respectively node number and edge number of the network. For a undirected complete graph, $0 < E < N^2$. Thus, the time complexity of Algorithm 1 is $O(VN^2 + N^3 \lg N)$. Although the algorithm 1 can help to achieve local maximum gain, but it will take a long time to search the nodes to be isolated.

3.2 Node Isolated Strategy Based on Maximum Node Degree

In essence, nodes isolated method is to change the network average degree. In algorithm 1, by isolating the nodes with the largest gain, the network average degree is reduced. Here we propose another node isolated strategy based on maximum node degree. At each step, we search and isolate the node with the largest degree of isolation. Meanwhile, the network performance gain is computed to ensure that every step of isolation brings positive benefits to the network. Algorithm 2 is designed for the strategy to isolate a node with maximum degree in the current network, called maximum degree isolation algorithm (MDIA).

Algorithm 2 MDIA

Input: current network graph G , current nodes total number N, importance matrix $IM(t)$, merged
  parameter set MP , initial network average information intensity μ_0

Output: The node v with maximum degree and its gain Q

1: Counting the degree of each node in the network k=count_deg(G)

2: Find the maximum degree and its subscripts [m, index] = max(k)

3: Initialize current network average information intensity $\mu \leftarrow 0$ and network average degree $k \leftarrow 0$,
  the new network average information intensity $\mu1 \leftarrow 0$

4: D=Dijkstra(G)

5: G_N=Remove(G, index)

6: D_N=Dijkstra(G_N)

7: **for** $i =1$ to N **do**

8:   $k = k + k[i]$

9: **for** $j =1$ to N **do**

10:   $\mu \leftarrow \mu + IM[i,j] / D[i,j]$, $\mu1 \leftarrow \mu1 + IM[i,j] / D\_N[i,j]$

11: **end for**

12: **end for**

13: $\mu \leftarrow \mu / N^2$, $\mu1 \leftarrow \mu1 / N^2$, $k = k / N$

14: Initialize transition probability β=MP[1] , ε=MP[2] , φ=MP[3] , γ=MP[4]

15: $\Delta\mu \leftarrow \mu1 - \mu$, $\Delta\delta \leftarrow [\varepsilon(\gamma + \varphi)(k - k1)] / [k\square k1\square\beta(\varepsilon + \gamma)]$

16: Cumulative information loss proportion λ=$(\mu0-\mu1)/\mu0$

17: Initialize transition probability $a1$=MP[5] , $b1$=MP[6] , $a2$=MP[7] , $b2$=MP[8] , $\lambda\_$max = MP[9]

18: **if** $\lambda < \lambda\_$max **then**

19:   $Q \leftarrow a1 \cdot \Delta\mu + b1 \cdot \Delta\delta$

20: **else**

21:   $Q \leftarrow a2 \cdot \Delta\mu + b2 \cdot \Delta\delta$

22: **end if**

In Algorithm 2, the node with maximum degree can be found directly through the degree matrix k and the time complexity of this step doesn't exceed $O(N)$. Thus, the time complexity of Algorithm 2 is $O(N^2)$. Compared with Algorithm 1, the time cost of algorithm 2 is significantly reduced. However, the gains acquiring from strategy in Subsect. 3.2 is different from strategy in Subsect. 3.1 and we will discuss the difference in simulations.

4 Numerical Simulation

In this section, the two isolation strategies are verified in simulation. To demonstrate the advantage of these two strategies, we compare them with node random isolation strategy.

4.1 Dataset and Experiment Setting

By using the complex network package developed by Muchnik [23], we generate a scale-free network and a small-world network to represent different types of network. The scale-free network and small-world network are selected for simulation. The two networks have the same number of nodes $N = 100$. The scale-free (SF) network have 768 edge and its average degree is 8. The small-world (SW) network have 800 edge and its average degree is 8.

Information importance matrix IM is set as a random matrix, and transition probability sets of two networks are the same for comparison under the same diffusion condition. Transition probabilities are respectively set as $\beta = 0.1$ $\varepsilon = 0.16$, $\varphi = 0.3$, $\gamma = 0.1$. Let Rv_1 and Rv_2 denote the basic reproductive number of SF network and SW network respectively. In this parameter setting, basic reproductive number $Rv_1 = Rv_2 = 1.25$, which indicates the two networks can't restore to the healthy state by themselves and the node isolation is needed. Different types of networks have different information transmission capabilities and security weights. Meanwhile, the maximum proportion of information loss tolerable are different between these two networks.

Suppose that scale-free network is concerned with information transmission capability, where the security is secondary. In contrast, information transmission capability of the small-world network is secondary. Therefore, the weight of average information intensity in scale-free network is larger than that in small-world network. Moreover, the robustness of networks concerned with information transmission capability is usually low for which the services can be operated better on the network. Based on the above, the gain parameter sets of scale-free network and small-world are respectively set as $GP_1 = \{0.4, 0.6, 0.8, 0.2, 0.1\}$ and $GP_2 = \{0.2, 0.8, 0.6, 0.4, 0.15\}$. The setting of parameters is to indicate that the two networks attach different importance to security.

4.2 The Effectiveness of Both Node Isolation Strategies

The two algorithms, MPGIA and MDIA, can isolate appropriate node according devised rules gradually. To verify the effectiveness of two algorithms on isolation, their gains are acquired in scale-free network and small-world network, and compared with random isolation algorithm (RIA). Figure 3(a) and Fig. 4(a) show the curve of cumulative gain over time in SF network and SW network respectively. To illustrate the influence of the maximum proportion of information loss on network performance gain, Fig. 3(b) and Fig. 4(b) gives the curve of network information loss proportion over time in these two networks simultaneously. It is noteworthy that time in simulations is unit time, which needs to be determined according to the transmission time between adjacent nodes of the network. Here we take 1(s) as unit time.

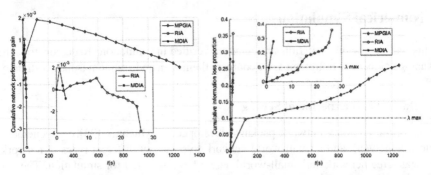

Fig. 3. The cumulative gain and information loss curves of three node isolated strategies in scale-free network (a) cumulative network performance gain (b) cumulative information loss proportion

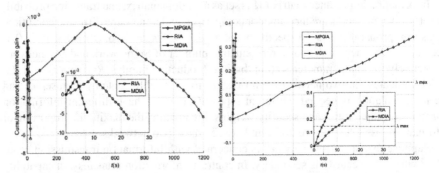

Fig. 4. The cumulative gain and information loss curves of three node isolated strategies in small-world network (a) cumulative network performance gain (b) cumulative information loss proportion

From Fig. 3 and Fig. 4, we can observe that when the same number of nodes are isolated, both of MPGIA and MDIA can get a better gain of network performance than RIA. When the cumulative information loss proportion doesn't exceed network maximum information loss proportion, the two strategies proposed in this paper can select more appropriate node to be isolated and get a greater network performance gain. In other words, both strategies can make a better trade-off between information transmission and information security, which are helpful to network managers.

4.3 The Strengths and Weakness Analysis of Both Node Isolation Strategies

The two strategies have their own strengths and weakness in running time and isolation effect. Here we execute MPGIA and MDIA in SW network mentioned above to illustrate the influence of node isolation on average information intensity and information security index. For comparison, the results of RIA are also given. When information security index $\delta = 1$, the simulation terminates. Figure 5(a) shows the curve of average information intensity over time and Fig. 5(b) shows the curve of information security index over time. To further illustrate the characteristics of isolated nodes, we give each node a serial number and acquire the numbered node set $V = \{v_1, v_2, \ldots, v_{100}\}$. In the range of network maximum information loss proportion $\lambda \leq 0.15$, the serial number of and degree of isolated nodes are recorded in Table 1. Simultaneously, cumulative information loss proportion, cumulative gain and finial network security index are also given in Table 1.

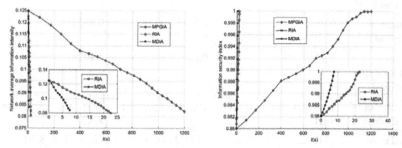

Fig. 5. The information intensity and security index curves of three node isolation strategies in small-world network (a) network average information intensity (b) information security index.

From Fig. 5(a) and Fig. 5(b), we can see that without considering of network information loss, MDIA makes the network achieve the largest security index with the fastest speed and the least number of nodes isolated. Thus, MDIA is suitable for infected network to quickly recover to health state regardless of network communication capability. From Table 1, we know in the range of network maximum information loss proportion, MPGIA can acquire the largest gain and a higher final network security index. However, its running time is relatively longer and the number of nodes isolated is relatively more than MPGIA. In general, MPGIA and MDIA first isolate nodes with high degree and the gain is greater than that acquired by RIA.

Table 1. The relevant indictors for three node isolated strategies

Relevant Indictors	Algorithm		
	MPGIA	MDIA	RIA
Cumulative information loss proportion	0.1446	0.1369	0.1429
Serial number of isolated nodes	$\{v_{90}, v_{95}, v_{49}, v_{26}, v_{81}, v_{89}\}$	$\{v_{95}, v_5, v_{26}, v_{49}\}$	$\{v_{42}, v_{91}, v_{50}, v_7, v_{67}, v_{85}\}$
Isolated node degree	$\{11, 12, 11, 11, 10, 6\}$	$\{12, 11, 11, 11\}$	$\{8, 9, 11, 8, 8, 9\}$
Cumulative network performance gain	0.006054	0.004235	0.004174
Final network security index	0.9887	0.9865	0.9867
Cumulative information loss proportion	0.1446	0.1369	0.1429
Serial number of isolated nodes	$\{v_{90}, v_{95}, v_{49}, v_{26}, v_{81}, v_{89}\}$	$\{v_{95}, v_5, v_{26}, v_{49}\}$	$\{v_{42}, v_{91}, v_{50}, v_7, v_{67}, v_{85}\}$
Isolated node degree	$\{11, 12, 11, 11, 10, 6\}$	$\{12, 11, 11, 11\}$	$\{8, 9, 11, 8, 8, 9\}$
Cumulative network performance gain	0.006054	0.004235	0.004174
Final network security index	0.9887	0.9865	0.9867

5 Conclusions

In this paper, we propose network average information intensity and network relative security index to quantitatively evaluate network information transmission capability and information security. In addition, the network performance gain function is devised to measure network performance gain when the network is adjusted. Based on these, two strategies are employed to properly isolate node and make a trade-off between information transmission capability and information security. The corresponding algorithms, MPGIA and MDIA, are devised to isolate node one by one. Compared with RIA, both of MPGIA and MDIA can get a better network performance gain at the same level of safety. Regardless of network information transmission, MDIA make network healthy with the fastest speed and the least number of nodes isolated. In consideration of network communication robustness, MPGIA can get the largest network performance and make a trade-off between network information transmission capability and network information security. But MPGIA takes more time than MDIA and RIA. Actually, not all nodes need to be attempted to isolate and we will optimize MPGIA with properly pruning strategies to reduce its costs on time.

Acknowledgment. The paper is supported by National Natural Science Foundation of China (No. 61573017, 61703420, 61873277). We declare that there is no conflict of interest regarding the publication of this paper.

References

1. Yu, H., Gibbons, P.B., Kaminsky, M., et al.: SybilLimit: a near-optimal social network defense against sybil attacks. IEEE/ACM Trans. Netw. **18**(3), 885–898 (2010)
2. Wang, G., Mohanlal, M., Wilson, C., et al.: Social turing tests: crowdsourcing sybil detection (2012). Eprint Arxiv https://arxiv.org/abs/1205.3856. Accessed 7 Dec 2012
3. Yang, C., Harkreader, R.C., Gu, G.: Empirical evaluation and new design for fighting evolving twitter spammers. IEEE Trans. Inf. Forensics Secur. **8**(8), 1280–1293 (2013)
4. Adewole, K.S., Anuar, N.B., Kamsin, A., et al.: Malicious accounts: dark of the social networks. J. Netw. Comput. Appl. **79**(1), 41–67 (2017)
5. Cheng, S.M., Vasileios, K., Chen, P.Y., et al.: Diffusion models for information dissemination dynamics in wireless complex communication networks. J. Complex Syst. **2013**, 1–13 (2013)
6. Zhao, J., Wu, J., Xu, K.: Weak ties: subtle role of information diffusion in online social networks. Phys. Rev. E **82**(1), 016105 (2010). https://doi.org/10.1103/PhysRevE.82.016105
7. Zhu, L., Wang, Y.: Rumor spreading model with noise interference in complex social networks. Phys. A: Stat. Mech. Appl. **469**, 750–760 (2016)
8. Huang, C.Y., Lee, C.L., Wen, T.H., et al.: A computer virus spreading model based on resource limitations and interaction costs. J. Syst. Softw. **86**(3), 801–808 (2013)
9. Yang, L.X., Yang, X., Tang, Y.: A bi-virus competing spreading model with generic infection rates. IEEE Trans. Netw. Sci. Eng. **5**, 2–13 (2017)
10. Xu, D., Xu, X., Xie, Y., et al.: Optimal control of an SIVRS epidemic spreading model with virus variation based on complex networks. Commun. Nonlinear Sci. Numer. Simul. **48**, 200–210 (2017)
11. Zhang, C.M.: Global behavior of a computer virus propagation model on multilayer networks. Secur. Commun. Netw. **2018**, 1–9 (2018)
12. Ali, J., Saeed, M., Rafiq, M., Iqbal, S.: Numerical treatment of nonlinear model of virus propagation in computer networks: an innovative evolutionary Padé approximation scheme. Adv. Differ. Equ. **2018**(1), 1–18 (2018). https://doi.org/10.1186/s13662-018-1672-1
13. Li, T., Wang, S., Li, B.: Research on suppression strategy of social network information based on effective isolation. Procedia Comput. Sci. **131**, 131–138 (2018)
14. Soini, M., Kukkurainen, J., Lauri, S.: Security and performance trade-off in KILAVI wireless sensor network. In: WSEAS International Conference on Computers. World Scientific and Engineering Academy and Society (WSEAS) (2010)
15. Lin, C., Jean, L.: Research article on multipath routing in multihop wireless networks: security, performance, and their tradeoff. Eurasip J. Wirel. Commun. Netw. **2009**(1), 1–13 (2009)
16. Lei, X., Jiang, C.X., Han, Z., et al.: Trust-based collaborative privacy management in online social networks. IEEE Trans. Inf. Forensics Secur. **14**(1), 48–60 (2019)
17. Chen, S., Jiang, J., Pang, S., et al.: Modeling and optimization of train scheduling network based on invulnerability analysis. Appl. Math. Inf. Sci. **7**(1), 113–119 (2013)

18. Gao, X., Keqiu, L.I., et al.: Invulnerability measure of a military heterogeneous network based on network structure entropy. IEEE Access **6**, 6700–6708 (2017). https://doi.org/10.1109/ACCESS.2017.2769964
19. Gan, C., Yang, X., Liu, W., Zhu, Q.: A propagation model of computer virus with nonlinear vaccination probability. Commun. Nonlinear Sci. Number Simul. **19**, 92–100 (2014)
20. Guo, H., Li, M.: Impacts of migration and immigration on disease transmission dynamics in heterogeneous populations. Discret. Contin. Dyn. Syst.-Ser. B **17**, 2413–2430 (2017)
21. Holme, P., Masuda, N.: The basic reproduction number as a predictor for epidemic outbreaks in temporal networks. PLoS ONE **10**(3), 1–15 (2015)
22. Li, T., Liu, X., Wu, J., et al.: An epidemic spreading model on adaptive scale-free networks with feedback mechanism. Phys. A: Stat. Mech. Appl. **450**, 649–656 (2016)
23. Muchnik, L.: Complex networks package for MATLAB. http://www.levmuchnik.net/Content/Networks/ComplexNetworksPackage.html (2013)

Author Index

Printed in the United States
By Bookmasters